Set Theory and the Continuum Problem

RAYMOND M. SMULLYAN

MELVIN FITTING

DOVER PUBLICATIONS, INC.
Mineola, New York

Copyright

Copyright © 2010 by Raymond M. Smullyan and Melvin Fitting
All rights reserved.

Bibliographical Note

This Dover edition, first published in 2010, is a revised and corrected republication of the work originally published in 1996 by Oxford University Press, New York. The authors have provided a new Preface for this revised edition.

Library of Congress Cataloging-in-Publication Data

Smullyan, Raymond M.
    Set theory and the continuum problem / Raymond M. Smullyan and Melvin Fitting. — [Rev. ed.].
      p. cm.
    Includes bibliographical references and index.
    ISBN-13: 978-0-486-47484-7
    ISBN-10: 0-486-47484-4
    1. Set theory. 2. Continuum hypothesis. I. Fitting, Melvin, 1942– II. Title.
QA248.S586 2010
511.3'22—dc22

                                                            2009036172

Manufactured in the United States by Courier Corporation
47484404
www.doverpublications.com

PREFACE TO THE REVISED 2010 EDITION

This book is intended to provide the reader with a complete foundation in modern set theory: how the subject is axiomatized, what ordinal and cardinal numbers are, what the role of the axiom of choice is, what constructible sets are, what forcing is. All the details are present, from initial axiomatic development to the relative consistency and independence of the continuum hypothesis. We have made use of our rather broad experience as logicians to organize things in what we hope will be seen as an elegant and insightful manner. The Preface to the first edition discusses the novelties of our approach in some detail—we do not repeat this here. Our coverage begins with Zermelo (and Cantor, of course), and ends with the work of Cohen. There are no new results, but there are new presentations of important results. After this book, the reader will have a full background in the subject and will be ready to move on to further and more current study. And whether or not further investigation of contemporary work in set theory is desired, the reader will have seen a thorough presentation of one of the great achievements of twentieth century mathematics.

The only changes made to this work, from the first edition, involve the correction of errors. While there were quite a few, most were minor. The most significant error concerned the proof that constructibility is absolute, in Chapter 14, and correcting this required the replacement of several theorems and proofs. Except for this, theorem and definition numbering are the same as in the original edition.

We want to thank the careful readers who reported errors to us. The following is a list (in alphabetical order): Evangelia Antonakos, Elliott Belbin, Victoria Gittman, Peter Gregory, and Sakama Tsuyoshi.

PREFACE

In the late nineteenth century Georg Cantor startled the mathematical world by proving that there are different orders of infinity. In particular he showed that for any set S of elements, the size of the power set $\mathcal{P}(S)$, of all *subsets* of S, is greater than the size of S. (His proof is given in Chapter 1 of this volume.) He then raised the following question: Given an infinite set S, does there exist a set X larger than S but smaller than the set $\mathcal{P}(S)$? He *conjectured* that there is no such intermediate set X, and this conjecture is known as the *generalized continuum hypothesis*. To this day, no one knows whether it is true or not (Gödel conjectured that it is false). But in 1938 Kurt Gödel made the amazing discovery that whether true or not, it is at least not *disprovable* from the accepted axioms of set theory (the Zermelo-Fraenkel axiom system, explained in this book) assuming these axioms themselves are consistent, and in 1963 Paul Cohen proved that the continuum hypothesis, even if true, is not *provable* either. So the continuum hypothesis is thus *independent* of these axioms (again assuming that the axioms are consistent, which we will henceforth do).

Gödel and Cohen proved similar results about the so-called *axiom of choice*—one form of which states that given any collection of non-empty sets, it is possible to simultaneously assign to each of the sets one of its members. The axiom of choice is certainly useful in modern mathematics, and most mathematicians regard it as self-evident, unlike the continuum hypothesis, which does not appear to be either self-evidently true or self-evidently false. Gödel showed that the axiom of choice is not disprovable from the Zermelo-Fraenkel axioms, and Cohen showed that the axiom is not provable either.

Later Gödel turned his attention to the more elegant system of *class-set* theory known as *NBG*, in place of the Zermelo-Fraenkel axioms, and this is the system we use here. We provide a general introduction to *NBG* and to the Gödel and Cohen proofs of relative consistency and independence of the axiom of choice and the continuum hypothesis. Our book is intended as a text for advanced undergraduates and graduates in mathematics and philosophy. It is close to self-contained, involving only a basic familiarity with the logical connectives and quantifiers. A semester course in symbolic logic is more than enough background.

About two-thirds of the book originated as class notes, used in teaching set theory over the years. We revised these chapters to make them more suitable as a text, and we added the remaining chapters. Now for some special features of the book.

Part I (Chapters 1–9) on basic set theory deals mainly with standard topics, but from a novel viewpoint. Standard topics include the axiomatic development of set theory, ordinal and cardinal numbers, and the investigation of various equivalent forms of the axiom of choice. (We do not cover the construction of the real number system—see

(Vaught 1995) or (Moschovakis 1994) for this.) Our main novelty lies in the use of Smullyan's double induction and double superinduction principles (in Chapters 3 and 4), which have hitherto appeared only in research papers, and which provide a unified approach to several of these topics, including Zermelo's well ordering theorem, Zorn's lemma, the transfinite recursion theorem, and the entire development of the theory of ordinals. It is high time that these should appear in a textbook. Closely connected with this is Cowen's theorem which also has not yet appeared in any textbook. Together they provide a particularly smooth and intuitive development of the ordinals.

We do as much as we can without using the axiom of substitution, which is not introduced until Chapter 6. In Chapter 4 we present the double superinduction principle and its applications to well ordering and various maximal principles such as Zorn's lemma. We present these principles in sharper forms than those usually given—forms which might be termed *local*, rather than global. For example, instead of proving the global statement that the axiom of choice implies that every set can be well ordered (Zermelo's theorem), we prefer the sharper local form that for any set S, if there is a choice function for S, then S can be well ordered. Thus the only choice function needed to well order S is one for S itself. In general, in dealing with maximal principles, we are interested in just what choice functions are needed to do a given job. Some of the results of these maximal principles may be new (for example, Theorem 6.2 of Chapter 4). The chapter concludes with the statement and proof of Cowen's theorem, which deserves to be better known.

Our treatment of ordinals (Chapter 5) is non-standard, in that our very definition of *ordinal number* involves quantifying over all *classes*, instead of all sets (x is an ordinal if and only if x belongs to all classes having certain closure properties) and it is therefore not immediate in the class-set theory that we are using that the class of all ordinals exists! Yet the definition leads to such a beautifully natural and elegant treatment of ordinals, if only it would work in *NBG*! Well, fortunately it *does* work by virtue of Cowen's theorem, from which the existence of the class of ordinals is immediate, and so this is the approach we happily take. This natural definition of *ordinal* was first published by Sion & Willmott (1962), and was independently thought of by Hao Wang (written communication) and by Smullyan. It is high time that this neat approach should be known!

Using the axiom of substitution, in Chapter 6 we establish various forms of the transfinite recursion theorem, but by quite a novel method: Our first form of this theorem (Theorem 5.1) appears to be new, and the proof is almost immediate from the double superinduction principle. From this theorem, the various other and more standard forms of the transfinite recursion theorem follow without further use of the Axiom of Substitution.

Chapters 7, 8, and 9 contain more or less standard material on rank, well foundedness, and cardinals. One general remark about style: Our approach is rather leisurely (particularly at the beginning) and free-wheeling, and is definitely *semantic* rather than syntactic. But we do indicate how it can all be formalized.

Part II (Chapters 10–15) is devoted to two of Gödel's proofs of the relative consistency of the continuum hypothesis. The first—and central—one is a modification of Gödel's 1938 proof, using later results such as the Mostowski-Shepherdson mapping theorem, reflection principles, and various contributions of Tarski, Vaught, Scott, Levy, Karp, and others. The second proof (given largely as a sequence of exercises) comes closer to Gödel's original proof, and is more involved, but reveals additional information of interest in its own right, information which was *implicit* in Gödel's proof, but which we bring out quite explicitly. These proofs combine four major themes—Mostowski-Shepherdson maps (Chapter 10), Reflection Principles (Chapter 11), Constructible sets (Chapter 12) (which is the central notion) and the notion of *absoluteness*. Our treatment of Mostowski-Shepherdson maps (also called *collapsing maps*) is considerably more thorough than usual, and brings to light interesting but less widely known results about rank, induction, and transfinite recursion, all generalized to well founded relational systems. We learned of this approach from Dana Scott. We also do more in Chapter 11 on Reflection Principles than is strictly necessary for Gödel's proof.

The four themes mentioned above unite in a remarkable way in the last chapters of Part II, leading to the basic result that the class of constructible sets satisfies all the axioms of class-set theory, as well as the axiom of choice and the generalized continuum hypothesis. Then, as explained in Chapter 15, all this can be formalized to show that if the formal axioms of class-set theory are consistent, then neither the axiom of choice nor the generalized continuum hypothesis can be disproved in the system.

Part III (Chapters 16–22) presents Paul Cohen's proofs of independence results using forcing. The approach it takes is novel and is, we believe, pedagogically an improvement on the usual versions. It is not hard to show that methods like those of Gödel that show the consistency of the continuum hypothesis will not allow one to construct a model of set theory that establishes its independence. But this fact applies only if a classical first-order model is desired. We introduce the notion of a *modal* model for set theory, and show there is one in which a version of the continuum hypothesis fails. We then show, by a simple argument, that this implies the *classical* independence.

Currently there are in the literature two main approaches to forcing. One version uses partially ordered sets of "forcing conditions," and follows Cohen's approach generally. Mathematically, such an approach can be hard for students to follow. In addition, certain countability assumptions are commonly made in order to present the proofs most naturally, then one shows how to avoid these assumptions. This is pedagogically confusing, at the least. The other standard version in the literature uses Boolean-valued models—in effect a kind of non-classical logic. This is mathematically easier to follow, but it is hard to see motivation for the specific models that are constructed. In fact, for motivation one generally returns to the Cohen-style approach.

We use the well-known modal logic $S4$. Partial orderings yield Kripke models for it, so we are in the Cohen tradition rather than in the Boolean-valued one. Even so, countability assumptions play no role. It is not that we can avoid them—they simply do not come up. Since we are following Cohen, details of the construction of particular models can be motivated fairly naturally. On the other hand, mathematical complexity is minimized in a way that is familiar to logicians generally—informal becomes formal; meta-language becomes language. Let us be more precise about this point, since it is a crucial one.

In most Cohen-style presentations one finds talk of *denseness*: a set S of forcing conditions is dense below p if every condition that strengthens p in turn has a strengthening that is in S. Then one finds assertions like: the set of conditions at which X is true is dense below p. There is much that must be established about which sets are dense, and all of this takes place at the meta-level, via arguments in English. But if one introduces a modal language, the assertion that X is true in a set that is dense below p becomes simply: $\Box \Diamond X$ is true at p. What had been meta-level arguments about sets of conditions become formal manipulations of the standard modal operators $\Box$ and $\Diamond$. Simple rules for these operators can be given once and for all. In this way the reader can concentrate on the key ideas of the independence proofs, without the details needing constant attention.

In a nutshell, we show various independence results using a natural $S4$ generalization of sets-with-rank (unramified forcing, in effect). Classical models are not constructed and countability assumptions are not made, or needed. Afterwards, countability assumptions are added, and it is shown how classical models can be produced. At this point, connections with other treatments are apparent, and the reader is prepared to consult the standard literature.

Our treatment is self-contained, and the ideas we need from modal logic, which are both standard and fairly minimal, are fully presented.

Acknowledgments We wish to thank Robert Cowen and Wiktor Marek for reading parts of the manuscript for this work, and making helpful comments.

CONTENTS

PREFACE TO THE REVISED 2010 EDITION   iii

PREFACE   v

I   AXIOMATIC SET THEORY   1

CHAPTER 1   GENERAL BACKGROUND   3
- §1   What is infinity? 3
- §2   Countable or uncountable? 4
- §3   A non-denumerable set 6
- §4   Larger and smaller 7
- §5   The continuum problem 9
- §6   Significance of the results 9
- §7   Frege set theory 11
- §8   Russell's paradox 11
- §9   Zermelo set theory 12
- §10   Sets and classes 13

CHAPTER 2   SOME BASICS OF CLASS-SET THEORY   15
- §1   Extensionality and separation 15
- §2   Transitivity and supercompleteness 17
- §3   Axiom of the empty set 18
- §4   The pairing axiom 19
- §5   The union axiom 21
- §6   The power axiom 23
- §7   Cartesian products 24
- §8   Relations 24
- §9   Functions 25
- §10   Some useful facts about transitivity 26
- §11   Basic universes 27

CHAPTER 3   THE NATURAL NUMBERS   29
- §1   Preliminaries 29
- §2   Definition of the natural numbers 32
- §3   Derivation of the Peano postulates and other results ... 33
- §4   A double induction principle and its applications .. 35
- §5   Applications to natural numbers 40
- §6   Finite sets 42
- §7   Denumerable classes 42
- §8   Definition by finite recursion 43

§9	Supplement—optional	44

CHAPTER 4 SUPERINDUCTION, WELL ORDERING AND CHOICE — 47
§1	Introduction to well ordering	47
§2	Superinduction and double superinduction	52
§3	The well ordering of g-towers	56
§4	Well ordering and choice	57
§5	Maximal principles	61
§6	Another approach to maximal principles	64
§7	Cowen's theorem	66
§8	Another characterization of g-sets	68

CHAPTER 5 ORDINAL NUMBERS — 71
§1	Ordinal numbers	71
§2	Ordinals and transitivity	75
§3	Some ordinals	76

CHAPTER 6 ORDER ISOMORPHISM AND TRANSFINITE RECURSION — 79
§1	A few preliminaries	79
§2	Isomorphisms of well orderings	80
§3	The axiom of substitution	82
§4	The counting theorem	83
§5	Transfinite recursion theorems	84
§6	Ordinal arithmetic	88

CHAPTER 7 RANK — 91
§1	The notion of rank	91
§2	Ordinal hierarchies	92
§3	Applications to the R_α sequence	93
§4	Zermelo universes	96

CHAPTER 8 FOUNDATION, ∈-INDUCTION, AND RANK — 99
§1	The notion of well-foundedness	99
§2	Descending ∈-chains	100
§3	∈-Induction and rank	101
§4	Axiom E and Von Neumann's principle	103
§5	Some other characterizations of ordinals	104
§6	More on the axiom of substitution	106

CHAPTER 9 CARDINALS — 107
§1	Some simple facts	107
§2	The Bernstein-Schröder theorem	108
§3	Denumerable sets	111
§4	Infinite sets and choice functions	111
§5	Hartog's theorem	112
§6	A fundamental theorem	114
§7	Preliminaries	116
§8	Cardinal arithmetic	118
§9	Sierpiński's theorem	121

II CONSISTENCY OF THE CONTINUUM HYPOTHESIS   125

CHAPTER 10   MOSTOWSKI-SHEPHERDSON MAPPINGS   127
§1   Relational systems 127
§2   Generalized induction and Γ-rank 129
§3   Generalized transfinite recursion 133
§4   Mostowski-Shepherdson maps 134
§5   More on Mostowski-Shepherdson mappings 136
§6   Isomorphisms, Mostowski-Shepherdson, well orderings 137

CHAPTER 11   REFLECTION PRINCIPLES   141
§0   Preliminaries 141
§1   The Tarski-Vaught theorem 145
§2   We add extensionality considerations 147
§3   The class version of the Tarski-Vaught theorems 148
§4   Mostowski, Shepherdson, Tarski, and Vaught 150
§5   The Montague-Levy reflection theorem 151

CHAPTER 12   CONSTRUCTIBLE SETS   155
§0   More on first-order definability 155
§1   The class L of constructible sets 156
§2   Absoluteness 158
§3   Constructible classes 163

CHAPTER 13 L IS A WELL FOUNDED FIRST-ORDER UNIVERSE   169
§1   First-order universes 169
§2   Some preliminary theorems about first-order universes 172
§3   More on first-order universes 174
§4   Another result 177

CHAPTER 14   CONSTRUCTIBILITY IS ABSOLUTE OVER L   179
§1   Σ-formulas and upward absoluteness 179
§2   More on Σ definability 181
§3   The relation $y = \mathcal{F}(x)$ 183
§4   Constructibility is absolute over L 189
§5   Further results 190
§6   A proof that L can be well ordered 191

CHAPTER 15   CONSTRUCTIBILITY AND THE CONTINUUM HYPOTHESIS   193
§0   What we will do 193
§1   The key result 194
§2   Gödel's isomorphism theorem (optional) 196
§3   Some consequences of Theorem G 198
§4   Metamathematical consequences of Theorem G 199
§5   Relative consistency of the axiom of choice 200
§6   Relative consistency of GCH and AC in class-set theory 201

III FORCING AND INDEPENDENCE RESULTS — 205

CHAPTER 16 FORCING, THE VERY IDEA — 207
§1 What is forcing? 207
§2 What is modal logic? 209
§3 What is $S4$ and why do we care? 213
§4 A classical embedding 215
§5 The basic idea 221

CHAPTER 17 THE CONSTRUCTION OF $S4$ MODELS FOR ZF — 223
§1 What are the models? 223
§2 About equality 229
§3 The well founded sets are present 233
§4 Four more axioms 235
§5 The definability of forcing 239
§6 The substitution axiom schema 242
§7 The axiom of choice 244
§8 Where we stand now 247

CHAPTER 18 THE AXIOM OF CONSTRUCTIBILITY IS INDEPENDENT — 249
§1 Introduction 249
§2 Ordinals are well behaved 249
§3 Constructible sets are well behaved too 251
§4 A real $S4$ model, at last 252
§5 Cardinals are sometimes well behaved 253
§6 The status of the generalized continuum hypothesis 256

CHAPTER 19 INDEPENDENCE OF THE CONTINUUM HYPOTHESIS — 259
§1 Power politics 259
§2 The model 259
§3 Cardinals stay cardinals 260
§4 CH is independent 262
§5 Cleaning it up 263
§6 Wrapping it up 266

CHAPTER 20 INDEPENDENCE OF THE AXIOM OF CHOICE — 267
§1 A little history 267
§2 Automorphism groups 268
§3 Automorphisms preserve truth 270
§4 Model and submodel 272
§5 Verifying the axioms 273
§6 AC is independent 278

CHAPTER 21 CONSTRUCTING CLASSICAL MODELS — 285
§1 On countable models 285
§2 Cohen's way 286
§3 Dense sets, filters, and generic sets 287
§4 When generic sets exist 289
§5 Generic extensions 291
§6 The truth lemma 294

§7	Conclusion	296

CHAPTER 22 FORCING BACKGROUND 299
§1 Introduction 299
§2 Cohen's version(s) 300
§3 Boolean valued models 301
§4 Unramified forcing 301
§5 Extensions 302

BIBLIOGRAPHY 305

INDEX 307

LIST OF NOTATION 313

PART I

AXIOMATIC SET THEORY

CHAPTER 1

GENERAL BACKGROUND

Infinity and the continuum problem

§1 What is infinity?

If you ask the average person what the word "infinite" means, the answer is likely to be: "endless," or "without end." This answer, though it somewhat captures the true picture, is not altogether satisfactory, since, for example, the circumference of a circle of radius 1 inch is certainly endless, in the sense of having no beginning or end, but one would not call it infinite (in the sense that a straight line extending infinitely far in at least one direction would be called infinite).

Now, the word "infinite" is used in mathematics in a very precise sense, and this is the sense we shall adopt. The first question is what are the sorts of things that can properly be said to be infinite or finite. The answer is that it is *sets of objects* to which these adjectives are applicable. And so, just what do we mean when we say that a set of objects is infinite or that it is finite? Before answering this question, some preliminaries are in order.

Let us first ask what it means for two sets to have the *same number of elements*, or to be of *the same size*. The clue here lies in the notion of a 1-1 correspondence, read "one to one correspondence." We say that a set A can be put into a 1-1 correspondence with a set B if it is possible to match each element of A with one and only one element of B in such a way that no element of B is left out and no two distinct elements of A are matched with the same element of B. For example, suppose you look into a theater and see that every seat is taken (and no one is sitting on anyone's lap) and that no one is standing. Then, without having to count either the number of seats or the number of people, you will know that the numbers are the same. The reason is that the set of people is in a 1-1 correspondence with the set of seats—each person corresponds to the seat on which he or she sits.

At this point we rely on the reader's familiarity with the natural numbers 0, 1, 2, ..., n, ..., that is, 0 and the positive whole numbers. (Later in this study we will see how the natural numbers can be explained in terms of the more basic notion of *set*, but for now, the reader's intuitive notion of "number" is sufficiently reliable.) Now, what does it mean to say that a set—say the set of fingers on my left hand—has exactly 5 members? It means that it can be put into a 1-1 correspondence with the set of positive whole numbers from 1 to 5. More generally, for any positive whole number n, to say that a set x has n elements is to say that x can be put into a 1-1 correspondence with the set of positive whole numbers from 1 to n. (Incidentally, the process of putting the elements of x into a 1-1 correspondence with the numbers from 1 to n is commonly called *counting*.) And so we have defined what it means for a set to have n elements, where n is a *positive* whole number. (The word "number" in this volume will always mean natural number, unless specified to the contrary.) As for the case $n = 0$, we say that a set has 0 elements if it has no elements at all (as, for example, if at a given time there is no one in a theater, we say that the set of people in the theater has *zero* elements, or that the set is *empty*).

Now that we have defined what it means for a set to have n elements, where n is any natural number, we can define what it means for a set to be finite or infinite.

Definition 1.1. A set S is said to be *finite* if there is a natural number n such that S has n elements. If there is no such natural number n, then S is said to be *infinite*.

This is the definition we shall adopt. Examples of finite sets abound. An obvious example of an *infinite* set is the set of all natural numbers themselves. Also the set of all *even* natural numbers is clearly infinite. As another example, consider a line segment one inch long. Although the *length* of the line is finite, the set of points on the line is infinite, since between any two distinct points there is a point (for example, the one halfway between them).

§2 Countable or uncountable?

The true father of the field known as *set theory* is the late nineteenth century mathematician Georg Cantor. A most fundamental problem about infinite sets that engaged his attention is this: are all infinite sets of the same size—that is, can any two infinite sets be put into a 1-1 correspondence with each other—or do they come in different sizes? What is your guess? (I have posed this question to non-mathematicians, and about half guess *yes* and half guess *no*.) The answer (as discovered by Cantor) will be given in this chapter.

An infinite set is called *denumerable* or *countably infinite* if it can be put into a 1-1 correspondence with the set $1, 2, 3, \ldots, n, \ldots$ of positive whole numbers. A set is called *uncountable* or *non-denumerable* if it is infinite but not denumerable. And so our question can be rephrased thus: is every infinite set denumerable or do there exist non-denumerable sets?

As I understand the history of the situation, Cantor spent 12 years trying to prove that every infinite set is denumerable and then in the thirteenth year discovered that the opposite is the case. What he did was to consider various sets which at first sight *seemed* too large to be denumerable, but then in each case he thought of an ingenious way of enumerating them (i.e., of putting them into a 1-1 correspondence with the set of positive integers).

I like to illustrate Cantor's constructions in the following manner. Imagine that you and I are immortal. I write down a positive whole number on a slip of paper and tell you: "Each day you have one and only one guess as to what the number is. If and when you guess it, you win a grand prize." Now, if you guess numbers at random, then it is perfectly possible that you will never get the prize. But if you guess systematically in the order $1, 2, 3, 4, \ldots, n, \ldots$, then obviously you will get the prize sooner or later.

Now, for another test a wee bit more difficult. This time I say: "I have now written down either a positive whole number or written down a negative whole number. Each day you have one and only one guess as to what it is. Again, if you ever guess it, you win a prize." Now do you have a strategy that will guarantee success sooner or later? Of course you do; you guess in the order 1, −1, 2, −2, 3, −3, ... and sooner or later you are bound to come to my number. (This shows that the set of all positive and negative whole numbers taken together, which at first sight appears to be "twice as large" as the set of positive whole numbers alone, is really denumerable after all!)

Now I give you a more difficult test. I write down *two* positive whole numbers on a slip of paper (maybe two different numbers, or maybe the same number written twice). Again you have one and only one guess each day as to what I wrote, and if you guess correctly you win a grand prize. However, you must guess *both* the numbers on the same day—you are not allowed to guess one of the numbers on one day and the other on another day, as this would be hardly any different from the first test. Is there *now* a strategy by which you can certainly win? At first thought, this may seem hopeless, since there are infinitely many possibilities for the first number that I wrote, and with each of the possibilities, there are infinitely many possibilities for the second number. But I can assure you that there is a strategy (a very simple one in fact) in which you can be sure to win the prize sooner or later. What strategy will work? I will leave this as a problem (call it *Problem 1*) which the reader is invited to try to solve. The solution will be given a bit later.

As a slight variant of this problem, suppose I require that you must not only name the numbers I have written, but also the order in which I wrote them (say, the first is written to the left of the second one). Now what strategy will work?

Next problem (Problem 2). I write down a positive fraction (a quotient of two positive whole numbers). On each day you have one and only one guess as to what the fraction is. What strategy will enable you to certainly name my fraction on some day or other? Or is there no such strategy? That is, is the set of positive fractions denumerable or not?

Now for a slightly harder problem (Problem 3). Is the set of all *finite* sets of positive whole numbers denumerable or not? That is, suppose I write down some *finite* set of positive integers (whole numbers). I don't tell you how many numbers are in the set, nor what is the highest number of the set. All you know is that the set is finite. Now do you feel that there is any chance of your certainly guessing my set?

At this point, let us discuss the solutions to these problems.

Solutions—For Problem 1, there is only one possibility in which the highest of the two numbers is 1—namely $\langle 1, 1 \rangle$; there are two possibilities in which the highest number is 2—namely, $\langle 1, 2 \rangle$ and $\langle 2, 2 \rangle$, and for each positive integer n, there are only n possibilities for the highest number being n—namely $\langle 1, n \rangle, \langle 2, n \rangle, \ldots, \langle n-1, n \rangle, \langle n, n \rangle$. And so you successively guess $\langle 1, 1 \rangle, \langle 1, 2 \rangle, \langle 2, 2 \rangle, \langle 1, 3 \rangle, \langle 2, 3 \rangle, \langle 3, 3 \rangle, \ldots, \langle 1, n \rangle, \langle 2, n \rangle, \ldots, \langle n, n \rangle, \ldots$. Sooner or later you are bound to guess my pair.

As for the variant of the problem in which you must guess also the order, you simply mention each pair of *distinct* numbers in both orders before proceeding to the next—that is you name the ordered pairs in the order: $\langle 1, 1 \rangle, \langle 1, 2 \rangle, \langle 2, 1 \rangle, \langle 1, 3 \rangle, \langle 3, 1 \rangle, \langle 2, 3 \rangle, \langle 3, 2 \rangle, \langle 3, 3 \rangle, \ldots, \langle 1, n \rangle, \langle n, 1 \rangle, \langle 2, n \rangle, \langle n, 2 \rangle, \ldots$.

Problem 2 is essentially the same as the last problem and differs only in that in writing a fraction $\frac{m}{n}$, where m, n are positive *whole* numbers, the numerator m is written *above* the denominator n, instead of to the left of it. Thus we can enumerate all the positive fractions in the order $\frac{1}{1}, \frac{1}{2}, \frac{2}{1}, \frac{1}{3}, \frac{3}{1}, \frac{2}{3}, \frac{3}{2}, \frac{3}{3}, \ldots, \frac{1}{n}, \frac{n}{1}, \frac{2}{n}, \frac{n}{2}, \ldots$. This result of Cantor—that the set of fractions (and consequently the set of rational numbers) is denumerable—was quite a shock to the mathematical world, when first announced.

For Problem 3, the key is that for each positive integer n, there are only finitely many sets whose highest member is n—namely, there are 2^{n-1} such sets. (Reason: each such set consists of n together with some subset of the numbers from 1 to $n-1$, and there are 2^{n-1} possible sets of numbers from 1 to $n-1$.) And so you first name the one and only set whose highest member is 1, then the two sets whose highest member is 2, then the four sets whose highest member is 3, then the eight sets whose highest member is 4, and so forth.

§3 A non-denumerable set

We have just seen that the collection of all *finite* sets of positive integers is denumerable. Now, what about the collection of *all* sets of positive integers—infinite sets as well as finite ones? Is this set denumerable or not? Well, Cantor showed that it is not! His argument is ingenious in its very simplicity and I would like to first illustrate it as follows.

Imagine that we have a book with denumerably many pages—the pages being consecutively numbered: page 1, page 2, ..., page n, ... (Thus the book has infinitely many pages, but these pages are in a 1-1 correspondence with the set of positive integers.) On each page is listed a set of positive integers. If *every* set of positive integers is listed in the book, then the book wins a blue ribbon. But the book *cannot* win the ribbon; there must be at least one set of positive integers that is *not* listed in the book! Why? Let us see.

We let S_1 be the set listed on page 1, S_2 the set listed on page 2, ..., S_n the set listed on page n, ... We wish to define a set S of positive integers which is different from every one of the sets $S_1, S_2, \ldots, S_n, \ldots$. Well, we first consider the number 1—whether it should go into our set S or not. We do this by considering the set S_1 listed on page 1 of the book. Either 1 belongs to the set S_1 or it doesn't. If it doesn't, then we shall include it in our set S, but if 1 *does* belong to S_1, then we *exclude* 1 from S. Thus whatever future decisions we make concerning the numbers 2, 3, ..., n, ..., we have secured the fact that $S \neq S_1$ because, of the two sets S and S_1, one of them contains 1 and the other doesn't. Next we consider the number 2. We put it into our set S just in case 2 does *not* belong to the set S_2, and this guarantees that $S \neq S_2$ (since one of them contains 2 and the other doesn't). And so on with every positive integer n. We thus take S to be the set of all positive integers n such that n does *not* belong to the set S_n. Then for every n, $S \neq S_n$ because, of the two sets S and S_n, one of them contains n and the other doesn't.

To make matters a bit more concrete, suppose, for example, that the first eight sets S_1–S_8 are the following:

* S_1 — The set of all numbers greater than 5
S_2 — The set of all even numbers
S_3 — The set of all prime numbers
* S_4 — The set of all odd numbers
S_5 — The set of all (positive whole) numbers
* S_6 — The empty set (the set with no numbers at all)
* S_7 — The set of all numbers divisible by 3
S_8 — The set of all numbers divisible by 4

I have starred those lines n such that n does *not* belong to S_n. (For example, 1 is not greater than 5, so 1 does not belong to S_1, so line 1 is starred. On the other hand, 2 is even, so 2 does belong to S_2, so line 2 is unstarred.) Then the numbers of the starred lines belong to S and those of the unstarred lines don't. Thus, of the first eight numbers, those that belong to S are 1, 4, 6 and 7. And so we see that $S \neq S_1$, because 1 belongs to S but not to S_1. As for S_2, $S \neq S_2$ because 2 doesn't belong to S, but does belong to S_2. As for 3, 3 belongs to S_3 but not to S, so $S \neq S_3$. And so forth.

What we have shown is that given any countably infinite (denumerable) sequence $S_1, S_2, \ldots, S_n, \ldots$ of sets of positive integers, there exists a set S of positive integers (namely, the set of all n such that n doesn't belong to S_n) such that S is different from each of the sets $S_1, S_2, \ldots, S_n, \ldots$. This means that *no* denumerable set of sets of positive integers contains every set of positive integers—in other words *the set of all sets of positive integers is non-denumerable*!

This is a special case of Cantor's theorem.

The more general case—For any set A, by the *power set of A*—symbolized $\mathcal{P}(A)$—is meant the set of all *subsets* of A. (A set B is called a subset of A if every member of B is also a member of A—as for example, the set of all even whole numbers is a subset of the set of all whole numbers.) We have just proved that if A is the set of positive integers, then $\mathcal{P}(A)$ cannot be put into a 1-1 correspondence with A. More generally, Cantor's theorem is that for *any* set A, it is impossible for $\mathcal{P}(A)$ to be put into a 1-1 correspondence with A. The proof is not much different from the proof for the special case that we have considered and runs as follows. Suppose we have a 1-1 correspondence that matches each element x of A with a subset S_x of A. Let S be the set of all elements x of A such that x does not belong to the set S_x. Then for every element x of A, $S \neq S_x$ because, of the sets S and S_x, one of them contains x and the other doesn't. And so there is *no* x in A that corresponds to S, and so in the purported 1-1 correspondence between A and the set of *all* subsets of A—the set S has been left out.

§4 Larger and smaller

Given two sets A and B, what does it mean to say that A is (numerically) *smaller* than B, or *larger* than B? It would be tempting to define A as being smaller than B, or B as being larger than A, if A can be put into a 1-1 correspondence with a *proper subset* of B—that is, with a subset of B which is not all of B (it leaves out at least one element of B). This definition would work fine for finite sets, but not for infinite sets, since, for example, the set of natural numbers can be put into a 1-1 correspondence with the set of natural numbers starting at 2, but we certainly wouldn't want to say that the set of natural numbers is smaller than itself! No, the correct definition is the following: we say that A is *smaller* than B—in symbols $A < B$—if A can be put into a 1-1 correspondence with some subset of B *and also A cannot* be put into a 1-1 correspondence with the whole of B (in other words, every 1-1 correspondence from A to part of B leaves out at least one element of B). We note that if A can be put into a 1-1 correspondence with a subset of B, then either A is smaller than B or A is the same size as B—in symbols, $A \leq B$.

Two interesting problems immediately arise. Suppose that A can be put into a 1-1 correspondence with a subset of B and B can be put into a 1-1 correspondence with a subset of A; does it necessarily follow that (all of) A can be put into a 1-1 correspondence with (all of) B? The answer is *yes*, and this has been proved in several ways, two of which we will study in this volume. (The result is known as the Schröder-Bernstein theorem.) A second and deeper problem is this. Two sets A and B are called *comparable* (comparable in size, that is), if either A can be put into a 1-1 correspondence with a subset of B, or B can be put into a 1-1 correspondence with a subset of A. Equivalently, A is *comparable* with B if and only if A is either smaller than B, larger than B, or the same size as B. If A and B are *finite* sets, then of course they are comparable. But is this necessarily true for infinite sets? Isn't it possible that A and B are such that there is no 1-1 correspondence between A and any subset of B, and also there is no 1-1 correspondence between B and any subset of A? Well, the statement that any two sets A and B *are* comparable is certainly accepted by most mathematicians, but it is equivalent to a somewhat controversial (though highly important) principle known as the *axiom of choice* (which in one form roughly says that, given any non-empty collection of non-empty sets, it is possible to simultaneously choose exactly one member from each of the sets). This equivalence is one of the topics we will study.

Cantor's theorem in full—We have shown that no set A can be put into a 1-1 correspondence with $\mathcal{P}(A)$, and so A is not of the same size as $\mathcal{P}(A)$. But A can be easily put into a 1-1 correspondence with a subset of $\mathcal{P}(A)$—namely, let any element x of A correspond to the set whose only element is x (this set is denoted $\{x\}$). And so we have:

Theorem 4.1 (Cantor's theorem). $A < \mathcal{P}(A)$ *for any set A. That is, A is smaller than $\mathcal{P}(A)$—i.e., A can be put into a 1-1 correspondence with a subset of $\mathcal{P}(A)$, but not with $\mathcal{P}(A)$.*

It of course follows from Cantor's theorem that for any infinite set A, the set $\mathcal{P}(A)$ is larger than A, and then $\mathcal{P}(\mathcal{P}(A))$ is in turn larger than $\mathcal{P}(A)$, then $\mathcal{P}(\mathcal{P}(\mathcal{P}(A)))$ is in turn larger than $\mathcal{P}(\mathcal{P}(A))$, and so we can generate an *infinite* sequence A, $\mathcal{P}(A)$, $\mathcal{P}(\mathcal{P}(A))$, $\mathcal{P}(\mathcal{P}(\mathcal{P}(A)))$, ... of sets—each one of a larger size than the preceding one. And so there is at least a denumerable infinity of different infinities (and, as we will later see, there are in fact many more than that!).

§5 The continuum problem

Two sets of the same size are also said to be of the same *cardinality*, or the same *power*. A set A that is larger than a set B is also said to be of higher cardinality or higher power than B (and then B is said to be of lower cardinality or lower power than A).

We let ω be the set of natural numbers (starting with 0). It of course has the same cardinality as the set of positive whole numbers (starting with 1)—since we can correspond to each natural number x the positive whole number $x + 1$, and this correspondence is clearly 1-1—so ω is a denumerable set. As we have seen, $\mathcal{P}(\omega)$ has higher power than ω and it is said to be of the power of the continuum, because it can be proved to be of the same size as the set of all real numbers, or as the set of all points on a straight line. Now, the continuum problem is this: does there exist a set x of higher cardinality than ω but of lower cardinality than $\mathcal{P}(\omega)$, or is the size of $\mathcal{P}(\omega)$ the "next" size after the size of ω? Cantor's continuum hypothesis (and it is only a *hypothesis*!) is that there is no such intermediate set x—in other words, any set that is larger than ω must be at least as large as $\mathcal{P}(\omega)$. And Cantor's *generalized* continuum hypothesis is that for *every* infinite set A, there can never be a set of higher cardinality than A but of lower cardinality than $\mathcal{P}(A)$. These hypotheses are, of course, only *conjectures* of Cantor. So far, no attempts have been the slightest bit successful in determining whether the continuum hypothesis is true or false! Another question *not to be confused with the truth or falsity of the continuum hypothesis* is whether it can be formally proved or disproved from the present day axioms of set theory. *This question is completely settled.*

We use the expression "present day axioms of set theory" to mean the system *ZF*—Zermelo-Fraenkel set theory—or the closely related system *NBG* of *class-set theory* (the system studied in this book), which are the systems currently in widespread use. Kurt Gödel (Gödel 1938, Gödel 1939) proved the celebrated result that the generalized continuum hypothesis is formally consistent with the axioms of *ZF* (assuming *ZF* is itself consistent, which we do throughout this book). And Paul Cohen (Cohen 1963, Cohen 1964, Cohen 1966) settled the matter in the other direction—he showed that the negation of the generalized continuum hypothesis (in fact even the negation of the special continuum hypothesis) is consistent in *ZF*. Thus the continuum hypothesis is *independent* of the axioms of *ZF*—i.e., it can neither be proved nor disproved in *ZF*. So the axioms of *ZF* are *not strong enough* to settle the continuum problem.

Another important independence result concerns the *axiom of choice*. Gödel has also shown that the axiom of choice is not disprovable in *ZF*, and Cohen has shown that it is not provable in *ZF*. So the axiom of choice is also independent of the axioms of *ZF*. The most remarkable result of Cohen is that even if we add the axiom of choice to the other axioms of set theory (i.e., to *ZF*) it is still not possible to prove the continuum hypothesis.

§6 Significance of the results

There has been a remarkable diversity of opinions concerning the significance of the independence results.

First of all we must realize that these results have been proved for the particular axiom systems *ZF* and *NBG* (though the arguments go through for many related systems). There are, however, very different *formal axiom systems* of set theory in which the arguments of Gödel and Cohen do *not* go through. But it is highly questionable whether these other systems really describe the notion of "set" as used by the working mathematician. For example, in Quine's systems *NF* (New Foundations) and *ML* (Mathematical Logic) (Quine 1937, Quine 1940) the axiom of choice is provably false. For one who regards all these alternative systems to be on an equal footing (and this includes many so-called *formalists*) the independence results may well seem insignificant; they could be construed as merely saying that the continuum hypothesis comes out positive in one system and negative in some other (equally good) system. On the other hand there are those who concede that *ZF* is the more "natural" set theory, but who cannot understand what could be meant by mathematical *truth* other than provability in *ZF*. They would therefore construe the independence results as saying that the continuum hypothesis is neither true or false. A slightly modified viewpoint is to the effect that the only propositions we can ever *know* to be true are those provable in *ZF*, hence that we can never *know* whether the continuum hypothesis is true or false.

The so-called *mathematical realist* or *Platonist* (and this seems to include a large number of working mathematicians) looks upon the matter very differently. We can describe the realist viewpoint as follows. There is a well defined mathematical reality of sets, and in this reality, the continuum hypothesis is definitely true or false. The axioms of *ZF* give a true but incomplete description of this reality. The independence results cast no light on the truth or falsity of the continuum hypothesis, nor do they in any way indicate that it is neither true nor false. Rather they highlight the inadequacy of our present day axiom system *ZF*. But it is perfectly possible that new principles of set theory may one day be found which, though not derivable from the present axioms, are nevertheless self-evident (as the axiom of choice is to most mathematicians) and which might settle the continuum hypothesis one way or the other. Indeed Gödel (1947)—despite his proof of the formal *consistency* of the continuum hypothesis—has conjectured that when such a principle is found, the continuum hypothesis will then be seen to be *false*. In this volume we adopt a realist viewpoint.

This book is divided into three parts. Part I contains the basics of axiomatic set theory. In Part II we give Gödel's proofs of the consistency of the axiom of choice and the continuum hypothesis. (Gödel proved these in two somewhat different ways (Gödel 1938, Gödel 1939) and (Gödel 1940), and both proofs are of considerable interest.) Part III is devoted to Paul Cohen's proofs of the consistency of the negation of the continuum hypothesis (even with the addition of the axiom of choice) and the consistency of the negation of the axiom of choice, as well as some related results.

Introduction to sets and classes

Before embarking on a rigorous development of class-set theory—which we begin in the next chapter—some brief historical background might be helpful.

One purpose of the subject known as *axiomatic set theory* is to develop mathematics out of the notions of logic (the logical connectives and quantifiers) together with the notion of an element being a *member* of a collection of elements. We presume the reader to be familiar with the basic symbols and concepts of first-order logic with identity: $\neg$ (not), $\wedge$ (and), $\vee$ (or, in the sense of at least one), $\supset$ (implies), $\equiv$ (if and only if), $\forall x$ (for every element x), $\exists x$ (there exists at least one x such that), and $=$ (equality or identity). We also use the symbol "$\in$" for "is a member of"—thus "$x \in A$" is read "x is a member of the collection A." For the moment, we will use the words *set*, *class*, *collection* synonymously; later, however, we will have to make certain technical distinctions between them.

§7   Frege set theory

One of the principal pioneers of formal set theory was Gottlob Frege. His system had, in addition to the axioms of logic, just one axiom of set theory—namely that given any property P, there exists a (unique) set A consisting of those and only those things that have the property P. Such a set is written "$\{x \mid P(x)\}$" and is read: "the set of all x's having the property P," or "the set of all x such that P of x." This principle of Frege is sometimes referred to as the *abstraction principle*, or the *unlimited* abstraction principle. It has the marvelous advantage of allowing us to obtain just about all the sets necessary for mathematics. For example, we can take some property $P(x)$ which doesn't hold for any x at all—such as $\neg(x = x)$—and form the set of all x having this property. We thus have the *empty* set whose usual name is "$\emptyset$." Thus $\emptyset = \{x \mid \neg(x = x)\}$.

Next, given any two sets a and b, we can form the set of all x such that $x = a$ or $x = b$. This set has a and b as its only elements and is denoted $\{a, b\}$. Thus $\{a, b\} = \{x \mid x = a \vee x = b\}$ and its existence is guaranteed by Frege's abstraction principle. The set $\{a, b\}$ is sometimes called the "unordered pair" of a and b.

We have already defined the *power set* $\mathcal{P}(a)$ of any set a as the set of all subsets of a. Then $\mathcal{P}(a) = \{x \mid x \subseteq a\}$ ($x \subseteq a$ means that x is a subset of a) and its existence is again guaranteed by Frege's abstraction principle.

For any set a, by $\cup a$ (read "the union of all the elements of a," or more briefly "union a") is meant the set of all elements of all elements of a. Thus $\cup a = \{x \mid (\exists y)(y \in a \wedge x \in y)\}$, and so its existence is guaranteed by Frege's principle.

As we shall see, the natural numbers can be defined as certain special sets, and the existence of the set ω of all natural numbers is a consequence of Frege's principle. But now we must pause for an important consideration.

§8   Russell's paradox

Despite the marvelous advantage of Frege's system of freely giving us all the sets we need, it has one serious drawback: It is logically inconsistent!

This startling observation was made by Bertrand Russell (and independently by Zermelo) who derived the following contradiction from Frege's unlimited abstraction principle. Given any set x, either x is a member of itself or it isn't. Call a set *ordinary* if it is not a member of itself and *extraordinary* if it is a member of itself. For example, the set of all chairs is not itself a chair, so it is an ordinary set. On the other hand, the set of all sets (which exists under Frege's principle) is itself a set, hence is one of its own members, hence is extraordinary. Whether extraordinary sets really exist or not, ordinary sets certainly do, and so we let O be the set of all ordinary sets. By Frege's principle, the set O exists since $O = \{x \mid \neg(x \in x)\}$. Now, is the set O ordinary or not? Either way leads to a contradiction. Suppose on the one hand that O is ordinary. Then, since *every* ordinary set belongs to O, then O belongs to O, making O extraordinary. This is a contradiction. On the other hand, suppose O is extraordinary. This means that O is a member of itself, but only ordinary sets are members of O, so we again have a contradiction.

More briefly, the argument is this. By definition of the set O, for every set x we have $x \in O \equiv \neg(x \in x)$. Since this is true for *every* x, then we can take O for x, and we thus have: $O \in O \equiv \neg(O \in O)$, which is a clear contradiction.

Thus, by Russell's famous argument, there cannot be such a thing as the set of all ordinary sets, yet the existence of such a set is implied by Frege's principle, and so Frege's principle leads to an inconsistency.

Remark Bertrand Russell later gave a popular version of his paradox in terms of a male *barber* of a certain town who shaved all and only those male inhabitants who didn't shave themselves. Did the barber shave himself or didn't he? If he didn't shave himself, then he failed to shave some man (namely himself) who didn't shave himself, thus violating the given conditions. On the other hand, if he shaved himself, then he again violated the conditions by shaving someone (namely himself) who shaves himself. The solution to this paradox is that there cannot exist any such barber—just as there cannot be any set that contains all and only those sets that are not members of themselves. More generally (and this takes care of both sets and barbers) given any relation $R(x, y)$ there cannot exist any element x that bears the relation R to all and only those y which do not bear the relation R to y—symbolically $\neg(\exists x)(\forall y)[R(x, y) \equiv \neg R(y, y)]$. (Indeed, this formula is a theorem of first-order logic.)

§9 Zermelo set theory

Frege was extremely upset by Russell's paradox, but more than need be, since his system, despite its inconsistency, has been subsequently modified to provide a most useful system indeed. Many of the main principles of modern day logic and set theory derive ultimately from the work of Frege.

It was Zermelo who replaced Frege's unsound abstraction principle by a less liberal, but presumably consistent, principle known as the *limited* abstraction principle, or *separation* principle—also known as *Aussonderungs*—which is this: Given any property P *and given any set* a there exists the set of all elements *of the set* a that have property P. Thus we cannot speak of the set of *all* x's having property P (as Frege did), but we can speak of the set of all x's *in* a that have property P. Thus we cannot (in general) talk of $\{x \mid P(x)\}$, but we can talk of $\{x \mid x \in a \wedge P(x)\}$. Now, this principle has never been known to lead to any contradiction and is indeed a principle in common use by the everyday mathematician (who speaks, e.g., of the set of all *numbers* having a given property P, or the set of all *points on a plane* having a given property P). Russell's paradox then disappears. We can no longer form the set of *all* ordinary sets, but given a set a in advance, we can form the set b of all ordinary elements *of the set* a. This leads to no paradox, but merely to the conclusion that b, though a subset of a, cannot be a member of a (for if it were, it couldn't be either ordinary or extraordinary without contradiction). It then follows that for any set a there is at least one subset of a that is not a member of a. (This result also follows from Cantor's theorem, since this theorem is to the effect that there are more subsets of a than elements of a.)

As a price for having given up Frege's *unlimited* abstraction principle, Zermelo had to take the existence of the sets $\emptyset$, $\{a, b\}$, $\cup a$, $\mathcal{P}(a)$ as separate axioms. He also took an axiom of *infinity*, which provided for the existence of the set ω of natural numbers. This constitutes *Zermelo* set theory. Later Fraenkel, and independently Skolem, added a powerful axiom known as the *axiom of substitution*, or the *axiom of replacement*, which roughly says that given any set x, one can form a new set by simply replacing each element of x by some element. (A more precise formulation of this axiom will be given later in this book.) The resulting system is known as *Zermelo-Fraenkel* set theory—abbreviated *ZF*.

One might also wish to state axioms providing for the existence of various *properties* of sets. Zermelo did *not* do this, and to that extent his system was not a completely *formal* axiom system in the modern sense of the term. It was Thoralf Skolem who proposed to identify *properties* with *first-order properties*, by which is meant conditions defined by *first-order formulas*—i.e., well formed expressions built from the set-membership symbol "$\in$," the variables $x, y, \ldots$ ranging over all sets, the connectives of propositional logic, and the quantifiers for the set variables $x, y, \ldots$. In Skolem's formulation, it was necessary to express Zermelo's separation principle as an *infinite* number of axioms, one for every first-order formula. (The system *ZF* when formulated along these lines of Skolem is sometimes referred to as *ZFS*.)

Now, Zermelo protested vigorously against this interpretation by Skolem. For Zermelo, properties were to be thought of as *all* meaningful conditions, not just those conditions given by first-order formulas (or higher order formulas, for that matter!). To which Skolem replied that Zermelo's notion of "property" was too vague to be satisfactory.

There are surely many realists who feel that Zermelo's system, despite its non-formal character, comes far closer to the true Cantorian set theory. Of course there is no harm in laying down formal axioms which force at least the first-order properties into the picture, but these are certainly not *all* the properties that there are. A still stronger system of set theory called *Morse-Kelley* set theory forces more properties into the picture—properties defined by quantifying over properties as well as sets, but this still does not provide for *all* the properties of sets that there are. In fact (by incompleteness results of Gödel) there is *no* formal axiomatization of properties that captures the entire picture, and this (we believe) is a key factor in the difficulty of deciding whether the continuum hypothesis is true or false.

§10 Sets and classes

Conceptually, Zermelo-Fraenkel set theory is a simple one, but technically it is in many ways quite awkward and inelegant. A far more attractive system was developed by Von Neumann, later revised by Robinson, Bernays, and Gödel and is now known as *NBG* (sometimes *VNB*). This is the main system that we study in this book. The basic idea here is that certain collections of things are called *classes* and certain collections are called *sets*. The term "class" is the more comprehensive one, since every set is also a class, but not every class is a set. Which classes are sets? Rather than attempt an absolute answer to this (which some authors have done with dubious success), we regard it as philosophically more honest to take these notions as only *relative* to any given *model* of the axioms of class-set theory. That is, a collection V is called a *model* of class-set theory if it satisfies the axioms of *NBG*, which will be given in the next several chapters. The elements of V are called the *sets* of the model and the subcollections of V are called the *classes* of the model. When the model V is fixed for the discussion, then the sets of the model are more briefly called "sets" and the classes of the model are simply called "classes." This is the procedure that we will adopt. And now we turn to a more formal development.

CHAPTER 2

SOME BASICS OF CLASS-SET THEORY

We presume familiarity with the notions of a *class* or *collection* of objects and the notion of an object x being a *member* or an *element* of a class A. We write $x \in A$ to mean that x is an element of A, or a member of A, or that x *belongs* to A (all these are synonymous). For example, the number 6 belongs to the class of all even numbers. We write $x \notin A$ to mean $\neg(x \in A)$ (it is not the case that x is in A). For any classes A and B we say that A is a *subclass* of B—in symbols, $A \subseteq B$—if every element of A is also an element of B. Thus:

Definition 0.1. $A \subseteq B =_{\text{df}} (\forall x)(x \in A \supset x \in B)$.

Warning! To say that A is a subclass of B is very different from saying that A is a *member* of B—a mistake commonly made by beginners! For example, the class E of all even natural numbers is certainly a subclass of the class of all natural numbers, but E is hardly a member of this larger class, since E is itself not a natural number. Thus in general, $A \subseteq B$ is not to be confused with $A \in B$.

§1 Extensionality and separation

Our first basic assumption about classes is that if A and B are classes which contain exactly the same members, then A and B are identical. Formally:

P_1 **[Axiom of extensionality]**

$$(\forall x)(x \in A \equiv x \in B) \supset A = B.$$

We can alternatively write this as: if $A \subseteq B$ and $B \subseteq A$ then $A = B$.

Separation—We now consider a class V—our so-called "universal class" which will be fixed for this volume. The elements of V will be called *sets*—each set being itself a class of elements of V—and V itself will be called the *class of all sets*. We do not specify what this class is; it can be any class which satisfies axioms A_1–A_8 (the axioms of *NBG*) which will be given. These axioms will provide V with a sufficiently rich structure so that virtually all of mathematics can be done within the universe V.

For the rest of this volume, "class" will mean *subclass* of V. (Any collection of objects which is not a subclass of V, though it may exist, will have no relevance to what we will be doing.)

We henceforth use capital letters, A, B, C, D, E, ... as standing for classes (sets included) and small letters, x, y, z, a, b, c, ... as standing for sets (elements of V). By a *first-order* property of sets we mean one defined by a formula in which we quantify only over *sets*. (Thus we allow $\forall x$, $\exists x$, for x a *set* variable, but we do not allow $\forall A$, $\exists A$, where A is a *class* variable.) We let $\varphi(A_1, \ldots, A_n, x)$ be a formula whose class variables are $A_1, \ldots, A_n$ (of course none of them are bound) and whose only free set variable is x. Then the axioms of *separation* or *class formation* are all sentences of scheme P$_2$ below (there are infinitely many axioms of separation—one for each formula $\varphi(A_1, \ldots, A_n, x)$).

P$_2$ [Separation]

$$(\forall A_1)\ldots(\forall A_n)(\exists B)(\forall x)[x \in B \equiv \varphi(A_1, \ldots, A_n, x)].$$

Intuitively, each axiom of P$_2$ says that given any subclasses $A_1, \ldots, A_n$ of V there exists the class B of all elements x of V that satisfies the condition $\varphi(A_1, \ldots, A_n, x)$. We remark that P$_1$ is to hold even when $n = 0$, i.e., even when there are no class variables in the formula.

We also remark that it is possible to replace the infinitely many axioms of P$_2$ by a finite number and derive the rest as consequences. We will do this in Chapter 13. An important consequence of the separation principle is the following.

Theorem 1.1. *Not every class is a set (i.e., not every subclass of V is a member of V).*

Proof This proof is a variant of Russell's paradox, but escapes being a paradox.

Call a set x *ordinary* if $x \notin x$. By P$_2$ (taking $\varphi(x)$ to be the formula $\neg(x \in x)$) there exists a class—call it "O"—whose elements are all and only those sets x that are ordinary. (The class O also happens to be unique, by P$_1$—the principle of extensionality.) Thus O is the class of all ordinary sets. Then for any *set* x we have

$$x \in O \equiv x \notin x. \tag{2.1}$$

If now O were a set, we could take O for x and have:

$$O \in O \equiv O \notin O.$$

However $O \in O \equiv O \notin O$ is a contradiction, hence O cannot be a set. And so the class of all ordinary sets, though a subclass of V, is not a set (not a member of V). ∎

Remark We see that Theorem 1.1 comes about as close to Russell's paradox as possible without itself being a paradox. The whole point is that the class O of all ordinary sets exists as a class, but escapes Russell's paradox by not being a set.

Actually, Theorem 1.1 is but a special case of the following theorem (whose proof is not much more complex than that of Theorem 1.1).

Theorem 1.2. *For any class A there is a subclass B of A such that B is not an element of A.*

Proof Given a class A, we now take for B the class of all *ordinary* elements of A, i.e., the class of all sets x such that $x \in A \wedge x \notin x$. This class B exists (by P$_2$) and is unique (by P$_1$). Then for any set x, we have the following equivalence:

$$x \in B \equiv (x \in A \wedge x \notin x). \tag{2.2}$$

Now, for any x which happens to be in A, $x \in A$ is true, and so for every $x \in A$:

$$x \in B \equiv x \notin x. \tag{2.3}$$

Remember, we are asserting (2.3) *only* for those x's which happen to be in A. And so if B were in A, then B would be one of those x's, and we would hence have $B \in B \equiv B \notin B$, which is impossible. And so B is not a member of A (though it is a subclass of A). ∎

§2 Transitivity and supercompleteness

Now we come to two crucial properties we shall later postulate about our universe V. First for two definitions.

Definition 2.1. A class A is called *transitive* (sometimes *complete*) if every element of A is itself a class of elements of A—in other words, every element of A is a subclass of A.

Another way of saying that A is transitive is that A contains with every element y all elements x of y as well—in other words A is transitive if and only if the following condition holds:

$$(\forall x)(\forall y)[(x \in y \land y \in A) \supset x \in A].$$

Definition 2.2. We will say that a class A is *swelled* if A contains with every element y all *subclasses* x of y as well—in other words, every subclass of every element of A is an element of A. Thus A is swelled iff [1] the following condition holds:

$$(\forall x)(\forall y)[(x \subseteq y \land y \in A) \supset x \in A].$$

Compare this with *transitivity*—"$x \subseteq y$" takes the place of "$x \in y$."

A class which is both transitive and swelled is called *supercomplete*. Thus if A is supercomplete, then for any element x of A, all *elements* of x and all *subclasses* of x are also in A.

The Axioms A_1–A_6

In this chapter we shall give six axioms, A_1–A_6, about our universe V which will ensure that V is what we term a *basic* universe. In the next chapter we will add A_7—the axiom of infinity—which will determine V to be a *Zermelo* universe (since all of Zermelo's theorems about sets will hold if "set" be understood as "element of V") and in Chapter 6 we add axiom A_8 (Fraenkel's axiom of *substitution*) which will determine V to be what we call a *Zermelo-Fraenkel* universe.

Rather than give all these axioms at once, we will introduce them one at a time and see what consequences can be drawn from those axioms already at hand. We begin with:

A_1 Every set is a class (i.e., every element of V is a subclass of V). In other words, V is *transitive*.

---

[1] We write "iff" for "if and only if"

A_2 Every subclass of a set is a set (i.e., every subclass of any element of V is an element of V). In other words, V is *swelled*.

And so by A_1 and A_2, V is supercomplete. Axiom A_2, together with Theorem 1.1 has the following important consequence.

Theorem 2.3. *V is not a set ($V \notin V$).*

Proof If V were a set, then by A_2, every subclass of V would be a set, contrary to Theorem 1.1. Thus the class V of all sets is not itself a set. ∎

Remark This theorem bears an amusing similarity to a saying of the ancient Chinese philosopher Lao Tse, who, when speaking of the Tao, said: "The Tao is that through which all things have come into being, hence the Tao is not a thing."

§3   Axiom of the empty set

A class A is called *empty* if it has no elements at all—i.e., if $(\forall x)(x \notin A)$—or equivalently, $\neg(\exists x)(x \in A)$. By the principle P_1 of extensionality there cannot exist more than one empty class, for if A_1, A_2 are both empty, then they contain exactly the same elements (namely, no elements at all), hence $A_1 = A_2$. Also, there exists at least one empty class by P_2, since we can take any property $P(x)$ that holds for no x at all (such as $\neg(x = x)$) and have a class consisting of all and only those sets x having this property. Such a class must be empty. Thus there is one and only one empty class—we denote it by "$\emptyset$."

Note 1. $\emptyset$ is a subclass of *every* class A.

What Note 1 says is that for any class A, the following holds: $(\forall x)(x \in \emptyset \supset x \in A)$. The reason this is true is that for any x it is *false* that $x \in \emptyset$, and since a false proposition implies any proposition, then $x \in \emptyset$ implies $x \in A$ (regardless of A), thus for all x, $x \in \emptyset \supset x \in A$, hence $\emptyset \subseteq A$.

Another way of looking at the situation is this. If you don't believe that every element of $\emptyset$ is also an element of A, just try to find an element of $\emptyset$ that *isn't* an element of A![2]

Indeed, given *any* property P it is true that *every* element of $\emptyset$ has the property P (because there isn't a single element of $\emptyset$ that doesn't have property P, since there are no elements of $\emptyset$ at all). One sometimes says it is *vacuously* true that all elements of $\emptyset$ have the property P. In particular, it is vacuously true that if x is any element of $\emptyset$, then all elements of x and all subclasses of x are also elements of $\emptyset$, and so we have:

Note 2. $\emptyset$ is (vacuously) supercomplete.

Consider now our universe V. For all we know so far, V might be empty, because if V were equal to $\emptyset$, V would satisfy axioms A_1 and A_2 (because $\emptyset$ does—$\emptyset$ is supercomplete). Our next axiom will change that.

A_3 The class $\emptyset$ is a set (i.e., $\emptyset \in V$).

---

[2] This cute way of looking at it is due to Halmos (1960).

And so V contains at least one element—namely $\emptyset$. Thus V is not empty.

Remark Suppose, instead of A_3, we had postulated the ostensibly weaker condition that V contains at least one element (not necessarily $\emptyset$). Could we then have derived A_3 as a theorem? Yes, we could, for suppose V contains an element x. Then x is also a class (by A_1), hence $\emptyset \subseteq x$ (by Note 1), and since every subclass of x is in V (by A_2), then $\emptyset \in V$, so $\emptyset$ is a set. And so in the presence of A_1 and A_2, the statements "$\emptyset \in V$" and "V is non-empty," are equivalent, and we could have taken either one for our axiom A_3.

§4 The pairing axiom

As far as axioms A_1, A_2 and A_3 are concerned, $\emptyset$ might be the *only* element of V (see Note 2 below), but we will soon fix that!

For any set a (element of V) we can form the class of all sets x such that $x = a$ (there is one such class by P_2 and only one by P_1); this class has a as its only element and is denoted by "$\{a\}$" (and read "singleton a" or "bracket a"). In particular, $\{\emptyset\}$ is the class whose only element is the empty set.

Note 1. The classes $\emptyset$ and $\{\emptyset\}$ are different classes—i.e., $\emptyset \neq \{\emptyset\}$, because $\{\emptyset\}$ is non-empty (it contains $\emptyset$ as an element), whereas $\emptyset$ has no elements at all. Hence by extensionality, $\emptyset \neq \{\emptyset\}$.

Note 2. The class $\{\emptyset\}$ is supercomplete.

Reasons Take any element x of $\{\emptyset\}$. Then x must be $\emptyset$. Hence all its elements (of which there are none) are vacuously elements of $\{\emptyset\}$, and all its subclasses (of which $\emptyset$ is the only one) are elements of $\{\emptyset\}$, and so $\{\emptyset\}$ is both transitive and swelled.

Note 2 explains our earlier remarks that as far as axioms A_1, A_2, A_3 are concerned, V could be $\{\emptyset\}$, since $\{\emptyset\}$ is supercomplete and contains $\emptyset$ as an element.

Another noteworthy fact about singletons is this.

Note 3. For any sets a and b, $a = b$ if and only if $\{a\} = \{b\}$.

Reasons If $a = b$, then $\{a\}$ and $\{b\}$ contain just the same elements (namely the element a), hence by extensionality, $\{a\} = \{b\}$.

Conversely, suppose $\{a\} = \{b\}$. Obviously $a \in \{a\}$, hence $a \in \{b\}$ (since $\{b\}$ is the same class as $\{a\}$), but b is the *only* element of $\{b\}$, hence $a = b$.

Pairing—For any sets a and b (whether the same or different) by $\{a, b\}$ we mean the class whose only elements are a and b—or equivalently the class of all x such that $x = a$ or $x = b$. (Such a class exists by P_2 and is unique by P_1.) Note that if a and b happen to be the same, then $\{a, b\} = \{a\}$—stated otherwise, $\{a, a\} = \{a\}$.

And so, for any sets a and b, $\{a, b\}$ exists as a *class* (subclass of V), but not necessarily as a *set* (element of V). Our next axiom is:

A_4 **[Pairing axiom]** For any sets a and b, the class $\{a, b\}$ is a set.

Note 4. Let A_4' be the following proposition: for any sets a and b there is a set c such that $a \in c$ and $b \in c$. Obviously A_4 implies A_4' (take c to be $\{a,b\}$). If we had A_4' in place of A_4, could we have derived A_4 as a theorem? (Yes, we could, but how? Of course we may use any of A_1, A_2, A_3. We will give the solution a bit later.)

The following is an obvious consequence of A_4:

Corollary 4.1. *For any set x, the class $\{x\}$ is a set.*

Discussion The corollary above has a drastic consequence: it not only implies V has at least two elements $\emptyset$ and $\{\emptyset\}$, but that V must have infinitely many elements!

Since $\emptyset$ is a set, so is $\{\emptyset\}$, hence so is $\{\{\emptyset\}\}$, hence so is $\{\{\{\emptyset\}\}\}$, and so forth. That is, we have an infinite sequence a_0, a_1, a_2, $\ldots$, a_n, $\ldots$, where $a_0 = \emptyset$ and for each n, $a_{n+1} = \{a_n\}$. (Thus a_n is $\emptyset$ enclosed in n brackets.) Furthermore, for $n \neq m$, a_n must be different from a_m, as can be shown by a rigorous mathematical induction argument, which we won't formally come to until the next chapter. But meanwhile, we can illustrate the proof by considering a special case—say $n = 3$ and $m = 5$. Well, suppose $\{\{\{\emptyset\}\}\} = \{\{\{\{\{\emptyset\}\}\}\}\}$. Then by Note 3, we would have $\{\{\emptyset\}\} = \{\{\{\{\emptyset\}\}\}\}$. Erasing another pair of brackets, we get $\{\emptyset\} = \{\{\{\emptyset\}\}\}$. Erasing another pair of brackets, we get $\emptyset = \{\{\emptyset\}\}$, which is impossible, since $\emptyset$ is empty and $\{\{\emptyset\}\}$ is not (it contains the element $\{\emptyset\}$). More generally, if $a_m = a_{m+n}$, where m is positive, then m applications of Note 3 yield $\emptyset = a_n$, hence a_n is empty, so $n = 0$.

And so we see that Axiom A_4 forces V to have infinitely many elements, but it does *not* force any element of V to be infinite. (Not until the next chapter will we have any axioms that force V to have an infinite element.)

At this point let us give the solution of Note 4 (for any reader who hasn't solved it). Well, suppose that instead of A_4 we are given that for any sets a and b there exists a set c to which they both belong. Then the class $\{a,b\}$ must be a subclass of such a set c, hence must be a set by A_2.

Ordered Pairs—We refer to $\{a,b\}$ as an *unordered* pair. Clearly $\{a,b\} = \{b,a\}$. We now wish to associate with each set a and each set b a set $\langle a,b \rangle$ called an *ordered* pair such that in general $\langle a,b \rangle$ will be different from $\langle b,a \rangle$, and such that for any sets a, b, c and d, we will have $\langle a,b \rangle = \langle c,d \rangle$ if and only if $a = c$ and $b = d$. There are several ways this can be done; we shall adopt the following.

Definition 4.2. By $\langle a,b \rangle$ we shall mean the set $\{\{a\}, \{a,b\}\}$.

Thus $\langle a,b \rangle$ has, in general, two members—one is $\{a\}$ and the other is $\{a,b\}$ (they are the same only in the degenerate case that $a = b$). We now wish to prove that $\langle a,b \rangle = \langle c,d \rangle$ iff $a = c$ and $b = d$. We first prove the following.

Lemma 4.3. *If $\{a,b\} = \{a,d\}$ then $b = d$.*

Proof Suppose $\{a,b\} = \{a,d\}$. Then, since $b \in \{a,b\}$, we have $b \in \{a,d\}$, hence $b = a$ or $b = d$. Suppose $b = a$. Then $\{a,b\} = \{a,a\} = \{a\}$, and since $d \in \{a,d\}$, then $d \in \{a,b\}$, hence $d \in \{a\}$, hence $d = a$, and we have $b = a$ and $d = a$, so $b = d$. Thus if $b = a$, then $b = d$. But if $b \neq a$, then again $b = d$ (since $b \in \{a,d\}$, hence $b = a$ or $b = d$). And so $b = d$. ∎

Theorem 4.4. *If $\langle a,b \rangle = \langle c,d \rangle$ then $a = c$ and $b = d$.*

Proof Assume $\langle a,b \rangle = \langle c,d \rangle$. We first consider the degenerate case that $a = b$. Then $\langle a,b \rangle = \langle a,a \rangle = \{\{a\},\{a,a\}\} = \{\{a\},\{a\}\} = \{\{a\}\}$. Then $\{\{a\}\} = \langle c,d \rangle$, which means that $\{\{a\}\} = \{\{c\},\{c,d\}\}$, hence $\{c,d\} \in \{\{a\}\}$, hence $\{c,d\} = \{a\}$, hence $c = a$ and $d = a$, and so $a = b = c = d$, hence of course $a = c$ and $b = d$.

Now for the more interesting case that $a \neq b$. Since $\{c\} \in \langle c,d \rangle$ then $\{c\} \in \langle a,b \rangle$, hence $\{c\} = \{a\}$ or $\{c\} = \{a,b\}$. However $\{c\} = \{a,b\}$ is not possible since $\{a,b\}$ contains two elements and $\{c\}$ contains only one, so $\{c\} = \{a\}$, and therefore $c = a$. Also $\{a,b\} \in \langle c,d \rangle$, hence $\{a,b\} = \{c\}$ or $\{a,b\} = \{c,d\}$, but $\{a,b\} \neq \{c\}$ (as we have seen), so $\{a,b\} = \{c,d\}$. But $a = c$, and so $\{a,b\} = \{a,d\}$, and therefore by the lemma above, $b = d$. Thus $a = c$ and $b = d$. ∎

Remark Another equally good schema is to define $\langle x,y \rangle$ to be $\{\{\emptyset, x\},\{\{\emptyset\}, y\}\}$—that is, Theorem 4.4 would also hold under that definition. Still another scheme (due to Norbert Wiener in 1914, and historically the first) is to define $\langle x,y \rangle$ to be $\{\{\emptyset, \{x\}\},\{\{y\}\}\}$.

n-**tuples** Having defined ordered pairs, we now define *n*-tuples ($n \geq 2$) by the following inductive scheme: We take $\langle x_1, x_2, x_3 \rangle$ to be $\langle \langle x_1, x_2 \rangle, x_3 \rangle$; we take $\langle x_1, x_2, x_3, x_4 \rangle$ to be $\langle \langle x_1, x_2, x_3 \rangle, x_4 \rangle$; etc. Thus for each $n \geq 2$, we take $\langle x_1, \ldots, x_n, x_{n+1} \rangle$ to be an ordered pair, namely $\langle \langle x_1, \ldots, x_n \rangle, x_{n+1} \rangle$. Also, for technical reasons, so that $\langle \ \rangle$ is always defined, we take $\langle x \rangle$ to be x.

EXERCISES

Exercise 4.1. Show that Theorem 4.4 is provable if ordered pairs are defined by:

(a) $\langle x,y \rangle =_{df} \{\{\emptyset, x\},\{\{\emptyset\}, y\}\}$;

(b) $\langle x,y \rangle =_{df} \{\{\emptyset, \{x\}\},\{\{y\}\}\}$.

§5 The union axiom

Given a bunch of sets, we can put all the elements of all of them together and form one big set called the *union* of the bunch of sets (or the result of *uniting* the sets). More formally, for any class A, by $\cup A$ (read "the union of all the elements of A") is meant the class of all elements of all elements of A. Thus x belongs to $\cup A$ if and only if x belongs to at least one element of A—i.e., if and only if $(\exists y)(x \in y \land y \in A)$. And so $\cup A$ exists (as a class) by P_2 and is unique by P_1. Our next axiom is:

A_5 **[Union axiom]** If x is a set, so is $\cup x$.

Note 1. We could replace A_5 by the apparently weaker statement A'_5: for any set x there is a set y that contains all elements of all elements of x.

Intersections For any class A, by $\cap A$ (read "the intersection of all the elements of A," or more briefly "intersection A") is meant the class of all elements that belong to *all* elements of A. Thus $x \in \cap A$ iff $(\forall y)(y \in A \supset x \in y)$. Again, $\cap A$ exists and is unique by P_2 and P_1.

Theorem 5.1.

(1) For any non-empty class A, $\cap A$ is a set. (In particular, if a is a non-empty set, then $\cap a$ is a set.)

(2) But $\cap \emptyset = V$!

Proof
(1) Suppose A is non-empty. Then at least one set x belongs to A. Then $\cap A \subseteq x$ (because every element of $\cap A$ belongs to *all* elements of A, hence to x). Thus $\cap A$ is a subclass of a set x, hence is a set by A_2.

(2) It is vacuously true that every element of V belongs to *every* element of $\emptyset$ (because there are no elements of $\emptyset$), and so every element of V belongs to $\cap \emptyset$, so $V \subseteq \cap \emptyset$. And of course $\cap \emptyset \subseteq V$, hence $\cap \emptyset = V$. ∎

Boolean operations For any classes A and B:

(1) $A \cup B$ means the class of all sets x such that $x \in A$ or $x \in B$.

(2) $A \cap B$ means the class of all sets x such that $x \in A$ and $x \in B$.

(3) $A - B$ means the class of all sets x such that $x \in A$ and $x \notin B$.

EXERCISES

Exercise 5.1. Why is Note 1 true?

Exercise 5.2. Prove that for any classes A and B:

(a) If $A \subseteq B$ then $\cup A \subseteq \cup B$.

(b) If $A \subseteq B$ then $\cap B \subseteq \cap A$.

Exercise 5.3. Which of the following statements are true?

(a) If A is transitive then $\cup A \subseteq A$.

(b) If $\cup A \subseteq A$ then A is transitive.

(c) If A is transitive then $\cup A$ is transitive.

(d) If every element of A is transitive then $\cup A$ is transitive.

Exercise 5.4. Show that for any sets x and y:

(a) $x \cup y = \cup \{x, y\}$

(b) $x \cap y = \cap \{x, y\}$.

Exercise 5.5. Let x be any set. Which of the following statements (if any) are necessarily true?

(a) $\cup\{x\} = x$

(b) $\cap\{x\} = x$.

Exercise 5.6. Show that for any classes A, B, C:

(a) $A \cap (B \cup C) = (A \cap B) \cup (A \cap C)$

(b) $A \cup (B \cap C) = (A \cup B) \cap (A \cup C)$

(c) $A - (A - B) = A \cap B$

(d) $B - (A - B) = \emptyset$

(e) $A \subseteq B$ iff $A \cap B = A$

(f) $A \subseteq B$ iff $A \cup B = B$

(g) If $B \subseteq A$, then $A - (A - B) = B$.

§6 The power axiom

For any *set* x by $\mathcal{P}(x)$ we mean the class of all subsets of x. (Since x is a set, any subclass of x is a set and is hence called a sub*set* of x.) Again $\mathcal{P}(x)$ exists as a class and is unique (by P_2 and P_1).

A_6 **[Power set axiom]** For any set x, $\mathcal{P}(x)$ is a set.

Remarks

(1) Again we could have taken the apparently weaker axiom A_6': for any set x there is a set y that contains all subsets of x as elements. Why would this have sufficed?

(2) Trivial as it is to say this, it is helpful to remember that the statements $x \in \mathcal{P}(y)$ and $x \subseteq y$ say exactly the same thing!

EXERCISES

Exercise 6.1. Show that for any sets x and y:

(a) $x \subseteq \mathcal{P}(\cup x)$

(b) $\cup(\mathcal{P}(x)) = x$

(c) If $x \subseteq y$ then $\mathcal{P}(x) \subseteq \mathcal{P}(y)$.

§7 Cartesian products

For any classes A and B, by $A \times B$ is meant the class of all ordered pairs $\langle a,b \rangle$ such that $a \in A$ and $b \in B$. Thus $x \in A \times B$ iff $(\exists a)(\exists b)(x = \langle a,b \rangle \wedge a \in A \wedge b \in B)$. Thus the class $A \times B$ exists and is unique (by separation and extensionality). The class $A \times B$ is called the *Cartesian product* of A and B.

Remark The term "Cartesian product" is named in honor of René Descartes, who invented the Cartesian co-ordinate system of analytic geometry, in which points on the plane are identified with ordered pairs of real numbers. (A point P is identified with the ordered pair $\langle a,b \rangle$, where a is the distance of P from the y-axis and b is the distance from the x-axis.) And so the entire plane is identified with Cartesian product $\mathbb{R} \times \mathbb{R}$, where $\mathbb{R}$ is the set of real numbers.

The word "product" does bear a close relation with arithmetical multiplication, since if A is a finite set with m members and B is a finite set with n members, then $A \times B$ does indeed have m times n members (e.g., the Cartesian product of a 3-element set with a 5-element set is a 15-element set). Analogous results will be seen to hold for infinite sets (once we have defined infinite cardinal numbers and multiplication on them).

Theorem 7.1. *If A and B are sets, so is $A \times B$.*

Proof Suppose A and B are sets. Then $A \cup B$ is a set (because $\{A, B\}$ is a set, but $\cup\{A, B\} = A \cup B$), hence $\mathcal{P}(A \cup B)$ is a set, and therefore $\mathcal{P}(\mathcal{P}(A \cup B))$ is a set. We now show that $A \times B$ is a subclass of the set $\mathcal{P}(\mathcal{P}(A \cup B))$ (and hence is a set by axiom A_2).

Well, suppose $x \in A \times B$. Then $x = \langle a, b \rangle$ for some $a \in A$ and some $b \in B$. Thus $x = \{\{a\}, \{a,b\}\}$. Since $a \in A$ and $b \in B$ then $a \in A \cup B$ and $b \in A \cup B$, hence $\{a\} \subseteq A \cup B$ and $\{a,b\} \subseteq A \cup B$, so $\{a\} \in \mathcal{P}(A \cup B)$ and $\{a,b\} \in \mathcal{P}(A \cup B)$, so $\{\{a\}, \{a,b\}\} \subseteq \mathcal{P}(A \cup B)$, so $\langle a,b \rangle \subseteq \mathcal{P}(A \cup B)$, so $x \subseteq \mathcal{P}(A \cup B)$, hence $x \in \mathcal{P}(\mathcal{P}(A \cup B))$. Thus every element x of $A \times B$ is an element of $\mathcal{P}(\mathcal{P}(A \cup B))$, so $A \times B \subseteq \mathcal{P}(\mathcal{P}(A \cup B))$. This concludes the proof. ∎

EXERCISES

Exercise 7.1. Determine whether the following statement is true or false: for any classes A, B, C, D:

$$(A \cup B) \times (C \cup D) = (A \times C) \cup (A \times D) \cup (B \times C) \cup (B \times D).$$

§8 Relations

By a (binary) *relation* R is meant a class of ordered pairs. Thus a relation R is nothing more nor less than a subclass of the Cartesian product $V \times V$. For any ordered pair $\langle x, y \rangle$, if $\langle x, y \rangle \in R$ we also write $R(x, y)$—and sometimes xRy—and we say that x *bears* the relation R to y, or that x *stands* in the relation R to y.

By the *domain*, $\mathsf{Dom}(R)$, of a relation R is meant the class of all x such that $\langle x, y \rangle \in R$ for at least one y. By the *range* of R, $\mathsf{Ran}(R)$, is meant the class of all y such that

$\langle x, y \rangle \in R$ for at least one x. $\mathsf{Dom}(R)$ and $\mathsf{Ran}(R)$ both exist as classes and are unique (for a given R) by P_2 and P_1. We note that $R \subseteq (\mathsf{Dom}(R) \times \mathsf{Ran}(R))$. We say that a relation R is *on* a class A if $\mathsf{Dom}(R)$ and $\mathsf{Ran}(R)$ are both subclasses of A. (This is equivalent to saying that R is a subclass of the Cartesian product $A \times A$.)

Lemma 8.1. *For any relation R:*

(1) $\mathsf{Dom}(R) \subseteq \cup(\cup R)$

(2) $\mathsf{Ran}(R) \subseteq \cup(\cup R)$.

Proof We will prove (2)—item (1) is, if anything, even simpler, and will be left to the reader.

Suppose $y \in \mathsf{Ran}(R)$. Then there is some x such that $\langle x, y \rangle \in R$. Thus $\{\{x\}, \{x, y\}\} \in R$. But also $\{x, y\} \in \{\{x\}, \{x, y\}\}$, hence $\{x, y\} \in \cup R$. But also $y \in \{x, y\}$, so $y \in \cup(\cup R)$. Thus every y in $\mathsf{Ran}(R)$ is in $\cup(\cup R)$, so $\mathsf{Ran}(R) \subseteq \cup(\cup R)$. ∎

By Lemma 8.1, and the fact that if R is a set, so is $\cup(\cup R)$ (by two applications of A_2) we have the following.

Theorem 8.2. *For any relation R, if R is a set, so are $\mathsf{Dom}(R)$ and $\mathsf{Ran}(R)$.*

EXERCISES

Exercise 8.1. Prove part (1) of Lemma 8.1.

§9 Functions

By a *function* or an *operation* or a *single-valued relation* is meant a relation R such that for any x there is at most one element y such that $\langle x, y \rangle \in R$. Thus if R is a function then for any x *in the domain of R* there is *exactly* one y (naturally in the range of R) such that $\langle x, y \rangle \in R$.

We often use the letters F, G, H, f, g, h for functions. For any function F and any element x in the domain of F, by $F(x)$ is meant that (unique) element y such that $\langle x, y \rangle \in F$. Thus $F(x) = y$ is synonymous with $\langle x, y \rangle \in F$. We sometimes refer to $F(x)$ as the result of *applying F* to x.

We say that a function F is *from* a class A *into* a class B if A is the domain of F and B *includes* (but is not necessarily identical with) the range of F. Thus if F is *into* B then the range of F is a subclass of B. If every element of B is in the range of F, then we say that F is *onto* B. Thus *into* B merely means that $\mathsf{Ran}(F) \subseteq B$, whereas *onto* B means that $\mathsf{Ran}(F) = B$.

Functions are also sometimes called *mappings* or *maps*, or sometimes a function is called a *many-one correspondence*. In the terminology of *mappings*, we say that x *maps* to y under F, or that F maps x to y, if $F(x) = y$. A function or map is called 1-1 (one-one), or is said to be a *1-1 correspondence* if for any x and y in the domain of F, if $x \neq y$ then $F(x) \neq F(Y)$—or equivalently, if $F(x) = F(y)$ then $x = y$. Thus for a 1-1 function, no two *distinct* elements map to the same element.

§10 Some useful facts about transitivity

Transitive classes will play a fundamental role in our study, so let us establish a few basic facts about them.

Let us first consider some sets that are transitive and some sets that are not. We already know that $\emptyset$ is transitive and $\{\emptyset\}$ is transitive. What about the set b consisting of the two elements $\emptyset$ and $\{\emptyset\}$—i.e., $b = \{\emptyset, \{\emptyset\}\}$; is b transitive or not? Well, all elements of $\emptyset$ are vacuously elements of b (since there are no elements of $\emptyset$) and all elements of $\{\emptyset\}$ (of which the only one is $\emptyset$) are elements of b. Thus, for each of the two elements of b, all of *their* elements are in b, so b is transitive.

Now let c be the set $\{\emptyset, \{\emptyset\}, \{\emptyset, \{\emptyset\}\}\}$. It has 3 elements and it is transitive, since all elements of $\emptyset$ are in c, all elements of $\{\emptyset\}$ are in c (since $\emptyset$ is in c) and all elements of $\{\emptyset, \{\emptyset\}\}$ are in c (since $\emptyset$ and $\{\emptyset\}$ are both in c). Thus c is transitive.

Suppose we remove the element $\{\emptyset\}$ from c—i.e., we consider the set $\{\emptyset, \{\emptyset, \{\emptyset\}\}\}$, which we will call d. This set d has just two elements—$\emptyset$ and $\{\emptyset, \{\emptyset\}\}$. Is d transitive? No, because one of its elements—the element $\{\emptyset, \{\emptyset\}\}$—contains an element—namely $\{\emptyset\}$—that is *not* an element of d. Thus d is not transitive. (A non-transitive class might be aptly described as a class with "holes" in it.) Now for some useful theorems on transitivity.

Theorem 10.1. *For any class A, the following two conditions are equivalent.*

(1) $\cup A \subseteq A$

(2) A is transitive.

Proof (1) Suppose $\cup A \subseteq A$. Now, suppose $x \in y$ and $y \in A$; we are to show $x \in A$. Well, since $x \in y$ and $y \in A$, then $x \in \cup A$, but since $\cup A \subseteq A$, then $x \in A$. This proves that A is transitive.

(2) Conversely, suppose A is transitive. Now suppose $x \in \cup A$. Then there is some y such that $x \in y$ and $y \in A$, hence $x \in A$ since A is transitive. Thus any x in $\cup A$ is also in A, so $\cup A \subseteq A$. ∎

Theorem 10.2. *If A is transitive, so is $\cup A$.*

Proof Suppose A is transitive. Then $\cup A \subseteq A$ by Theorem 10.1. Hence $\cup(\cup A) \subseteq \cup A$ by Exercise 5.2 of §5. Hence $\cup A$ is transitive by Theorem 10.1. ∎

The following theorem is particularly important.

Theorem 10.3. *If every element of A is transitive, then $\cup A$ is transitive.*

Proof Suppose every element y of A is transitive. To say that $\cup A$ is transitive is to say that every element x of $\cup A$ is also a subset of $\cup A$. Well, suppose $x \in \cup A$. Then x is an element of some element y of A. But y is transitive (by hypothesis). Hence $x \subseteq y$. And of course $y \subseteq \cup A$ (since $y \in A$), and so $x \subseteq \cup A$. ∎

Theorem 10.4. *For any set a, the following two conditions are equivalent:*

(1) $a \subseteq \mathcal{P}(a)$

(2) a *is transitive.*

Proof (1) Suppose $a \subseteq \mathcal{P}(a)$. Then any element of a is an element of $\mathcal{P}(a)$, hence is a subset of a. Thus a is transitive.

(2) Conversely, suppose a is transitive. Then any element of a is a subset of a, hence an element of $\mathcal{P}(a)$, which means $a \subseteq \mathcal{P}(a)$. ∎

Remark By Theorem 10.4 and Theorem 10.1, we see that for any *set a*, the following *three* conditions are equivalent.

(1) $\cup a \subseteq a$.

(2) $a \subseteq \mathcal{P}(a)$.

(3) a is transitive.

Theorem 10.5. *For any set x, if x is transitive then $\mathcal{P}(x)$ is transitive.*

Proof Suppose x is transitive. Then $x \subseteq \mathcal{P}(x)$, by Theorem 10.4, so $\mathcal{P}(x) \subseteq \mathcal{P}(\mathcal{P}(x))$ (obvious!). Hence $\mathcal{P}(x)$ is transitive by Theorem 10.4 again. ∎

EXERCISES

Exercise 10.1.

(a) Is the set $\{\{\emptyset\}\}$ transitive?

(b) Show that if we remove any element x from V, the resulting class $V - \{x\}$ is *not* transitive.

§11 Basic universes

We shall call a class A a *basic universe* if the axioms A_1–A_6 all hold for A, reading "element of A" for "set," and "subclass of A" for "class." Thus A is a basic universe if and only if the following six conditions hold.

(1) A is transitive.

(2) A is swelled.

(3) $\emptyset \in A$.

(4) For any x and y in A, $\{x, y\} \in A$.

(5) For any x in A, $\cup x \in A$.

(6) For any x in A, $\mathcal{P}(x) \in A$.

In the next chapter we will add another axiom A_7 (an "axiom of infinity") characterizing V as what is called a *Zermelo* universe, and we will see that arithmetic can then be done within V.

CHAPTER 3

THE NATURAL NUMBERS

§1   Preliminaries

We are all intuitively familiar with the natural numbers, 0, 1, 2, 3, ... Peano gave a famous set of postulates for them. He took as undefined the notion of "natural number," the notion of "zero," and the notion of "successor" (the successor n^+ of n is $n + 1$). The Peano postulates are the following:

P_1 0 is a natural number.

P_2 If n is a natural number, so is n^+.

P_3 For any natural number n, $n^+ \neq 0$. [This is paraphrased by saying that 0 has no predecessor.]

P_4 For any n and m, if $n^+ = m^+$ then $n = m$. [Stated otherwise, no two *distinct* natural numbers have the same successor.]

P_5 **Principle of mathematical induction**   For any set A, the following two conditions are jointly sufficient that A contain every natural number:

  (1) $0 \in A$.
  (2) For every natural number n, $n \in A$ implies $n^+ \in A$.

Remarks on P_5   For students who encounter mathematical induction for the first time, and who find it difficult of comprehension, we like to illustrate it by the following example. Imagine you are immortal, and also are now living in the good old-fashioned days when the milkman delivered milk to your door. The milkman is going to come to your door today (the 0th day) to receive instructions. So you leave a note in your empty bottle saying "If ever you leave milk one day, be sure and leave it the next day as well." Now, it is perfectly possible for the milkman *never* to leave milk without violating your instructions. (The only way he can violate your instructions is by leaving milk one day and failing to leave it the next.) So the milkman can go several thousand days without leaving milk, and you cannot justly accuse him of disobeying your orders. Now suppose after several thousand days the milkman, on a pure whim, decides to leave milk one day. Then he must also leave it the next, and the next after that, and so forth. However your note never guarantees that he will leave milk at all. So instead of writing *that* note, you write the following better note: "If ever you leave milk one day, be sure and leave it the next day as well. Also, leave milk today!" *This* note will guarantee delivery on the 0th day (today), and also delivery tomorrow, and also the day after tomorrow, etc.[1]

Concerning the Peano postulates, as Bertrand Russell correctly pointed out, these postulates give the key properties of the natural numbers which mathematicians need to develop number theory, but the postulates do not tell you what the natural numbers actually are. For example, if we interpreted "natural number" to mean any natural number greater than or equal to 100 and "0" to mean 100, and "successor" as usual, all of conditions P_1–P_5 would still hold. Indeed, any denumerable sequence $\{a_0, a_1, a_2, \ldots, a_n, \ldots\}$ without repetitions (where $a_0, a_1, \ldots$ are perfectly arbitrary objects) will serve as a model of the Peano postulates, if we interpret "0" to mean a_0, and the "successor" of the object a_i as the next term a_{i+1} of the sequence. Thus it is indeed true that the Peano postulates make no attempt to tell us what the natural numbers actually are.

What *is* a natural number anyhow? What, for example, is the natural number 3? As Russell also helpfully pointed out, one of the reasons it took so long to discover a "definition" of the natural numbers, is that the names of the natural numbers— e.g. "3"—are used both as adjectives and nouns, and this distinction was not sooner realized. For example, when one says of a given set S that it has 3 elements, one is using "3" as an *adjective* (i.e. to qualifiy the set). But when one says "3+5 = 8," one is using "3" as a noun—one is talking about a certain entity 3, an entity 5, performing the operation of "addition" on them, and obtaining the entity 8. Now, the Frege-Russell idea is to *first* define 3 as an adjective—i.e. to define what it means for a set S to have 3 objects—and *then* to define the noun 3 in terms of the adjective 3.

It is easy to define in terms of pure logic and set theory what it means for a set S to have exactly 3 elements. S has 3 elements if and only if there exists an x, y, z in S such that x is distinct from y, x is distinct from z, y is distinct from z, and for any w in S, $w = x$ or $w = y$ or $w = z$. In symbols,

$$S \text{ has 3 elements} \quad \text{iff} \quad (\exists x)(\exists y)(\exists z)\{x \in S \land y \in S \land z \in S$$
$$\land x \neq y \land x \neq z \land y \neq z$$
$$\land (\forall w)[w \in S \supset (w = x \lor w = y \lor w = z)]\}.$$

We remark that Poincaré (and others) objected to this definition as a sort of "cheat" in that in setting it up, one *counted* the number of variables (x, y, z)! Therefore, reasoned Poincaré, the intuitive arithmetical notion of 3 is really prior to the logical set-theoretic notion of 3 given by the above definition. Now Poincaré is probably right if by "prior" is meant "psychologically prior"—indeed for one who did not already know how to count up to 3 (and therefore did not know that the set of symbols "x", "y", "z" contains exactly 3 variables), such a one could not recognize the correctness of the definition. However Russell never claimed that logical or set theoretic notions are *psychologically* more immediate than arithmetic ones, but only that in a purely *logical* sense, arithmetic is reducible to logic and set theory. More specifically,

---

[1] A cute variant, suggested by the computer scientist Alan Tritter, illustrates what might be called the *recursion* approach. The note consists of just one sentence: "Leave milk today and read this note again tomorrow."

the definition above is formed purely in the language of logic and set theory, yet completely characterizes the property of threeness in the exact sense that a given set S satisfies this logical-set-theoretic definition precisely in case S does have exactly 3 elements.

Having now defined what it means for a set to have 3 elements, how does Russell define the natural number 3 as an entity? Actually Russell worked within a different system—the system of type theory—so the closest definition to Russell's which can be given in our present framework is to define the number 3 as the *class* of all sets S which satisfy the definition above. This, then, would make the natural number 3 a *class* rather than a set, and for many reasons this is undesirable. We therefore seek other definitions of the natural numbers which will make them *sets* (elements of our universe V rather than only subclasses).

Zermelo constructed the natural numbers as follows. He took for 0 the empty set $\emptyset$. For 1 he took the set $\{\emptyset\}$, whose only element is 0. For 2 he took the set $\{1\}$—i.e. $\{\{\emptyset\}\}$ whose only element is 1, and so forth (that is, $n + 1$ is the unit set whose only element is n). Thus the notation for the Zermelo integer n is "0" enclosed in n pairs of braces.

This scheme of Zermelo is quite simple, and would work fine for purposes of the arithmetic of finite numbers. But it does not nicely generalize to the theory of the transfinite ordinal numbers, which we will need for subsequent developments. And therefore we shall not use Zermelo's scheme, but rather the following scheme of von Neumann.

Von Neumann defined the natural numbers in such a way that each natural number turns out to be the set of all lesser numbers. This means that 0 must be the empty set $\emptyset$ (since there are no natural numbers less than 0). Then 1 must be the set of all natural numbers less than 1—this set has 0 as its only member. So $1 = \{0\}$—i.e. $\{\emptyset\}$. So far 0, 1 are for von Neumann the same as for Zermelo. But with 2, the situation changes. For Zermelo, $2 = \{1\}$, but for von Neumann, $2 = \{0, 1\}$—i.e. $\{\emptyset, \{\emptyset\}\}$; $3 = \{0, 1, 2\}$—i.e. $\{\emptyset, \{\emptyset\}, \{\emptyset, \{\emptyset\}\}\}$, etc.

Let us note first that the von Neumann natural number n is a set which contains exactly n elements (unlike the Zermelo n, which contains only one element, unless $n = 0$). Thus, e.g. $7 = \{0, 1, 2, 3, 4, 5, 6\}$, which contains 7 elements. This fact is technically useful (indeed we can define a set S to have 7 elements iff it can be put into a 1-1 correspondence with 7!)

We shall now follow only the von Neumann development, and the natural numbers will be understood accordingly. Let us note with great care their scheme of generation. How, e.g., does 7 relate to 6? Well, $6 = \{0, 1, 2, 3, 4, 5\}$ and $7 = \{0, 1, 2, 3, 4, 5, 6\}$. So 7 consists of the elements of 6 *together with 6 itself thrown in as a new element*! In other words, $7 = 6 \cup \{6\}$. (Recall that to adjoin an element y to a set x, the result is $x \cup \{y\}$, not $x \cup y$!) Likewise $1 = 0 \cup \{0\}$, $2 = 1 \cup \{1\}$, ..., $n + 1 = n \cup \{n\}$.

The operation $x \cup \{x\}$ thus turns out to be a crucial one, hence we introduce the following notation.

Definition 1.1. For any set x, by x^+ we mean $x \cup \{x\}$.

Thus the natural numbers are the sets $0 = \emptyset, 0^+, 0^{++}, \ldots$.

§2 Definition of the natural numbers

Given any natural number n, I can define n for you in a finite number of steps. For example, if you asked me to define 4, I would do so by the following chain of definitions.

$$\begin{aligned} 0 &=_{df} \emptyset \\ 1 &=_{df} 0^+ \\ 2 &=_{df} 1^+ \\ 3 &=_{df} 2^+ \\ 4 &=_{df} 3^+ \end{aligned}$$

So for each n, I can give you a definition of n. But I still have not defined for you what it means for an arbitrary set x to be a natural number!

Of course, I realize that you already intuitively know what I mean by the class of objects $\{0, 0^+, 0^{++}, \ldots\}$. But I have not yet given you a definition of this class *in terms of the notions of logic and set theory*! And it is of crucial importance to do this.

How can this be done? The answer is both simple and profound. The underlying idea is due to Frege, and this idea has been of enormous importance in many branches of mathematics. First for a basic preliminary definition.

Definition 2.1. We call a class A *inductive* if it satisfies the following two conditions:

(1) $0 \in A$;

(2) for all x, if $x \in A$ then $x^+ \in A$.

The class V is obviously inductive (why?). Now, we wish to define "natural number" in such a way that the Peano postulates will hold. We also want the "0" of the Peano postulates to be $\emptyset$, and the *successor* of n to be $n \cup \{n\}$. This leads to the following definition.

Definition 2.2. We call a set x a *natural number* if x belongs to every inductive set.

We shall shortly see that this definition works. We let ω be the class of natural numbers. From axioms A_1–A_6 defining a basic universe, it is impossible to infer that ω is a set. (We shall later see an example of a basic universe in which ω, though of course a subclass, is not a member.) And so now we add the following axiom.

A_7 ω is a set ($\omega \in V$).

A_7 is called the *axiom of infinity*. A basic universe (a class satisfying axioms A_1–A_6) which also satisfies A_7 will be called a *Zermelo universe*. For the rest of this volume, V will be assumed to be a Zermelo universe. A basic universe, though necessarily infinite, doesn't necessarily contain any infinite element, whereas a Zermelo universe contains the infinite element ω.

Remarks A_7 of course implies that V contains an inductive set—namely ω. Consider now the ostensibly weaker axiom A_7': There exists an inductive set. Is A_7' really weaker than A_7? No, it is not. Suppose we take A_7' as an axiom in place of A_7. We can then derive A_7 as follows. By A_7' there is some inductive set b. Then all natural numbers belong to b (they all belong to all inductive sets) so $\omega \subseteq b$. Since the class ω is thus a subclass of b and b is a set, then ω is a set (by A_2). Thus we have A_7.

Some authors take A_7' as an axiom instead of A_7.

Next, something rather amusing. Suppose that instead of A_7 or A_7' we take the following axiom A_7'': Not every set is a natural number. Could we then derive A_7 or A_7' as a consequence? Yes, we could, for if A_7' were false, there would be no inductive sets, hence every set would vacuously belong to *all* inductive sets, hence every set would be a natural number! Thus we could take the curious statement A_7'' as an alternative to A_7 or A_7'.

§3 Derivation of the Peano postulates and other results

Postulates P_1, P_2, P_5 are utterly trivial consequences of our definition. Obviously 0 belongs to every inductive set (by definition of an inductive set), so 0 is a natural number. This proves P_1. As for P_2, obviously if x belongs to *every* inductive set, then each of these sets must also contain x^+, so x^+ also belongs to every inductive set. Thus if x is a natural number, so is x^+. As for P_5, a natural number x by definition belongs to every inductive set, hence any inductive set A must contain all the natural numbers.

Postulate P_3 is also quite trivial. The reason $n^+ \neq 0$ is that n^+ is non-empty (since n is an element), but 0 is empty, so $n^+ \neq 0$.

Postulate P_4 is a bit more tricky to verify. We shall first prove a few other theorems about the natural numbers.

Theorem 3.1. *Every natural number is transitive.*

Proof We prove this by mathematical induction. We show that 0 is transitive, and for any n, if n is transitive so is n^+. This means then that the set of all transitive natural numbers is inductive, hence must contain every natural number, and hence every natural number must be transitive.

(1) Obviously, 0 is vacuously transitive, i.e., since $x \in 0$ is always false, then $(x \in 0) \supset (x \subseteq 0)$ is always true, so every element x of 0 is a subset of 0. (If you don't believe this, just try to find an element of 0 which is not.)

(2) Suppose n is transitive. We must show n^+ is transitive. Let $x \in n^+$. We must show $x \subseteq n^+$. Since $x \in n^+$, then $x \in n$ or $x = n$ (why?). If $x \in n$, then $x \subseteq n$ (because n is transitive). If $x = n$, then certainly $x \subseteq n$. So in either case, $x \subseteq n$. But $n \subseteq n^+$ (why?), so $x \subseteq n^+$.

This concludes the proof. ∎

Theorem 3.2. *Every natural number n is ordinary—i.e. $n \notin n$.*

Proof Again by induction.

(1) Since 0 is empty, then $0 \notin 0$, so 0 is ordinary.

(2) We must show $(n \notin n) \supset (n^+ \notin n^+)$. This is equivalent to showing $(n^+ \in n^+) \supset (n \in n)$. So suppose $n^+ \in n^+$. Thus $n^+ \in n$ or $n^+ = n$. If $n^+ \in n$ then $n^+ \subseteq n$ (since n is transitive), and if $n^+ = n$, then of course $n^+ \subseteq n$, so in either case $n^+ \subseteq n$. But $n \in n^+$, and since $n^+ \subseteq n$, then $n \in n$. This proves $(n^+ \in n^+) \supset (n \in n)$, and hence that $(n \notin n) \supset (n^+ \notin n^+)$, and this completes the induction. ∎

Theorem 3.3. *Not both $n \in m$ and $m \in n$.*

Proof Suppose $n \in m$ and $m \in n$. Since $m \in n$, then $m \subseteq n$ (because n is transitive). Hence we have $n \in m \subseteq n$, and so $n \in n$. This contradicts Theorem 3.2. ∎

Now we are in a position to easily prove the Peano postulate P_4. Suppose $n^+ = m^+$. Since $n \in n^+$ and $m \in m^+$, we then have $n \in m^+$ and $m \in n^+$. Thus $n \in m \vee n = m$, and also $m \in n \vee m = n$. If $n = m$ were false, then from $n \in m \vee n = m$ we would have $n \in m$, and from $m \in n \vee m = n$ we would have $m \in n$. Thus if $n \neq m$ we have both $n \in m$ and $m \in n$, contrary to Theorem 3.3. Thus $n = m$.

Theorem 3.4. *Every element of a natural number is a natural number.*

Proof Vacuously, all elements of 0 are natural numbers. Now suppose n is a natural number such that all its elements are natural numbers. Then all elements of n^+ are natural numbers, since every element of n^+ is either an element of n, or is n itself. By induction it follows that for every natural number n, all elements of n are natural numbers. ∎

EXERCISES

Exercise 3.1. [Principle of complete induction] Suppose that A is a class of natural numbers such that for every natural number n, if A contains all numbers less than n, then A contains n as well. Prove that A then contains all natural numbers. [Hint: Show by induction on n that for every n, A contains all numbers less than n.]

Exercise 3.2. Given a natural number n, which of the following statements are necessarily true?

(a) $n \subseteq \cup n$ (Recall that "$\cup$" is the union symbol.)

(b) $\cup n \subseteq n$

(c) $\cup n = n^+$

(d) $\cup n^+ = n$

(e) $\cup n = n$.

Exercise 3.3. Which of the following statements are true?

(a) $\cup \omega = \omega$

(b) $\cup \omega^+ = \omega$.

§4 A double induction principle and its applications

It should be intuitively obvious that for any natural numbers n and m, either $n \subseteq m$ or $m \subseteq n$ (indeed, if these are distinct, the smaller is a subset of the larger). We shall soon prove this rigorously. Our proof will use a certain principle which we call a "double induction principle." This principle will be subsequently extended to the transfinite, and will play a basic role in our entire treatment of further topics (e.g. the proof of Zorn's lemma and the well ordering theorem, our development of the theory of transfinite ordinals, and the transfinite recursion theorem).

We shall first consider a special case of this "double induction" principle: Suppose R is a relation on the natural numbers satisfying the following two conditions.

D_1 $R(n, 0)$ holds for every natural number n.

D_2 For any natural numbers m and n, if $R(m, n)$ and $R(n, m)$, then $R(m, n^+)$.

Does it necessarily follow that $R(m, n)$ holds for *all* natural numbers m and n? Yes, it does! We invite the reader to try proving this as an exercise before turning to the more general situation (which is really no more difficult to prove!) which now follows.

First for a definition which generalizes the notion of inductive. Instead of working with the particular set ω and the successor function S (where $S(x) = x^+$) we consider an arbitrary class A and an arbitrary function g.

Definition 4.1. We shall say that A is *inductive under g*, or more briefly *g-inductive* if the following two conditions hold:

(1) $\emptyset \in A$,

(2) for every x in A, the set $g(x)$ is in A. (We paraphrase this by saying that A is *closed* under g).

Thus *inductive*, in the sense defined in §2, is inductive under the successor function.

Definition 4.2. We shall say that A is *minimally* inductive under g if A is inductive under g and if no *proper* subclass of A is inductive under g—in other words, any subclass of A that is inductive under g contains all elements of A.

If a set M is minimally inductive under g, then we have the following generalized induction principle: To show that a given property P holds for all elements of M it suffices to show that $\emptyset$ has the property P and that for any element x in M, if x has property P, so does $g(x)$ (for then the set of all elements of M having the property is inductive under g, hence contains all elements of M). This method of proof might aptly be called *proof by g-induction* and generalizes the ordinary principle of mathematical induction. We note that the Peano postulates P_1, P_2 and P_5 are jointly to the effect that ω is minimally inductive under the successor function.

Now for our double induction principle.

Theorem 4.3 (Double induction principle). *Suppose M is minimally inductive under g and that R is a relation satisfying the following two conditions:*

D_1 $R(x, 0)$ *holds for every x in M;*

D_2 *for all x, y in M, if $R(x,y)$ and $R(y,x)$ then $R(x,g(y))$.*

Then $R(x,y)$ holds for all x,y in M.

Proof Assume the hypothesis. Call an element x of M *left normal* if $R(x,y)$ holds for all y in M, and *right normal* if $R(y,x)$ holds for all y in M.

Step 1) We show that each right normal element is also left normal. So suppose x is right normal. We show by g-induction on y that $R(x,y)$ holds for all y in M—i.e., we show that the set of y in M for which $R(x,y)$ holds is inductive under g.

(1) By D_1, $R(x,0)$ holds.

(2) To show $R(x,y) \supset R(x,g(y))$, suppose $R(x,y)$. Also $R(y,x)$ (since x is right normal). Thus $R(x,y)$ and $R(y,x)$, hence $R(x,g(y))$ by D_2. Thus we have $R(x,y) \supset R(x,g(y))$.

By (1) and (2), the set of y such that $R(x,y)$ is inductive under g, and since M is minimally inductive under g, then this set is all of M. Thus $R(x,y)$ holds for all y in M, so x is left normal.

Step 2) Next we show by g-induction that every x in M is right normal (and hence also left normal by Step 1).

(1) Clearly 0 is right normal by D_1.

(2) Suppose x is right normal. Then it is also left normal (by Step 1), hence for every y in M, $R(x,y)$, and since $R(y,x)$, then $R(y,g(x))$ (by D_2), thus $g(x)$ is right normal. Thus if x is right normal, so is $g(x)$, which completes the induction. Thus every element x of M is right normal, hence $R(y,x)$ holds for every y. ∎

Definition 4.4. We shall say that a function g from sets to sets is *progressing* if $x \subseteq g(x)$ for every x in the domain of g.

An important example of a progressing function is the successor function, since obviously $x \subseteq x^+$.

Definition 4.5. Two sets x and y are called *comparable* (with respect to inclusion) if either $x \subseteq y$ or $y \subseteq x$. A class A is called a *nest* if every two elements of A are comparable.

We recall that we are aiming to prove that ω is a nest—i.e., that every two natural numbers are comparable. More generally, we will prove that if M is a set which is minimally inductive under g and if g is progressing, then any two elements of M are comparable. Better yet (and this will be important when we come to the subject of *well ordering*) we will prove that the same hypothesis implies that for any elements x and y of M, either $g(x) \subseteq y$ or $y \subseteq x$ (the former of course implies $x \subseteq y$, since $x \subseteq g(x)$). This will follow from our double induction principle (Theorem 4.3) by virtue of the following lemma (which will have other applications in the next chapter).

Lemma 4.6 (Progressing function lemma). *Let g be a progressing function, and take $R(x,y)$ to be the relation: $g(x) \subseteq y \vee y \subseteq x$. Then:*

(1) for every set y in the domain of g, $R(y,0)$;

(2) for any sets x and y in the domain of g, $(R(x,y) \wedge R(y,x)) \supset R(x,g(y))$.

§4. A DOUBLE INDUCTION PRINCIPLE AND ITS APPLICATIONS

Proof

(1) Obviously $0 \subseteq y$, hence *either $g(y) \subseteq 0$ or $0 \subseteq y$*, so $R(y, 0)$ holds.

(2) Suppose $R(x, y)$ and $R(y, x)$. Thus:

   (a) $g(x) \subseteq y$ or $y \subseteq x$
   (b) $g(y) \subseteq x$ or $x \subseteq y$.

We are to prove $R(x, g(y))$—i.e., that $g(x) \subseteq g(y)$ or $g(y) \subseteq x$.

If $g(x) \subseteq y$ then $g(x) \subseteq g(y)$ (since $y \subseteq g(y)$) and $R(x, g(y))$ then holds. If $g(y) \subseteq x$, then of course $R(x, g(y))$ holds (i.e., $g(x) \subseteq g(y) \vee g(y) \subseteq x$). If neither $g(x) \subseteq y$ nor $g(y) \subseteq x$ then by (a) and (b), both $y \subseteq x$ and $x \subseteq y$, hence $x = y$, hence $g(x) = g(y)$, hence $g(x) \subseteq g(y)$, so again $R(x, g(y))$ holds. This completes the proof. ∎

If now in Theorem 4.3 we take $R(x, y)$ to be the relation $g(x) \subseteq y$ or $y \subseteq x$, the lemma at once yields the following.

Theorem 4.7. *If M is minimally inductive under g, where g is progressing, then M is a nest in which for all x and y in M, either $g(x) \subseteq y$ or $y \subseteq x$.*

Remark The fact that $g(x) \subseteq y$ or $y \subseteq x$ implies that $x \subseteq y$ or $y \subseteq x$, because $x \subseteq g(x)$.

Definition 4.8 (Least and greatest elements). An element x of a class A is said to be the *least* element of A (with respect to the relation of inclusion) if x is a subset of every element of A, and x is said to be the *greatest* element of A if every element of A is a subset of x.

A class A is said to be *well ordered under inclusion* if A is a nest and if every non-empty subclass of A has a least element.

This notion is but a special case of the more general notion of being well ordered under an arbitrary relation. We study this generalization in the next chapter. Our present aim is to show that if M is minimally inductive under a progressing function g, then M is well ordered under inclusion. Several other interesting things will be proved along the way.

We shall use the notation $x \subset y$ to mean that x is a *proper* subset of y ($x \subseteq y$ but not $y \subseteq x$). We write $x \subset y \subset z$ to mean $x \subset y$ and $y \subset z$, and we write $x \subseteq y \subseteq z$ to mean $x \subseteq y$ and $y \subseteq z$.

The following simple lemma will be used in this chapter and the next.

Lemma 4.9. *Suppose that N is closed under g, g is progressing and for any elements x and y of N, either $g(x) \subseteq y$ or $y \subseteq x$. Then for any elements x, y of N the following three conditions hold:*

(1) *(The sandwich principle) If $x \subseteq y \subseteq g(x)$ then either $x = y$ or $y = g(x)$.*

(2) *If $x \subset y$ then $g(x) \subseteq y$.*

(3) *If $x \subseteq y$ then $g(x) \subseteq g(y)$.*

Proof To begin with, since g is progressing, then N must be a nest (since $g(x) \subseteq y$ implies $x \subseteq y$, and so $g(x) \subseteq y \lor y \subset x$ implies $x \subseteq y$ or $y \subseteq x$).

(1) By hypothesis, either $g(x) \subseteq y$ or $y \subseteq x$, hence $y \subset g(x)$ and $x \subset y$ cannot both hold. Therefore $x \subseteq y \subseteq g(x)$ implies $y = g(x)$ or $x = y$.

(2) Suppose $x \subset y$. By hypothesis either $y \subseteq x$ or $g(x) \subseteq y$, but $y \subseteq x$ is false (since $x \subset y$), hence $g(x) \subseteq y$.

(3) Suppose $x \subseteq y$. If $x = y$, then $g(x) = g(y)$, in which case $g(x) \subseteq g(y)$. If $x \neq y$ then $x \subset y$, hence $g(x) \subseteq y$ (by 2), hence $g(x) \subseteq g(y)$ (since g is progressing). And so in either case $g(x) \subseteq g(y)$. ∎

Remarks Conclusions (1) and (2) above do not require that g be progressing, but (3) does.

The term "sandwich principle" might be aptly applied to (1), if we think of $x \subset y \subset z$ as "y is *sandwiched* between x and z." Conclusion (1) then says that for no x and y in N is it the case that y is sandwiched between x and $g(x)$.

From Theorem 4.7 and Lemma 4.9 we of course have the following.

Theorem 4.10. *Suppose that M is minimally inductive under g where g is progressing. Then for any elements x and y of M:*

(1) $x \subseteq y \subseteq g(x)$ implies $x = y$ or $y = g(x)$.

(2) $x \subset y$ implies $g(x) \subseteq y$.

(3) $x \subseteq y$ implies $g(x) \subseteq g(y)$.

Definition 4.11 (Bounded sets). We shall say that a class B is *bounded* by a set x if every element of B is a subset of x. We also say that B is a bounded subset of a class A if B is a subset of A and is bounded by at least one element of A.

If B is bounded by a set x then B must be a set, because B is then a subclass of $\mathcal{P}(x)$, hence is a set, since V is supercomplete.

Theorem 4.12. *Suppose M is minimally inductive under g, where g is progressing. Then every non-empty bounded subset of M has a greatest element.*

Proof Assume the hypothesis. We show by g-induction on x that every non-empty subset of M that is bounded by x has a greatest element.

For $x = \emptyset$, the only non-empty set bounded by $\emptyset$ is $\{\emptyset\}$, which has $\emptyset$ as its greatest element.

Now suppose x is such that every non-empty subset of M bounded by x has a greatest element. Then consider any non-empty subset B of M bounded by $g(x)$. We are to show that B has a greatest element. If $g(x) \in B$, then of course $g(x)$ is the greatest element of B, so we suppose that $g(x) \notin B$. Then every element of B is a *proper* subset of $g(x)$. Let y be any element of B. Since y is a *proper* subset of $g(x)$, then $g(x) \subseteq y$ doesn't hold, hence $y \subseteq x$ (since either $g(x) \subseteq y$ or $y \subseteq x$). Thus every element of B is a subset of x, which means that B is bounded by x, and therefore has a greatest element by the induction hypothesis. This completes the g-induction. ∎

§4. A DOUBLE INDUCTION PRINCIPLE AND ITS APPLICATIONS

Definition 4.13 (Fixed points). An element x is called a *fixed point* of a function f if $f(x) = x$.

Theorem 4.14. *If M is minimally inductive under g, where g is progressing, then for any x in M, if x is a fixed point of g, then x is the greatest element of M.*

Proof Suppose that x is an element of M such that $g(x) = x$. To show that x is the greatest element of M, we show by g-induction that every element of M is a subset of x. Well, obviously $\emptyset$ is a subset of x. Now suppose $y \subseteq x$. Then $g(y) \subseteq g(x)$ (by Theorem 4.10), but since $g(x) = x$, then $g(y) \subseteq x$. This completes the induction. ∎

Now for a key lemma, that will be used in this chapter and the next.

Lemma 4.15. *Suppose that N and g satisfy the following conditions:*

C_1 *N is closed under g and g is progressing;*

C_2 *For any x and y in N, either $g(x) \subseteq y$ or $y \subseteq x$;*

C_3 *For any x in N, if $g(x) = x$, then x is the greatest element of N.*

Then for any non-empty subclass A of N and any element x of N, if x is a proper subset of all elements of A, and is the greatest such element, then $g(x)$ is in A and is the least element of A.

Proof Assume the hypothesis. Let A be any non-empty subclass of N and let L be the class of all elements y such that y is a proper subset of every element of A. Let x be the greatest element of L. We are to show that $g(x)$ is the least element of A. Now, since x is a proper subset of every element of A, then by Lemma 4.9, conclusion (2), $g(x)$ is a subset of every element of A. It remains to be shown that $g(x)$ itself is an element of A.

Since x is a *proper* subset of every element of A and A contains at least one element, then x cannot be the greatest element of N, and therefore by hypothesis, $x \neq g(x)$, and so x is a *proper* subset of $g(x)$. Therefore, since x is the greatest element of L, $g(x)$ is not an element of L, and so there is at least one element y of A such that $g(x)$ is not a proper subset of y. But $g(x)$ is a subset of y (as we have shown), hence $g(x) = y$, and so $g(x) \in A$. This concludes the proof. ∎

Now we have

Theorem 4.16. *If M is minimally inductive under g, where g is progressing, then M is well ordered under inclusion.*

Proof Suppose M is minimally inductive under g, where g is progressing. Then by Theorems 4.7 and 4.14, the conditions C_1, C_2 and C_3 of the hypothesis of Lemma 4.15 hold, reading "M" for "N." Since M is a nest, it remains to show that every non-empty subclass A of M has a least element.

Let L be the class of all elements x such that x is a proper subclass of all elements of A. If $\emptyset \in A$, then $\emptyset$ is obviously the least element of A, so we now consider the case that $\emptyset \notin A$. Then $\emptyset$ is a *proper* subset of every element of A, hence $\emptyset \in L$, and hence L is non-empty. Now, L is bounded by all elements of A, and since A is non-empty, L is bounded by at least one element of A, hence by Theorem 4.12, L has a greatest element x. Therefore $g(x)$ is the least element of A, by Lemma 4.15. ∎

* **A more general situation (optional)** The next theorem generalizes some earlier results.

* **Theorem 4.17.** *Suppose that N is closed under g, where g is progressing, and that b is an element of N and that no proper subclass of N containing b is closed under g. Then:*

(1) For all x and y in N, either $g(x) \subseteq y$ or $y \subseteq x$.

(2) Every bounded subset of N has a greatest element.

(3) g has no fixed point, unless possibly the greatest element of N (if there is one).

(4) N is well ordered under inclusion.

(5) b is the least element of N.

EXERCISES

Exercise 4.1. Prove Theorem 4.17, and explain how it generalizes earlier results.

Exercise 4.2. Show that if A is inductive under g and A is a set, then some subset of A is minimally inductive under g.

§5 Applications to natural numbers

We now apply results from §4 to the natural numbers, easily establishing facts about their ordering.

Definition 5.1. For any natural numbers m and n, we say that m is *less than n*—in symbols $m < n$—if $m \subset n$ (m is a proper subset of n) and that m is *less than or equal to n*—in symbols $m \leq n$—if $m \subseteq n$ (m is a subset of n). If $m < n$, we also write $n > m$ (read "n is greater than m"), and if $m \leq n$, we also write $n \geq m$ (read "n is greater than or equal to m").

Since ω is minimally inductive under the successor function, then from Theorem 4.7 we immediately have the following.

Theorem 5.2. *Any two natural numbers m and n are comparable—in fact either $m^+ \subseteq n$ or $n \subseteq m$.*

Remark Central to our proof of Theorem 5.2 was the double induction principle (Theorem 4.3). There is an older double induction principle (Smullyan 1965) that yields an alternative proof of Theorem 5.2, which we discuss in the supplement of this chapter.

§5. APPLICATIONS TO NATURAL NUMBERS

From Theorem 4.10 we get the following.

Theorem 5.3. *For any natural numbers m and n:*

(1) If $m \leq n \leq m^+$, then $m = n$ or $n = m^+$.

(2) $m < n$ implies $m^+ \leq n$.

(3) $m \leq n$ implies $m' \leq n'$.

From Theorem 4.12 we have the following.

Theorem 5.4. *Every bounded non-empty set of natural numbers has a greatest element.*

From Theorem 4.16 we have the following.

Theorem 5.5 (Well ordering principle). *Every non-empty set of natural numbers has a least element.*

Now additional results on the natural numbers follow easily.

Theorem 5.6. *For any natural numbers m and n, $m < n$ iff $m \in n$.*

Proof
(1) Suppose $m \in n$. Then $m \subseteq n$ (since n is transitive), but if $m = n$, we would have $m \in m$, contrary to Theorem 3.2. Thus $m \leq n$ and $m \neq n$, so $m < n$.
(2) Conversely suppose $m < n$. Then $m^+ \leq n$ (by Theorem 5.3), but $m \in m^+$, hence $m \in n$. ∎

Remark Theorem 5.6 is to the effect that every natural number is the set of all lesser natural numbers.

Theorem 5.7 (Trichotomy principle). *For any natural numbers m and n, either $m \in n$ or $m = n$ or $n \in m$. Stated otherwise, given any two distinct natural numbers, one of them is an element of the other.*

Proof By virtue of Theorem 5.6, the theorem is equivalent to the statement that for any numbers m and n, either $m \subset n$ or $m = n$ or $n \subset m$, which is true since any two natural numbers are comparable. ∎

Theorem 5.8. *For any natural numbers m and n, if $m \leq n^+$ then either $m \leq n$ or $m = n^+$.*

Proof Suppose $m \leq n^+$. If $m \leq n$ is false, then $n^+ \leq m$, and hence $m = n^+$ (since $m \leq n^+$). ∎

EXERCISES

Exercise 5.1. Suppose the natural number n is in A and that for any number x in A, x^+ is also in A. Prove that A contains every number $\geq n$.

Exercise 5.2. Suppose f is a function from ω into A.

(a) Show that if for all n, $f(n) \subseteq f(n^+)$, then for all n and m, if $n \leq m$ then $f(n) \subseteq f(m)$.

(b) Show the same result with "$\subset$" in place of "$\subseteq$."

§6 Finite sets

For any natural number n, a class A is said to have n elements, or to be an n-element class, if A can be put into a 1-1 correspondence with n. (For $n \neq 0$, this is equivalent to saying that A can be put into a 1-1 correspondence with the set $\{1, \ldots, n\}$—i.e., the set $n^+ - \{0\}$.)

A class A is called *finite* if there exists some natural number n such that A has n elements; otherwise A is called *infinite*.

We leave the proofs of the following facts as exercises. (In each case the proof is by induction on the number of elements of the class or set under consideration.)

Facts (1) Every finite class is a set.

(2) Every subset of a finite set is finite.

(3) Every non-empty finite set of natural numbers has a greatest element.

EXERCISES

Exercise 6.1. Prove the three facts above.

Exercise 6.2. Prove that the union of any two finite sets is finite.

Exercise 6.3. Prove that the union of any finite set of finite sets is finite.

Exercise 6.4. Suppose M is minimally inductive under g where g is progressing. Prove that for every x in M, the set of all $y \in M$ such that $y \subseteq x$ is finite. (Use g-induction on x.)

Exercise 6.5. Suppose M is minimally inductive under g, where g is progressing. Suppose also that M contains an element x such that $x = g(x)$. Show that M is finite.

§7 Denumerable classes

A class is called *denumerable* if it can be put into a 1-1 correspondence with ω. A class is called *non-denumerable* if it is neither denumerable nor finite. In Theorem 4.1 of Chapter 1 we proved Cantor's theorem—a special case of this theorem is that $\mathcal{P}(\omega)$ is non-denumerable.

Is every denumerable class a set? Our axioms for a Zermelo universe are not strong enough to settle this. (Later, when we add Fraenkel's axiom of substitution, the answer will be affirmative.)

EXERCISES

Exercise 7.1. Show that every set of natural numbers is either finite or denumerable.

§8 Definition by finite recursion

Let c be an element of a class A, and g be a function from A into A. Does there necessarily exist a function f from numbers to elements of A such that $f(0) = c$ and for every n, $f(n^+) = g(f(n))$ (in other words $f(0) = c$, $f(1) = g(c)$, $f(2) = g(g(c))$, ...)? An obvious induction argument on n shows that there cannot be more than one such function f, but how do we know that there is at least one? Unfortunately some articles and textbooks have given the fallacious argument that f is defined on 0 (since $f(0) = c$) and if f is defined on n, then f is also defined on n^+ (since $f(n^+) = g(f(n))$), hence by mathematical induction, f is defined on every natural number n. The counter to that fallacious argument is: *What* function f? How do we know that there is such an f? That's just the thing we are trying to prove!

The correct way to go about it is as follows. First show by induction on n that for every n, there is a *unique* function f_n defined on the set of integers from 0 to n such that $f_n(0) = c$ and for each number $i < n$, $f_n(i^+) = g(f_n(i))$. Next, show that for each n, $f_n \subseteq f_{n^+}$—in other words, that for every number $k \leq n$, $f_n(k) = f_{n^+}(k)$. (This can be shown by showing that $f_n(0) = f_{n^+}(0)$, which is obvious, and that for any $k < n$, if $f_n(k) = f_{n^+}(k)$, then $f_n(k^+) = f_{n^+}(k^+)$.) It then follows that $f_n(n) = f_{n^+}(n)$. We then take our desired function f (defined on all n) to be the set of all ordered pairs $\langle n, f_n(n) \rangle$ (which is clearly a function), and so for each n, $f(n) = f_n(n)$. It is then easily seen that $f(0) = c$ and $f(n^+) = g(f(n))$.

We leave the details of this proof to the reader as Exercise 8.2. It yields the following.

Theorem 8.1. *For any function g from A into A and any set $c \in A$ there is a unique function f from numbers to elements of A such that:*

(1) $f(0) = c$;

(2) for every n, $f(n^+) = g(f(n))$.

Sequences It is helpful to think of Theorem 8.1 in terms of *sequences*. A finite sequence $a_0, a_1, \ldots, a_n$ is nothing more than a function whose domain is the set of natural numbers $\leq n$—the function f that assigns to each $i \leq n$ the element a_i. A denumerable sequence $a_0, a_1, \ldots, a_n, a_{n+1}, \ldots$ is a function f whose domain is ω and which assigns to each natural number n the element a_n. In terms of sequences, Theorem 8.1 says that for any element c and any function g there exists the infinite sequence $c, g(c), g(g(c)), \ldots, g^n(c), g^{n+1}(c), \ldots$, where for each natural number n, $g^{n+1}(c)$ is $g(g^n(c))$ (thus, e.g., $g^3(c) = g(g(g(c)))$).

Stated otherwise, Theorem 8.1 says that for any function g and any set c, there is an infinite sequence $a_0, a_1, \ldots, a_n, a_{n+1}, \ldots$, such that $a_0 = c$ and for any n, $a_{n^+} = g(a_n)$.

EXERCISES

Exercise 8.1. Show that if A is a non-empty set of natural numbers having no greatest element, then A is denumerable.

Exercise 8.2. Fill in the details of the proof of Theorem 8.1.

Exercise 8.3. Prove the following strengthening of Theorem 8.1. For any class A and any element c in A and any function $g(x, y)$ from $\omega \times A$ into A, there is a unique function f from ω into A such that $f(0) = c$ and for any number n, $f(n^+) = g(n, f(n))$. Why is this a strengthening of Theorem 8.1?

Exercise 8.4.

(a) Prove that there is a unique function $A(x, y)$ from $\omega \times \omega$ into ω such that for any natural numbers x and y, $A(x, 0) = x$ and $A(x, y^+) = A(x, y)^+$. (This is the usual *addition* function. Writing $x + y$ for $A(x, y)$, we have $x + 0 = x$ and $x + y^+ = (x + y)^+$.)

(b) Prove there is a unique function $M(x, y)$ from $\omega \times \omega$ into ω such that $M(x, 0) = 0$ and $M(x, y^+) = M(x, y) + y$. (Writing $x \cdot y$ for $M(x, y)$, we have $x \cdot 0 = 0$ and $x \cdot y^+ = (x \cdot y) + y$. This is the usual multiplication function. All arithmetic arises out of set theory.)

Exercise 8.5. Suppose that $a_0, a_1, \ldots, a_n, a_{n+1}, \ldots$ is an infinite sequence such that for each n, $a_n \subseteq a_{n+1}$. Let c be the set $\{a_0, a_1, \ldots, a_n, a_{n+1}, \ldots\}$ (i.e., the set of all elements a_i, where i is any natural number). Prove that c is a nest.

Exercise 8.6. Suppose g is a function and $a_0, a_1, \ldots, a_n, a_{n+1}, \ldots$ is an infinite sequence such that $a_0 = \emptyset$, and for each n, $a_{n+1} = g(a_n)$. Prove that the set $\{a_0, a_1, \ldots, a_n, a_{n+1}, \ldots\}$ is the set that is minimally inductive under g.

§9 Supplement—optional

As remarked in §5, Theorem 5.2 can be alternatively derived from another double induction principle, which is the following.

Theorem S_1 *Suppose that M is minimally inductive under g and that R is a relation satisfying the following two conditions:*

D'_1 $R(x, 0)$ and $R(0, x)$ hold for every x in M.

D'_2 For every x and y in M, $[R(x, y) \wedge R(x, g(y)) \wedge R(g(x), y)] \supset R(g(x), g(y))$.

Then $R(x, y)$ holds for all x, y in M.

Proof Exercise 9.1. ∎

We will call a function g *slowly progressing* if g is progressing and if for all x in the domain of g, the set $g(x)$ has at most one more element than x.

Lemma S_2 *Suppose that g is slowly progressing and $R(x, y)$ is the relation $x \subseteq y$ or $y \subseteq x$. Then R satisfies conditions D'_1, D'_2 of the theorem above.*

Proof Exercise 9.1 again. ∎

§9. SUPPLEMENT—OPTIONAL

From Theorem S_1 and Lemma S_2 it follows that if M is minimally inductive under a *slowly* progressing function g, then M is a nest. To get the additional conclusion that for any x and y in M, either $g(x) \subseteq y$ or $y \subseteq x$, the following suffices.

Lemma S_3 *Suppose g is slowly progressing. Then:*

(1) $x \subset y \subset g(x)$ can never hold.

(2) If N is a nest closed under g, then for any elements x and y of N, either $g(x) \subseteq y$ or $y \subseteq x$.

Proof Exercise 9.1 once more. ∎

It then follows that Theorem 4.7 holds for the special case that g is *slowly* progressing. This, of course, is a weaker result, but it suffices to yield Theorem 5.2, since the successor function is *slowly* progressing (n^+, in fact, contains *exactly* one more element than n—namely n itself).

EXERCISES

Exercise 9.1. Prove Theorem S_1, Lemma S_2 and Lemma S_3. (Proofs can be found in (Smullyan 1965).)

CHAPTER 4

SUPERINDUCTION, WELL ORDERING AND CHOICE

Part I—Superinduction and Well Ordering

§1 Introduction to well ordering

As remarked in the last chapter, the notion of well ordering under inclusion is but a special case of a more general notion of well ordering under an arbitrary relation, and this more general notion plays a key role in set theory. To define this, we first make some preliminary definitions.

We recall that a *relation* R is a class of ordered pairs, and the notations $R(x, y)$ and xRy are synonymous, and both mean that the ordered pair $\langle x, y \rangle$ is a member of R. By the relation $=$ of identity, we mean the class of all ordered pairs $\langle x, y \rangle$ such that x is identical to y—in other words, the class of all ordered pairs $\langle x, x \rangle$. By the relation $\subseteq$ of set inclusion, we mean the class of all ordered pairs $\langle x, y \rangle$ such that x is a subset of y. By the *field* of a relation R is meant the class of all elements of either the domain or range of R. Thus the field of R is $\mathsf{Dom}(R) \cup \mathsf{Ran}(R)$.

A relation R is called *reflexive* if xRx holds for every x in the field of R. An obvious example is ordinary identity ($x = x$ always holds). Another obvious example is set inclusion, since $x \subseteq x$ holds for every set x.

A relation R is called *anti-symmetric* if for all x and y, $(xRy \wedge yRx) \supset x = y$. (Equivalently stated, if x and y are *distinct* elements, then xRy and yRx cannot both hold.) The relation $\subseteq$ of set inclusion is obviously anti-symmetric, since if $x \subseteq y$ and $y \subseteq x$ then $x = y$. (We might note that the term "anti-symmetric" comes about as follows. A relation R is called *symmetric* if xRy always implies yRx. Thus *anti-symmetric* is sort of the opposite, in that we can never have both xRy and yRx, unless $x = y$.)

A relation R is called *transitive* if for all x, y and z, $(xRy \wedge yRz) \supset xRz$. The relation $\subseteq$ of set inclusion is transitive, since if $x \subseteq y$ and $y \subseteq z$, then $x \subseteq z$.

By a *partial ordering* is meant a relation that is reflexive, anti-symmetric and transitive. The standard notation for a partial ordering is the usual less-than-or-equal-to symbol "$\leq$," and this is the notation we will adopt. Thus a relation $\leq$ is a partial ordering iff the following three conditions hold (for all x, y and z in the field of the relation).

O_1 $x \leq x$

O_2 $(x \leq y \wedge y \leq x) \supset x = y$

O_3 $(x \leq y \wedge y \leq z) \supset x \leq z$.

The relation $\subseteq$ of set inclusion is thus a partial ordering. In the terminology of partial ordering, if $\leq$ is a partial ordering, then if $x \leq y$, we say that x is *less than or equal* to y (it is understood that this is with respect to $\leq$). We write $x < y$—read "x is less than y"—if $x \leq y$ and $x \neq y$.

As with the natural numbers, if $x < y$, we also write $y > x$ (read "y is greater than x") and if $x \leq y$, we also write $y \geq x$ (read "y is greater than, or equal to x"). Also, if $x < y$, we say that x *precedes* y (in the partial ordering $\leq$).

As with the special case of set inclusion, two elements x and y of the field of a partial ordering $\leq$ are called *comparable* if either $x \leq y$ or $y \leq x$. A partial ordering is called a *linear* ordering if every two elements of its field are comparable. Thus a linear ordering is a partial ordering $\leq$ satisfying the following additional condition (for every x and y in the field of $\leq$).

O_4 $x \leq y \vee y \leq x$.

For any relation R and any class A, by the *restriction* of R to A—symbolized $R \upharpoonright A$—is meant the class of all ordered pairs $\langle x, y \rangle$ *of elements x, y of A* such that xRy. Thus $R \upharpoonright A = R \cap (A \times A)$. Now, we say that a class A is *partially ordered*, respectively *linearly ordered*, under a relation R if A is a subclass of the field of R and the restriction of R to A is respectively a partial ordering, or a linear ordering. Thus, for example, the class V is partially ordered under inclusion. An important example of a linear ordering emerged in the last chapter: the set ω of natural numbers is *linearly* ordered under $\subseteq$, by Theorem 4.7 of Chapter 3.

For any class A of elements of the field of a partial ordering $\leq$, an element x of A is called the *least (first, smallest)* element of A (with respect to $\leq$) if $x \leq y$ for every y in A. (There cannot be more than one least element by anti-symmetry.) An element x of A is called the *greatest* element of A if $y \leq x$ for every $y \in A$. And now by a *well ordering* is meant a linear ordering $\leq$ satisfying the following additional condition.

O_5 Every non-empty subclass of the field of $\leq$ has a least element.

We say that a class A is *well ordered* under a relation R if the restriction $R \upharpoonright A$ of R to A is a well ordering.

Proposition 1.1. *If A is well ordered under R, so is every subclass of A.*

Proof Obvious. ∎

We say that a class A *can be well ordered* if there exists a relation R under which A is well ordered.

Proposition 1.2. *If a class B can be well ordered, then any class A that can be put into a 1-1 correspondence with a subclass of B can be well ordered.*

More specifically, suppose that R is a well ordering of B and that F is a 1-1 function that maps each element x of A to an element x' of B. Let $\leq$ be the class of all ordered pairs $\langle x, y \rangle$ of elements of A such that $x'Ry'$ (and thus $x \leq y$ iff $x'Ry'$). Then A is well ordered under this relation $\leq$.

Proof Let A' be the range of F (the set of all elements x', where $x \in A$). Then A' is well ordered under R (Proposition 1.1).

For any element $a \in A$, we have $a'Ra'$ (since R is reflexive), hence $a \leq a$, so $\leq$ is reflexive.

For any elements a, b, c of A, suppose $a \leq b$ and $b \leq c$. Then $a'Rb'$ and $b'Rc'$, hence $a'Rc'$ (since R is transitive), hence $a \leq c$. Thus $\leq$ is transitive.

Now suppose $a \leq b$ and $b < a$ (a and b are elements of A). Then $a' \leq b'$ and $b' \leq a'$, hence $a' = b'$ and then, since the correspondence is 1-1, $a = b$. Thus the relation $\leq$ is anti-symmetric.

Next, for every element a and b of A, either $a'Rb'$ or $b'Ra'$ (since R is a linear ordering), hence either $a \leq b$ or $b \leq a$, so $\leq$ is a linear ordering.

Finally, suppose C is any non-empty subclass of A. Let C' be the class of all elements x' such that $x \in C$. Then C' contains a least element (with respect to R), and this element is b' for some $b \in C$. Then for any $x \in C$, $x' \in C'$, and so $x'Rb'$, and therefore $x \leq b$. Thus b is the least element of C (with respect to $\leq$). Thus $\leq$ is a well ordering of A. ∎

Discussion Since the set ω of natural numbers can be well ordered, it follows from the proposition above that every denumerable set can be well ordered. The set $\mathbb{Q}$ of rational numbers is denumerable, as shown in Chapter 1, hence $\mathbb{Q}$ can be well ordered. Now $\mathbb{Q}$ is certainly *not* well ordered under the usual relation $\leq$ among the rationals, but the point is that it is possible to arrange the rationals in a different order which *is* a well ordering—if we enumerate the rationals as indicated in Chapter 1, and we define $x \leq y$ to mean that $x = y$ or x comes earlier than y in the enumeration, then under *this* relation, $\mathbb{Q}$ is well ordered.

Successor elements and limit elements Consider now a *linear* ordering $\leq$ on a class A. For any elements x and y of A, we say that y is the *successor* of x, or that x is the *predecessor* of y if $x < y$ but for no z in A is it the case that $x < z < y$ holds. Our justification for using the word "the" for successor or predecessor, instead of "a," is given by the following.

Proposition 1.3. *In a linear ordering, no element can have more than one successor, nor more than one predecessor.*

Proof

(1) Suppose that y and z are both successors of x. Then $x < y$ and $x < z$. If $y < z$ then $x < y < z$, contrary to the fact that z is the successor of x. Likewise we cannot have $z < y$ (as this would imply $x < z < y$). Hence $y = z$.

(2) The proof that an element cannot have more than one predecessor is similar and is left to the reader. ∎

Proposition 1.4. *If $\leq$ is a well ordering of A, then every element x of A that is not the greatest element of A has a successor.*

Proof The successor of x is the *least* element of A that is greater than x. ∎

Definition 1.5. An element x of A is called a *limit* element (under the well ordering $\leq$) if x is neither the least element of A nor a successor element.

Thus any element x of A belongs to exactly one of the following three categories:

(1) x is the least element of A.

(2) x is a successor element.

(3) x is a limit element.

Notation If $\leq$ is a well ordering of a class A then for any element x of A, by $L_<(x)$ we shall mean the class of all elements of A less than x, and by $L_\leq(x)$ we shall mean the class of all elements less than or equal to x. Thus $y \in L_<(x)$ iff $y < x$, and $y \in L_\leq(x)$ iff $y \leq x$.

Proposition 1.6. *In a well ordering, if x is not the least element, then x is a limit element if and only if $L_<(x)$ has no greatest element.*

Proof
(1) If $L_<(x)$ has a greatest element y, then x is the successor of y, hence x is not a limit element. Therefore, if x is a limit element then $L_<(x)$ cannot have a greatest element.
(2) If x is a successor element then the predecessor of x is the greatest element of $L_<(x)$. Therefore if $L_<(x)$ has no greatest element then x cannot be a successor element, hence if it is not the least element of the ordering, it must be a limit element. ∎

As an obvious corollary we have the following.

Proposition 1.7. *If x is a limit element of a well ordering then $L_<(x)$ has infinitely many elements.*

Transfinite induction principles One importance of well ordered classes is that they permit proofs by what is termed *transfinite induction*.

Theorem 1.8 (Transfinite induction principle 1). *Let A be well ordered under $\leq$. Let P be a property satisfying the following condition: For every $x \in A$, if P holds for every $y < x$, then P holds for x. Then P holds for every element of A.*

Proof If P failed to hold for some element of A, then there must be a least element x of A for which P fails to hold. Then for every $y < x$, P holds for y. This violates the hypothesis. ∎

Theorem 1.9 (Transfinite induction principle 2). *Let A be well ordered under $\leq$ and let P satisfy the following three conditions:*

(1) P holds for the least element of A.

(2) For any x having a successor $S(x)$, if P holds for x, then P holds for $S(x)$.

§1. INTRODUCTION TO WELL ORDERING

(3) For any limit element y, if P holds for all $x < y$, then P holds for y.

Then P holds for every element of A.

Proof Again, if P failed to hold for some element of A, then there must be a least element x of A for which P fails to hold. By (1), x cannot be the least element of A. By (2), x cannot be a successor element. By (3), x cannot be a limit element. Therefore P holds for all elements of A. ∎

Well ordering and choice We know every denumerable set can be well ordered, but what about non-denumerable sets? Can *every* set be well ordered? Zermelo (1904) proved the remarkable result that if we assume the axiom of choice, then every set can be well ordered.

A few words now about the axiom of choice, which we informally stated in Chapter 1 §4. More formally, this axiom (in one of its several equivalent forms) is that given any non-empty set A of non-empty sets, there is a function f that maps each element x of A to an element of x—in other words, for each x in A, $f(x) \in x$. (The function f, so to speak, "chooses" an element from each element of A.) Bertrand Russell gave a neat popular illustration of this axiom: Suppose we have an infinite number of pairs of socks. Using the axiom of choice, it is possible to simultaneously pick one sock from each pair. But now if instead, we have an infinite number of pairs of shoes, one doesn't need the axiom of choice to pick one shoe from each pair; simply pick the left shoes. Stated in more purely mathematical terms, given an infinite set of *unordered* pairs, one needs the axiom of choice to pick one element from each pair, but given an infinite set of *ordered* pairs, one doesn't need the axiom of choice to pick one element from each of the ordered pairs—simply pick the first element of each pair.

Zermelo's famous *well ordering theorem* is that the axiom of choice implies every set can be well ordered. This is one of the many interesting things that will be proved in this chapter.

EXERCISES

Exercise 1.1. Show our definition of well ordering is stronger than necessary, by showing the following: If $\leq$ is a *partial* ordering satisfying the additional condition that every non-empty subclass of the field of $\leq$ has a least element, then $\leq$ is a *linear* ordering, and hence is a well ordering.

Exercise 1.2. For a *linear* ordering $\leq$ on a class A, a subclass L of A is called a *lower section* of A (with respect to $\leq$, understood) if every element of L is less than every element of A not in L. A lower section L of A is called a *proper lower section* if it is non-empty and not the whole of A.

Prove that a sufficient condition for a linear ordering $\leq$ of A to be a *well* ordering of A is that, for every proper lower section L of A, there is a least element x of A not in L.

Exercise 1.3. Prove that a linear ordering $\leq$ of A is a well ordering of A if and only if the following three conditions hold:

(a) A itself has a least element a_0.

(b) Every element of A except the greatest (if there is one) has a successor.

(c) For any proper lower section L of A, if L has no greatest element, then there is a least element of A not in L.

Exercise 1.4. By a *backward path* of a linear ordering is meant a finite or denumerable sequence of elements such that each term of the sequence is less than the preceding term.

Prove that in a *well* ordering, all backward paths must be finite.

Exercise 1.5. The axiom of choice is sometimes stated in the following form: If A is a non-empty set of pairwise disjoint non-empty sets (i.e., if no element of A is empty, and if no two *distinct* elements of A have any element in common) then there exists a set C that consists of elements of elements of A and that contains exactly one element of each element of A (i.e., for each $x \in A$, the set $C \cap x$ contains exactly one element).

Prove that this form of the axiom of choice (sometimes called the axiom of *selection*) is equivalent to the previously given form.

Exercise 1.6. Another form of the axiom of choice is that for any non-empty set A, there is a function f such that for every proper subset x of A, $f(x) \in x$.

Show that this is equivalent to the other two forms.

Exercise 1.7. Using the axiom of choice, prove that every infinite set S can be put into a 1-1 correspondence with a proper subset of S. (Hint: do the next exercise first.)

Exercise 1.8. Using the axiom of choice, show that every infinite set has a denumerable subset.

§2 Superinduction and double superinduction

We now turn to a principle—a transfinite analog of the double induction principle of the last chapter—that will be basic to our entire study. It provides a unified treatment of Zermelo's well ordering theorem and several other major results of the next several chapters.

By a *chain* we shall mean a nest that is also a set. We shall say that a class A is *closed under chain unions* if for every chain C of elements of A, the set $\cup C$ (the union of C) is also in A.

And now we define a class A to be *superinductive* under a function g if A is both inductive under g and closed under chain unions. Thus A is superinductive under g iff the following three conditions hold:

(1) $\emptyset \in A$;

(2) for all $x \in A$, $g(x) \in A$;

(3) A is closed under chain unions—i.e., for any chain C of elements of A, $\cup C \in A$.

§2. SUPERINDUCTION AND DOUBLE SUPERINDUCTION

We call A *minimally superinductive* under g if A is superinductive under g and no proper subclass of A is superinductive under g. If A is minimally superinductive under g, then we have available the following method of proof, which we term *proof by superinduction*. Let P be a property that we wish to show holds for all elements of A. Then it suffices to show three things:

(1) P holds for $\emptyset$;

(2) for any x in A, if P holds for x, then P holds for $g(x)$;

(3) for any chain C of elements of A, if P holds for every element of C, then P holds for $\cup C$.

These three conditions suffice, for they imply that the class A_0 of elements of A for which P holds is superinductive under g, hence $A_0 = A$, since A is minimally superinductive under g. Thus P holds for all elements of A.

Double superinduction The double induction principle of the last chapter will henceforth be referred to as D.I.P. (Theorem 4.3, Chapter 3). Using it, we proved that if M is minimally inductive under a progressing function g, then M is well ordered under inclusion. We now wish to prove that any class that is minimally *superinductive* under a progressing function g is well ordered under inclusion. This result is basic to our entire study and to prove it we need in place of D.I.P. the following *double superinduction* principle—D.S.P.

Theorem 2.1 (Double superinduction principle). *Assume M is minimally superinductive under g. Then the following three conditions are sufficient for a relation $R(x, y)$ to hold for all x, y in M:*

D_1 *(Same as in D.I.P.) $R(x, 0)$ holds for all x in M.*

D_2 *(Same as in D.I.P.) For all x, y in M, $[R(x, y) \wedge R(y, x)] \supset R(x, g(y))$.*

D_3 *For any $x \in M$ and any chain C of elements of M, if $R(x, y)$ holds for every y in C, then $R(x, \cup C)$.*

Proof Assume the hypothesis. As in the proof of D.I.P. we define an element x of M to be *left* normal if $R(x, y)$ holds for all $y \in M$, and *right* normal if $R(y, x)$ holds for all $y \in M$.

Step 1 As in the proof of D.I.P. we show that every right normal element is also left normal. So suppose x is right normal. We show by superinduction on y that $R(x, y)$ holds for all $y \in M$, and hence that x is left normal. Well, $R(x, 0)$ holds by D_1. Now suppose that $R(x, y)$. Then also $R(y, x)$ (since x is right normal). Hence $R(x, g(y))$ by D_2. Thus $R(x, y) \supset R(x, g(y))$. (So far, we have only repeated part of the argument of the proof of D.I.P.). Now suppose that C is a chain of elements of M such that $R(x, y)$ holds for every y in C. Then $R(x, \cup C)$ by D_3. This completes the superinduction, and so $R(x, y)$ holds for all y in M, which means that x is left normal.

Step 2 We now show by superinduction on x that every x in M is right normal. Well 0 is, by D_1. Now suppose x is right normal. Then x is also left normal by Step 1. Now take any y in M. Then $R(y,x)$ and $R(x,y)$ both hold, hence $R(y,g(x))$ by D_2. Thus $g(x)$ is right normal. Finally suppose C is a chain of right normal elements. Let x be any element of M. Then $R(x,y)$ holds for every $y \in C$ (since every y in C is right normal). Hence $R(x,\cup C)$ by D_3, so $\cup C$ is right normal. This completes the superinduction, so every element y of M is right normal, and therefore $R(x,y)$ holds for all x and y in M. ∎

Applications to g-towers We shall call a class M a g-*tower* if M is minimally superinductive under g and g is progressing. We now wish to prove that every g-tower is a nest (linearly ordered under inclusion)—and more strongly, that for any elements x and y of a g-tower M, either $g(x) \subseteq y$ or $y \subseteq x$. To do this it suffices to show the following.

Lemma 2.2. *Let g be a progressing function and $R(x,y)$ be the relation $g(x) \subseteq y \vee y \subseteq x$. Then R satisfies conditions D_1, D_2 and D_3 of the hypothesis of Theorem 2.1.*

Proof Conditions D_1 and D_2 were already proved in Chapter 3 (Lemma 4.6). As for D_3, suppose that C is a chain (or any set, for that matter) such that $R(x,y)$ holds for every $y \in C$. Thus for every y in C, either $g(x) \subseteq y$ or $y \subseteq x$. If $y \subseteq x$ holds for every y in C, then $\cup C \subseteq x$, hence $R(x, \cup C)$. If it is not the case that $y \subseteq x$ for every $y \in C$, then there is at least one element $y \in C$ for which $y \subseteq x$ doesn't hold, and for such element y, it must be that $g(x) \subseteq y$ (since $R(x,y)$), and hence $g(x) \subseteq \cup C$ (since $y \subseteq \cup C$), hence again $R(x, \cup C)$. This proves D_3, which completes the proof of the lemma. ∎

From Theorem 2.1 and the lemma we now have the following.

Theorem 2.3. *If M is a g-tower then M is a nest, and moreover for any elements x and y of M, either $g(x) \subseteq y$ or $y \subseteq x$.*

From Theorem 2.3 above and Lemma 4.9 of Chapter 3 we have the following.

Theorem 2.4. *For any g-tower M and any elements x and y of M the following conditions hold:*

(1) (Sandwich principle) $x \subset y \subset g(x)$ cannot hold—alternatively, $x \subseteq y \subseteq g(x)$ implies $x = y$ or $y = g(x)$.

(2) $x \subset y$ implies $g(x) \subseteq y$.

(3) $x \subseteq y$ implies $g(x) \subseteq g(y)$.

Greatest elements and fixed points We now wish to show that Theorem 4.14 of the last chapter holds good replacing "inductive under g" by "superinductive under g"—i.e., we show the following.

Theorem 2.5. *If M is a g-tower, then for any element x of M, if $x = g(x)$ then x is the greatest element of M.*

§2. SUPERINDUCTION AND DOUBLE SUPERINDUCTION

Proof Suppose $x = g(x)$. We show by superinduction on y that $y \subseteq x$ for every y in M.

(1) Obviously $\emptyset \subseteq x$.

(2) Suppose $y \subseteq x$. Then $g(y) \subseteq g(x)$ (by Theorem 2.4). Hence $g(y) \subseteq x$ (since $g(x) = x$). Thus $y \subseteq x$ implies $g(y) \subseteq x$.

(3) Now suppose C is a chain of elements of M such that every y in M is a subset of x. Then obviously $\cup C \subseteq x$.

This completes the superinduction. ∎

Theorem 2.6. *If M is a g-tower and a set, then $\cup M \in M$ and $\cup M$ is the greatest element of M and is the unique fixed point of g.*

Proof Assume the hypothesis. Since M is a nest and a set, then M is a chain, and since M is closed under chain unions, $\cup M \in M$. Then clearly $\cup M$ is the greatest element of M. Thus $\cup M$ cannot be a proper subset of $g(\cup M)$, hence $\cup M = g(\cup M)$. By Theorem 2.5, M has no other fixed point. ∎

Lemma 2.7. *For any set S:*

(1) If every element of S is closed under a function g, so is $\cap S$.

(2) If every element of S is closed under chain unions, so is $\cap S$.

Proof Exercise 2.1 below. ∎

Theorem 2.8. *If S is superinductive under a progressing function g and S is a set, then some element x of S is a fixed point of g.*

Proof Assume the hypothesis. Let M be the intersection of all subsets of S that are superinductive under g. M is minimally superinductive under g (Lemma 2.7). Also M is a set. Then by Theorem 2.6, $\cup M$ is a fixed point of g, and $\cup M \in S$ (since $\cup M \in M$). ∎

We shall call a function g *strictly progressing* if for every x in the domain of g, x is a *proper* subset of $g(x)$. Thus a strictly progressing function is a progressing function that has no fixed point.

Theorem 2.8 obviously yields the following.

Theorem 2.9. *If A is superinductive under a strictly progressing function g, then A is not a set. Stated otherwise, no set can be superinductive under a strictly progressing function.*

Theorem 2.8 also has as a corollary the following, which will be needed in §5 and §6.

Theorem 2.10. *Suppose that g is progressing and S is a non-empty set that is closed under g and closed under chain unions. Then any element b of S is a subset of some element $x \in S$ such that $g(x) = x$.*

Proof Assume the hypothesis. If $b = \emptyset$, the case reduces to Theorem 2.8, so we assume $b \neq \emptyset$. Let S_b be the set of all elements of S that include b, together with $\emptyset$. It is obvious that S_b is also closed under chain unions. We now define the function g' on S_b as follows: for any $x \in S_b$ other than $\emptyset$, take $g'(x) = g(x)$, and take $g'(\emptyset) = b$. Then g' is also progressing, S_b is closed under g', and $\emptyset \in S_b$, so S_b is superinductive under g'. Then by Theorem 2.8, S_b contains an element x such that $g'(x) = x$. Now, for such an x, $x \neq \emptyset$, since $g'(\emptyset) = b$, hence $b \subseteq x$. Also, since $x \neq \emptyset$, $g'(x) = g(x)$, and so $g(x) = x$. Thus $b \subseteq x$ and $g(x) = x$. ∎

EXERCISES

Exercise 2.1. Prove Lemma 2.7.

§3 The well ordering of g-towers

Theorem 3.1. *Every g-tower is well ordered under inclusion. More specifically, if M is a g-tower then M is linearly ordered under inclusion and for any non-empty subclass A of M, if L is the set of all elements x that are proper subsets of all elements of A, then the following three conditions hold:*

(1) If L is empty, then $\emptyset$ is the least element of A.

(2) If L is non-empty and contains a greatest element x, then $g(x)$ is the least element of A.

(3) If L is non-empty and contains no greatest element, then $\cup L$ (the union of L) is the least element of A.

Proof Assume the hypothesis. By Theorem 2.3, M is a nest (linearly ordered under $\subseteq$).

(1) If L is empty, $\emptyset \notin L$, hence $\emptyset$ is not a proper subset of every element of A, but $\emptyset$ is a subset of every element of A, hence $\emptyset$ must be one of the elements of A, and is then obviously the least element of A.

(2) This follows by Lemma 4.15 of Chapter 3, since by Theorem 2.3 and Theorem 2.5 the conditions C_1, C_2 and C_3 of the hypothesis of the lemma hold, reading "M" for "N" (i.e., *(i)* M is closed under g and g is progressing; *(ii)* $g(x) \subseteq y$ or $y \subseteq x$ for every x and y in M; *(iii)* the only possible fixed point of g is the greatest element of M).

(3) Now suppose that L is non-empty and has no greatest element. Then $\cup L$ is not a member of L (for if it were, then it *would* be the greatest element of L). But $\cup L \in M$ (since L is a chain). Now, for each y in A, $\cup L \subseteq y$, since each element of L is a subset of y. However $\cup L \notin L$, which means there is at least one y in A such that $\cup L$ is not a *proper* subset of y, and hence for such an element y, it must be that $\cup L = y$. Thus $\cup L \in A$. ∎

We recall that in a well ordering, for any element x we write $L_<(x)$ for the class of all elements less than x. Thus in a g-tower M, for any $x \in M$, $L_<(x)$ is the set of all elements of M that are proper subsets of x.

Corollary 3.2. *Suppose that M is a g-tower and y is an element of M other than $\emptyset$. Then $L_<(y)$ is non-empty and exactly one of the following conditions holds.*

C_1 *$L_<(y)$ has a greatest element x and $g(x) = y$.*

C_2 *$L_<(y)$ has no greatest element and $\cup L_<(y) = y$.*

Proof Since $y \neq \emptyset$ then y is not the least element, so $\emptyset \in L_<(y)$, hence $L_<(y)$ is non-empty. The rest follows from Theorem 3.1, taking $A = \{y\}$. (The least element of $\{y\}$ is obviously y.) ∎

Theorem 3.3. *If M is a g-tower then M is well ordered under inclusion in such a way that the following conditions hold.*

(1) $\emptyset$ is the least element of M.

(2) If x is any element of M other than the greatest, then the successor of x is $g(x)$.

(3) If x is a limit element, then x is the union of the set of all lesser elements ($x = \cup L_<(x)$).

Proof
(1) Obvious.
(2) Suppose $x \in M$ and x is not the greatest element of M. Then $x \neq g(x)$ (Theorem 2.5), hence $x \subset g(x)$. By Theorem 2.4, there is no element y such that $x \subset y \subset g(x)$, and hence $g(x)$ is the successor of x.
(3) Suppose x is a limit element. Then x is not a successor element, hence by (2) there is no $y \in M$ such that $x = g(y)$ (for if there were, x would be the successor of y), hence there is no $y \in L_<(x)$ such that $x = g(y)$. Then condition C_1 of Corollary 3.2 doesn't hold, hence condition C_2 does, and so $\cup L_<(x) = x$. ∎

§4 Well ordering and choice

We now turn to the proof of Zermelo's well ordering theorem; we shall prove several other things along the way.

Slow Well Orderings and Slow g-Towers

Definition 4.1. We will say that a class N is *slowly* well ordered under inclusion if N is well ordered under inclusion in such a way that the following three conditions hold:

S_1 $\emptyset$ is the least element of N;

S_2 each successor element contains exactly one more element than its predecessor;

S_3 each limit element is the union of the set of its lesser elements.

Definition 4.2. We define a function g to be *slowly progressing* if g is progressing and for every x in the domain of g, the set $g(x)$ contains at most one more element than x (and thus either $g(x) = x$ or $g(x) - x$ contains exactly one element). And we shall call M a *slow g-tower* if M is a g-tower and g is slowly progressing.

Theorem 4.3. *If M is a slow g-tower then M is slowly well ordered under $\subseteq$.*

Proof Suppose that M is a slow g-tower. Then conditions S_1 and S_3 of Definition 4.1 hold by Theorem 3.3. As for condition S_2, suppose that y is the successor of x. Then x is not the greatest element of M, and by (2) of Theorem 3.3, $y = g(x)$. Hence y contains at most one more element than x. But since y is the successor of x, then x is a proper subset of y, so y contains at least one more element than x. Hence y contains exactly one element not in x. ∎

We wish to show that if M is a slow g-tower, then $\cup M$ can be well ordered. More generally, we show the following.

Theorem 4.4. *If N is slowly well ordered under inclusion then $\cup N$ can be well ordered.*

Proof Suppose N is slowly well ordered under $\subseteq$. Let $A = \cup N$. For any element $a \in A$, let $F(a)$ be the first (least) element of N that contains a. Then for any elements a and b of A, we define $a \leq b$ iff $F(a) \subseteq F(b)$. We will show that A is well ordered under this relation $\leq$. By Proposition 1.2 it suffices to show that the function F is 1-1. So suppose that $F(a) = F(b)$; we must show that $a = b$.

First of all, $F(a)$ is non-empty, since $a \in F(a)$; hence $F(a)$ cannot be the least element of N (by condition S_1 of Definition 4.1). If $F(a)$ were a limit element, it would be the union of all lesser elements (condition S_3 of Definition 4.1), but no lesser element contains a, hence the union of them cannot contain a. Hence $F(a)$ is not a limit element, so it is a successor element. Let x be the predecessor of $F(a)$. Then $a \notin x$, but also $F(a)$ contains exactly one element not in x (condition S_2 of Definition 4.1), hence this element must be a. Thus $F(a) = x \cup \{a\}$. Hence $F(b) = x \cup \{a\}$. Since $F(b)$ is the least element that contains b, then $b \notin x$, hence $b \in \{a\}$, so $b = a$. Thus F is a 1-1 function, and so by Proposition 1.2, A is well ordered under $\leq$. ∎

Discussion Suppose N is slowly well ordered under $\subseteq$ and $A = \cup N$. For any elements a and b of A, let us say that *a comes into A before b* if the least element of N containing a does not contain b. Then in the well ordering $\leq$ of the proof above, $a \leq b$ iff either $a = b$, or a comes into A before b. Thus $a < b$ iff a comes into A before b. Informally speaking, we are "ordering" the elements of A according to the order in which they come into A, and the key feature of *slow* well orderings which makes things work is that only one new element comes into A at a time.

We should note that we could alternatively define $a \leq b$ to mean that every element of N that contains b also contains a (Exercise 4.1 below). The particularly interesting thing to note is that under the ordering $\leq$, the successor of any element a of A is the least element of $A - F(a)$ (Exercise 4.1 again). Also F is a 1-1 correspondence from A onto the class of all successor elements of M—that is, every successor element is $F(a)$ for some $a \in A$ (Exercise 4.1 yet again).

§4. WELL ORDERING AND CHOICE

From Theorems 4.3 and 4.4 we get the following key result.

Theorem 4.5. *If M is a slow g-tower then $\cup M$ can be well ordered.*

From this result and Theorem 2.6 we get the following.

Theorem 4.6. *For any set S, if there exists a slowly progressing function g from all subsets of S to subsets of S whose only fixed point is S, then S can be well ordered.*

Proof Assume the hypothesis. Then $\mathcal{P}(S)$ is closed under g, and of course $\mathcal{P}(S)$ contains $\emptyset$ and is closed under chain unions, so $\mathcal{P}(S)$ is superinductive under g. Let M be the intersection of all subsets of $\mathcal{P}(S)$ that are superinductive under g. Then M is minimally superinductive under g (Lemma 2.7), and so M is a slow g-tower. By Theorem 2.6 $\cup M \in M$ and $\cup M$ is a fixed point of g. But S is the only fixed point of g, hence $\cup M = S$. Then, since $\cup M$ can be well ordered (Theorem 4.5), S can be well ordered. ∎

Choice functions and well orderings For any *set* S, a *choice function* for S is a function C that assigns to each non-empty subset x of S an element of x. Thus if C is a choice function for S, then $C(x) \in x$ for each non-empty subset x of S. For any non-empty subset x of S, we will refer to $C(x)$ as the *C-chosen element of x*. The axiom of choice, in the form that we shall use, is that every non-empty set has a choice function.

Theorem 4.7. *For any set S, if there is a choice function for S, then S can be well ordered.*

Proof Suppose that C is a choice function for S. For any subset x of S, whether empty or not, we define $g(x)$ as follows: If x is a *proper* subset of S, then $S - x$ is non-empty and we take $g(x)$ to be the result of adjoining the C-chosen element of $S - x$ to x—i.e., $g(x) = x \cup \{C(S - x)\}$. For $x = S$, we take $g(S) = S$. Thus $g(S) = S$ and for any proper subset x of S, $g(x)$ adds to x some element of S not in x. Clearly g is slowly progressing and S is the only fixed point of g. Thus g satisfies the hypotheses of Theorem 4.6, and so S can be well ordered. ∎

Discussion Let us briefly review the main points involved in the proof of Theorem 4.7 and see how the choice function is related to the well ordering. Given a choice function C for a set S, we let g be that function from $\mathcal{P}(S)$ into $\mathcal{P}(S)$ such that $g(S) = S$ and for any proper subset x of S, $g(x)$ is the result of adjoining the C-chosen element of $S - x$ to x. Then g is slowly progressing and the set $\mathcal{P}(S)$ is superinductive under g. We take the intersection of all subsets of $\mathcal{P}(S)$ that are superinductive under g and this set M is minimally superinductive under g, the greatest element of M is S, and $S = \cup M$. We then *well order* S by the scheme $a \leq b$ iff $F(a) \subseteq F(b)$. Thus $a < b$ iff the least set containing a doesn't contain b.

Now let us see just how the choice function C is related to the well ordering $<$. Well, the least element a_0 of M—the first element to come into S—is the C-chosen element of S. The next element a_1 is the C-chosen element of $S - \{a_0\}$, next comes the chosen element of $S - \{a_0, a_1\}$, and so forth. For any element a other than the greatest, the next element after a is the C-chosen element of the set of elements of S that have not yet come into S—i.e., the successor of a is the C-chosen element of $S - F(a)$.

Of course Theorem 4.7 implies the following.

Theorem 4.8 (Zermelo's well ordering theorem). *The axiom of choice implies that every set can be well ordered.*

Remark These days Zermelo's well ordering theorem is usually proved via a result known as *Zorn's lemma*, which we will look at in the next section. We prefer the proof above in that it elucidates more clearly just how the choice function comes into the well ordering. Another proof that also does this will be indicated in Chapter 6, but it requires the axiom of substitution.

The converse of Zermelo's well ordering theorem is quite trivial.

Theorem 4.9. *If every set can be well ordered, then the axiom of choice holds. More specifically, for any set S, if S can be well ordered then there is a choice function for S.*

Proof Suppose S is well ordered under $\leq$. Then for any non-empty subset x of S, take $C(x)$ to be the least element of x (under the well ordering $\leq$). ∎

Since every denumerable set can be well ordered, it follows from this theorem that every denumerable set has a choice function.

EXERCISES

Exercise 4.1. For the well ordering $\leq$ of $\cup N$ described in the proof of Theorem 4.4, show the following facts:

(a) For any a and b of $\cup N$, $a \leq b$ iff every element of N that contains b also contains a.

(b) $x \in F(a) \supset x \leq a$ (for each $x \in N$, $a \in \cup N$).

(c) For any $a \in \cup N$, if a is not the greatest element of $\cup N$ then its successor is the least element of $\cup N - F(a)$.

(d) Every successor element of N is $F(a)$ for some $a \in N$.

Exercise 4.2. Show that for any set S, if there exists a choice function for $\mathcal{P}(S)$, then there exists a choice function for S.

Exercise 4.3. Show that if there exists a choice function for S then there exists a choice function for $\cup S$.

Part II—Maximal principles

§5 Maximal principles

Many interesting principles have been derived from the axiom of choice—principles known as *maximal principles*, sometimes referred to as forms of *Zorn's lemma*, sometimes as *Kuratowski principles* or *Hausdorff principles*. These principles were discovered independently by (among others) Kuratowski (1922), Hausdorff (1914) and Zorn (1935). The principles have in fact all been proven *equivalent* to the axiom of choice.

Definition 5.1. An element x of a class A is called a *maximal* element of A if x is not a proper subset of any element of A.

A maximal element need not be the greatest element, for it could happen that x and y are maximal elements, but neither $x \subseteq y$ nor $y \subseteq x$, and so neither x nor y would be the greatest element. Of course if A is a nest, then any maximal element of A must be the greatest element of A.

We will say that a set S is of type M if every element of S can be extended to (i.e., is a subset of) a maximal element of S. Obviously any non-empty set of type M must have at least one maximal element. What we are terming *maximal principles* are various sufficient conditions for a set to be of type M. We shall now look at some maximal principles that are consequences of the axiom of choice (and are in fact equivalent to it).

Kuratowski's maximal principle Every set closed under chain unions is of type M.

Theorem 5.2. *The axiom of choice implies Kuratowski's maximal principle.*

We will in fact prove the following sharper result.

Theorem 5.3. *If S is a set closed under chain unions and if there is a choice function for S, then S is of type M.*

Proof Suppose S is closed under chain unions and that C is a choice function for S. For each $x \in S$, let x^* be the set of all elements y of S such that x is a proper subset of y. Obviously x^* is empty if and only if x is a maximal element of S. We now define a progressing function g from S into S as follows: for any $x \in S$, if x is a maximal element of S, we take $g(x) = x$, and if x is not a maximal element of S, we take $g(x)$ to be the C-chosen element of x^*. Thus if x is maximal then $g(x) = x$, but if x is not maximal, then x is a proper subset of $g(x)$. By Theorem 2.10, any element b is a subset of some element x of S such that $x = g(x)$. Such an element x is then a maximal element of S.
∎

The Tukey-Teichmüller maximal principle Kuratowski's maximal principle is what we will need for subsequent chapters, but there are several other interesting ones that we will look at.

A class A is said to be of *finite character* if for every set x, $x \in A$ if and only if all finite subsets of x are in A.

Sometimes the term "finite character" is applied to properties of sets rather than to classes of sets. A property P of sets is said to be of finite character if for every set x, x has property P if and only if all finite subsets of x have property P.

A well-known property of finite character is that of being a *nest*: If x is a nest, then every finite subset (in fact every subset) of x is a nest. Conversely, if every finite subset of x is a nest, then every 2-element subset $\{a,b\}$ of x is a nest, which means that for any elements a, b of x, either $a \subseteq b$ or $b \subseteq a$, and hence x is a nest.

Another well-known example is from mathematical logic—the property of *consistency*. If a set of formulas is inconsistent; that is, if a contradiction can be derived from it, then the contradiction can be derived from some *finite* subset (since a proof uses only finitely many formulas), and of course if some finite subset of it is inconsistent, then the set itself is inconsistent. Thus a set of formulas is inconsistent if and only if some finite subset of it is inconsistent. Equivalently, a set is consistent if and only if all finite subsets are consistent, and so consistency is a property of finite character. A well-known result in mathematical logic is that any consistent set can be extended to (i.e., is a subset of) a maximal consistent set. The only feature of consistency needed for the proof is that consistency is a property of finite character. And so this result is but a special case of what is known as *Tukey's lemma*, or *Teichmüller's lemma*, which is that every set of finite character is of type M. We will show that Kuratowski's maximal principle implies Tukey's lemma, and hence that the axiom of choice implies Tukey's lemma. This is immediate from (2) of the following lemma.

Lemma 5.4. *Suppose A is of finite character. Then:*

(1) A contains, with each element x, all subsets of x as well (A is swelled);

(2) A is closed under chain unions.

Proof
(1) Suppose $x \in A$ and $y \subseteq x$. Then every finite subset of y is also a finite subset of x, and hence is an element of A (since $x \in A$ iff all finite subsets of x are in A). Thus all finite subsets of y are in A, hence $y \in A$.

(2) Suppose that C is a chain of elements of A (C is a *set*). To show that $\cup C \in A$ it suffices to show that every finite subset of $\cup C$ is an element of A. Well, consider any finite subset $\{y_1, \ldots, y_n\}$ of $\cup C$. For each $i \leq n$, y_i is an element of some c_i of C. Let c be the greatest of the sets $c_1, \ldots, c_n$. Then each of the sets $y_1, \ldots, y_n$ is an element of c, so $\{y_1, \ldots, y_n\} \subseteq c$. Thus every finite subset of $\cup C$ is a subset of some element of C, hence a subset of some element of A, and hence is an element of A, by (1). Then, since A is of finite character, $\cup C \in A$. ∎

From this lemma we have the following.

Proposition 5.5. *Kuratowski's maximal principle implies Tukey's lemma.*

From this and Theorem 5.3 we have the following.

Theorem 5.6. *The axiom of choice implies Tukey's lemma.*

Remark By virtue of Theorem 5.3 we can say more specifically that if S is of finite character and there is a choice function for S, then S is of type M. However, we can do better: as will be shown (Corollary 6.4) if S is of finite character, then the existence of a choice function for $\cup S$ is enough to guarantee that S is of type M. (This is a stronger result by virtue of Exercise 4.3). In particular, this means that if S is a set of subsets of a *denumerable* set D, then without assuming any choice functions, if S is of finite character then S is of type M, because $\cup S$ is then denumerable, and the existence of a choice function for $\cup S$ is then automatic.

Zorn's lemma One form of Zorn's lemma is that if A is a set partially ordered under a relation $\leq$, then the set S of all subsets of A that are linearly ordered under $\leq$ is of type M.

Now, the property of being linearly ordered under $\leq$ is obviously a property of finite character (a subset B is linearly ordered under $\leq$ iff every 2-element subset of B is linearly ordered under $\leq$), and so the following is immediate.

Proposition 5.7. *Tukey's lemma implies Zorn's lemma.*

We thus now have the following.

Theorem 5.8. *The axiom of choice implies Zorn's lemma.*

Hausdorff's maximal principle The Hausdorff maximal principle is: for every non-empty set A, the set S of all chains of elements of A is of type M (every chain of elements of A is a subset of a maximal chain of elements of A).

This principle is simply the special case of Zorn's lemma in which $\leq$ is the relation $\subseteq$ of set inclusion, and so we immediately have the following.

Proposition 5.9. *Zorn's lemma implies Hausdorff's principle.*

We next show the following.

Proposition 5.10. *Hausdorff's principle implies Kuratowski's principle.*

Proof Assume Hausdorff's principle. Now suppose that S is a non-empty set closed under chain unions. Consider any element b. Then by Hausdorff's principle, b is an element of some maximal chain C of elements of S (since the singleton $\{b\}$ is trivially a chain, and hence is a subset of a maximal chain of elements of S). By hypothesis, S is closed under chain unions, so $\cup C \in S$. Then $b \subseteq \cup C$ and $\cup C$ is a maximal element of S, because if $\cup C$ were a proper subset of some element x of S, then C would be a proper subset of the chain $C \cup \{x\}$, contrary to the fact that C is a maximal chain of elements of S. Thus $\cup C$ is a maximal element of S. ∎

By Propositions 5.5, 5.7, 5.9 and 5.10 we now have the following.

Theorem 5.11. *Kuratowski's principle, Tukey's lemma, Zorn's lemma and Hausdorff's principle are all equivalent.*

All these maximal principles are implied by the axiom of choice. They are in fact equivalent to the axiom of choice, by virtue of the following theorem, whose proof we will sketch.

Theorem 5.12. *Hausdorff's maximal principle implies the axiom of choice.*

Proof (Sketch) Assume Hausdorff's maximal principle. Given a non-empty set S, let Σ be the set of all choice functions of all non-empty subsets of S. The set Σ is non-empty (since every finite subset of S obviously has a choice function) and Σ can be seen to be closed under chain unions. (For two functions f and g, to say that $f \subseteq g$ is to say that the domain of f is a subset of the domain of g and that for every element x in the domain of f, $f(x) = g(x)$.) Then by Hausdorff's maximal principle, Σ has a maximal element f, and f is then easily seen to be a choice function for the entire set S. ∎

Remark Of course Theorem 5.12 provides another proof of Proposition 5.10, but the first one is simpler and more direct.

EXERCISES

Exercise 5.1. Fill in the details of the proof of Theorem 5.12.

§6 Another approach to maximal principles

In this section we obtain some sharpenings of earlier results.

We call a class B an *extension* of a class A if $A \subseteq B$. We shall say that B is an *immediate extension* of A if B is an extension of A, and B contains exactly one more element than A. Now, here is another principle that on the surface is weaker than Kuratowski's maximal principle, but is actually equivalent to it.

Principle E If S is closed under chain unions, then every element b of S is a subset of some element of S that has no immediate extension in S.

Obviously, any *maximal* element of a set S has no immediate extension in S (in fact no proper extension at all in S), and so Principle **E** is a trivial consequence of Kuratowski's principle. However, if S is swelled (every subset of every element of S is an element of S), then an element x of S has no immediate extension in S if and only if x is a maximal element of S, and so Principle **E** implies that for every *swelled* set closed under chain unions, every element b of S is a subset of some maximal element of S. But also, by Lemma 5.4, any set of finite character is both swelled and closed under chain unions. Therefore, Principle **E** implies Tukey's lemma, which we know is equivalent to Kuratowski's principle, and so we have the following.

Theorem 6.1. *Principle E is equivalent to Kuratowski's maximal principle.*

The following is the main result of this section, and appears to be incomparable in strength with Theorem 5.2 (since the hypothesis and conclusion are both weaker).

Theorem 6.2. *Suppose that S is closed under chain unions and that there is a choice function C for $\cup S$. Then any element b of S is a subset of an element of S that has no immediate extension in S.*

§6. ANOTHER APPROACH TO MAXIMAL PRINCIPLES

Proof Assume the hypothesis. Let $A = \cup S$. For any element x of S, let $E(x)$ be the set of all elements $a \in A$ such that $x \cup \{a\}$ is an immediate extension of x. (We note that $E(x)$ is a subset of $\cup S$, whereas in the proof of Theorem 5.2, x^* is a subset of S.) We now define a progressing function g from S into S as follows. If x has no immediate extension in S, we take $g(x) = x$, whereas if x does have an immediate extension in S, then $E(x)$ is non-empty and we take $g(x)$ to be the result of adjoining to x the C-chosen element of $E(x)$ ($g(x)$ is then $x \cup \{C(E(x))\}$). Thus we have a progressing function g from S into S whose only fixed points are those elements of S that have no immediate extensions in S. By Theorem 2.10, any element b of S is a subset of some fixed point x of g, and such an element x has no immediate extension in S. ∎

We recall that if S is swelled, then any element of S having no immediate extension in S is a maximal element of S. Thus we have the following.

Corollary 6.3. *If S is swelled and closed under chain unions, and if $\cup S$ has a choice function, then S is of type M.*

Remark Would the corollary hold if we left out "swelled?" We posed this question to Professor Herman Rubin, who informed us that it would not.

By Lemma 5.4 we have the following further result.

Corollary 6.4. *If S is of finite character and $\cup S$ has a choice function, then S is of type M.*

Since there is a choice function for any denumerable set D (since D can be well ordered) we have the following.

Corollary 6.5. *If S is of finite character and $\cup S$ is denumerable, then S is of type M.*

Remark Since consistency is a property of finite character, it follows from this corollary that in a language in which the set of sentences is denumerable, any consistent set is a subset of some maximally consistent set. This result is due to Lindenbaum—see (Tarski 1930)—who proved it by a different method which is easily generalizable to an alternative proof of Corollary 6.5. Briefly, Lindenbaum's proof is as follows. Suppose that S is of finite character and that $\cup S$ is a denumerable set D. Let $d_1, \ldots, d_n, d_{n+1}, \ldots$ be any enumeration of all the elements of D. Now let b be any element of S. Using Theorem 8.1 from Chapter 3, we can generate a denumerable sequence of elements of S as follows. Take $b_0 = b$. Then, assuming b_n has been defined, we take b_{n+1} to be $b_n \cup \{d_n\}$, providing $b_n \cup \{d_n\}$ is an element of S; otherwise we take $b_{n+1} = b_n$. We let C be the set $\{b_0, b_1, \ldots, b_n, b_{n+1}, \ldots\}$. Then C is a chain (Exercise 8.5 of Chapter 3), hence $\cup C \in S$. Then it is not difficult to show that $\cup C$ is a maximal element of S (Exercise 6.1 below). This is the Lindenbaum construction.

EXERCISES

Exercise 6.1. In the Lindenbaum construction described above, prove that $\cup C$ is a maximal element of S.

Part III—The existence of minimally superinductive classes

§7 Cowen's theorem

Given a function g, we shall call a set x a *g-set* if x belongs to every class that is superinductive under g. Now, this definition of g-set involves quantifying over all *classes*, hence it does not follow from the separation principle (P_2 of Chapter 2) that there is such a thing as the class of all g-sets. In (Smullyan 1967) there is a proof that there is such a class, but the proof involves both the axiom of substitution (which will be introduced in Chapter 6) and the axiom of choice (the latter, incidentally is eliminable—see Exercise 5.3 of Chapter 6). Now, the only g-sets that will interest us are those in which g is progressing, and Cowen (1971) has shown that for a progressing function g, the existence of the class of all g-sets is provable without using the axiom of substitution. He did this by giving an equivalent characterization of g-sets that involves quantifying only over *sets*, and hence the existence of the class of all g-sets follows from the separation principle. This characterization will be very useful for several results in subsequent chapters, and Cowen's theorem, which we prove in this section, will play a fundamental role in further developments.

Throughout this section, g will be a progressing function fixed for the discussion. "Superinductive" will mean superinductive under g; "closed" will mean closed under g.

Definition 7.1. For any sets y and x, we will say that y is *closed* (under g) *relative to* x if for any $z \in y \cap \mathcal{P}(x)$, $g(z) \in y$. (Thus $(z \in y \wedge z \subseteq x) \supset g(z) \in y$.)

Definition 7.2. We shall say that a set S is *special for* x, or more briefly *x-special* if $\emptyset \in S$, S is closed relative to x, and S is closed under chain unions.

Definition 7.3. By M_x we mean the intersection of all x-special subsets of $\mathcal{P}(x)$.

We let M be the class of all x such that $x \in M_x$. Cowen's theorem is that M is minimally superinductive under g (and hence is the class of all g-sets). We now turn to the proof of this.

Lemma 7.4. $\mathcal{P}(x)$ *is x-special.*

Proof Obvious. ∎

Proposition 7.5. $\emptyset \in M$.

Proof By Lemma 7.4, $\mathcal{P}(\emptyset)$ is $\emptyset$-special, and $\mathcal{P}(\emptyset) = \{\emptyset\}$, and so $\{\emptyset\}$ is $\emptyset$-special, and is the only special subset of $\{\emptyset\}$ (since the only other subset of $\{\emptyset\}$ is $\emptyset$, which is empty). Therefore $M_\emptyset = \{\emptyset\}$, and since $\emptyset \in \{\emptyset\}$, then $\emptyset \in M_\emptyset$, and so $\emptyset \in M$. ∎

Lemma 7.6. *If* $x \subseteq y$ *then* $M_x \subseteq M_y$.

Proof Suppose $x \subseteq y$. We will first show that $M_y \cap \mathcal{P}(x)$ is x-special.
  (1) Since $\emptyset \in M_y$ and $\emptyset \in \mathcal{P}(x)$ then $\emptyset \in M_y \cap \mathcal{P}(x)$.
  (2) Suppose $z \in M_y \cap \mathcal{P}(x)$ and $g(z) \subseteq x$. Since $x \subseteq y$ then also $g(z) \subseteq y$. Since $z \in M_y$ and $g(z) \subseteq y$, then $g(z) \in M_y$. Thus $g(z) \in \mathcal{P}(x)$ and $g(z) \in M_y$, so $g(z) \in M_y \cap \mathcal{P}(x)$. Thus $M_y \cap \mathcal{P}(x)$ is closed relative to x.

(3) Since M_y and $\mathcal{P}(x)$ are both closed under chain unions, so is $M_y \cap \mathcal{P}(x)$.

By (1), (2) and (3), $M_y \cap \mathcal{P}(x)$ is x-special, and of course $M_y \cap \mathcal{P}(x) \subseteq \mathcal{P}(x)$, and so $M_x \subseteq M_y \cap \mathcal{P}(x)$. Therefore $M_x \subseteq M_y$. ∎

Proposition 7.7. *M is closed under chain unions.*

Proof Let C be a chain of elements of M. For each element $x \in C$, we have $x \subseteq \cup C$, hence $M_x \subseteq M_{\cup C}$ (by Lemma 7.6), but also $x \in M_x$ (since $x \in M$), hence $x \in M_{\cup C}$, and so $C \subseteq M_{\cup C}$, and therefore $\cup C \in M_{\cup C}$ (since $M_{\cup C}$ is closed under chain unions). Thus $\cup C \in M$. ∎

To show that M is closed under g is a bit more of a problem, and we use the following somewhat intricate lemma.

Lemma 7.8 (Main lemma). *If $x \subseteq y$ then $M_y \subseteq M_x \cup (\mathcal{P}(y) - \mathcal{P}(x))$.*

Proof Suppose $x \subseteq y$. Let $S = M_x \cup (\mathcal{P}(y) - \mathcal{P}(x))$. We first show that S is y-special.
(1) Since $\emptyset \in M_x$, then $\emptyset \in S$.
(2) Suppose $z \in S$ and $g(z) \subseteq y$. We are to show that $g(z) \in S$. If $g(z) \notin \mathcal{P}(x)$ then $g(z) \in \mathcal{P}(y) - \mathcal{P}(x)$ (since $g(z) \subseteq y$), hence $g(z) \in S$. Now suppose $g(z) \in \mathcal{P}(x)$—i.e., $g(z) \subseteq x$. This is the more interesting case. Since g is progressing, $z \subseteq g(z)$, and so $z \subseteq x$, and so $z \in \mathcal{P}(x)$. Then it is false that $z \in \mathcal{P}(y) - \mathcal{P}(x)$, but $z \in S$, so $z \in M_x$. Thus $z \in M_x$ and $g(z) \subseteq x$, so $g(z) \in M_x$, hence $g(z) \in S$. Thus $(z \in S \wedge g(z) \subseteq y)$ implies $g(z) \in S$, and so S is closed relative to y.
(3) Next we show that S is closed under chain unions. Suppose C is a chain of elements of S. All elements of S are subsets of y, because all elements of M_x are subsets of x, and hence of y, and of course all elements of $\mathcal{P}(y) - \mathcal{P}(x)$ are subsets of y. Since all elements of C are subsets of y, so is $\cup C$, hence $\cup C \in \mathcal{P}(y)$. If $\cup C \notin \mathcal{P}(x)$ then $\cup C \in \mathcal{P}(y) - \mathcal{P}(x)$, hence $\cup C \in S$. Now suppose $\cup C \in \mathcal{P}(x)$. Then for every element z of C, $z \subseteq \cup C \subseteq x$, so $z \subseteq x$. Then $z \in \mathcal{P}(y) - \mathcal{P}(x)$ is false, so $z \in M_x$. So if $\cup C \in \mathcal{P}(x)$ then $C \subseteq M_x$, hence $\cup C \in M_x$, and then $\cup C \in S$. Thus, whether $\cup C \in \mathcal{P}(x)$ or not, $\cup C \in S$. Thus S is closed under chain unions, which concludes the proof that S is y-special. Also $S \subseteq \mathcal{P}(y)$ (since every element of M_x is a subset of x, and hence of y, and $\mathcal{P}(y) - \mathcal{P}(x) \subseteq \mathcal{P}(y)$), and so $M_y \subseteq S$. ∎

Proposition 7.9. *For each z, $M_z \subseteq M$.*

Proof Suppose $x \in M_z$. Let $y = x \cup z$. Since $z \subseteq y$, then $M_z \subseteq M_y$ (Lemma 7.6), and hence $x \in M_y$. Also $x \subseteq y$, hence $M_y \subseteq M_x \cup (\mathcal{P}(y) - \mathcal{P}(x))$ (Lemma 7.8), so $x \in M_x \cup (\mathcal{P}(y) - \mathcal{P}(x))$. But $x \notin \mathcal{P}(y) - \mathcal{P}(x)$, hence $x \in M_x$. Thus $x \in M$, and so $M_z \subseteq M$.
∎

Proposition 7.10. *M is closed under g.*

Proof Suppose $x \in M$. Then $x \in M_x$. Since $x \subseteq x \cup g(x)$, $M_x \subseteq M_{x \cup g(x)}$ (Lemma 7.6), hence $x \in M_{x \cup g(x)}$. Also $g(x) \subseteq x \cup g(x)$, and so $g(x) \in M_{x \cup g(x)}$. Also $M_{x \cup g(x)} \subseteq M$ (Proposition 7.9), hence $g(x) \in M$. Thus M is closed under g. ∎

We now know that M is superinductive (by Propositions 7.5, 7.10 and 7.7). To show that M is minimally superinductive we prove the following.

Lemma 7.11. *For any superinductive class A and any element x, $M_x \subseteq A$.*

Proof Suppose A is superinductive. We show that $A \cap \mathcal{P}(x)$ is x-special.
 (1) Since $\emptyset \in A$ and $\emptyset \in \mathcal{P}(x)$, then $\emptyset \in A \cap \mathcal{P}(x)$.
 (2) Suppose $y \in A \cap \mathcal{P}(x)$ and $g(y) \in \mathcal{P}(x)$. Since $y \in A$ then $g(y) \in A$, hence $g(y) \in A \cap \mathcal{P}(x)$. Thus $A \cap \mathcal{P}(x)$ is closed relative to x.
 (3) Since A and $\mathcal{P}(x)$ are both closed under chain unions, so is $A \cap \mathcal{P}(x)$.
 Thus $A \cap \mathcal{P}(x)$ is x-special, and of course $A \cap \mathcal{P}(x) \subseteq \mathcal{P}(x)$, and so $M_x \subseteq A \cap \mathcal{P}(x)$, hence $M_x \subseteq A$. ∎

Now we have

Proposition 7.12. *M is minimally superinductive under g.*

Proof As already remarked, M is superinductive by Propositions 7.5, 7.10 and 7.7.
 Now, suppose A is any superinductive class. For any $x \in M$, $x \in M_x$ and $M_x \subseteq A$ (Lemma 7.11), hence $x \in A$. Thus $M \subseteq A$. And thus M is minimally superinductive. ∎

We thus have the following.

Theorem 7.13 (Cowen). *For any progressing function g, there is a class M whose elements are precisely those elements x which belong to all classes that are superinductive under g. Also, for any x, $x \in M$ if and only if x belongs to every subset of $\mathcal{P}(x)$ that contains $\emptyset$, is closed under g relative to x, and is closed under chain unions.*

§8 Another characterization of g-sets

Let us say that a class A is *g-ordered* if A is well ordered under inclusion in such a way that the following three conditions hold:

(1) the least element is $\emptyset$;

(2) each successor element y is $g(x)$, where x is the predecessor of y;

(3) each limit element is the union of the set of lesser elements.

Theorem 3.3 says that every g-tower is g-ordered.

Theorem 8.1. *If g is progressing, then for any set x, x is a g-set if and only if x is an element of some g-ordered set.*

Proof
 (1) Let M be the class of all g-sets. Then M is a g-tower, hence M is g-ordered by Theorem 3.3. Then for any $x \in M$, the set $L_{\leq}(x)$ (the set of all elements of M that are subsets of x) is obviously g-ordered, and x is a member of $L_{\leq}(x)$. Thus every $x \in M$ is a member of some g-ordered set.

(2) If y is g-ordered, then by an obvious transfinite induction on the g-ordering of y, every element of y must be an element of M (because $\emptyset$ is an element of M, and if $x \in y$ is an element of M other than the greatest element of y, then the successor of x is $g(x)$, which is an element of M, and for any limit element z of y, if each element of $L_<(z)$ is in M, so is z, since $z = \cup(L_<(z))$ and M is closed under chain unions). Thus every element of a g-ordered set y is an element of M. ∎

CHAPTER 5

ORDINAL NUMBERS

§1   Ordinal numbers

Natural numbers actually play two different roles. On the one hand we use them to specify the *size* of a finite set, as when we say "there are nine planets." But we also use them to express *order*, as for example "Pluto is the ninth planet from the sun." To determine how many members a finite set has, we count it, and when we do, we arrange its members in a sequence and say "first," "second," and so forth, thus introducing an *order*. This is the second role played by the natural numbers. Now, Cantor discovered that these two roles split apart when considering infinite sets. On the one hand we have the *cardinal numbers*, which measure *sizes* of sets, and which we study in Chapter 9. For now, we will study the *ordinal* numbers, which are used to measure well orderings. These ordinal numbers play an absolutely fundamental role in just about all the chapters that follow, and are crucial for the understanding of the independence proofs of Gödel and Cohen, at which we are aiming.

This chapter defines the ordinal numbers and establishes some of their basic properties. Their use as measures of well orderings will be dealt with in the next chapter, and their use in independence proofs will be amply apparent in Parts II and III of this volume.

We let σ be the successor function of Chapter 3 — i.e., for any set x, the set $\sigma(x) = x \cup \{x\}$. We also write x^+ for $\sigma(x)$. In this chapter, σ will be the only progressing function in which we are interested, and the terms "inductive," "superinductive," "tower" shall now be understood always with reference to σ. (Thus, e.g., "superinductive" means superinductive under σ.)

We defined x to be a *natural number* if x belongs to every inductive set. The notion of "ordinal number"—more briefly "ordinal"—is an extension of the notion of "natural number." Various equivalent definitions of "ordinal" exist in the literature. It turns out that under any of these definitions, x is an ordinal number if and only if x belongs to every *superinductive class*. We shall take this property as our very definition of "ordinal number," since it seems a particularly natural extension of the notion of "natural number" and a technical development using this definition is smooth and simple. We shall subsequently show the equivalence of this definition to some of the standard ones.

Definition 1.1. x is an ordinal number if x belongs to every superinductive class. We use Greek letters $\alpha, \beta, \gamma, \ldots$, as variables over ordinals.

Although our definition of *ordinal* involves quantifying over all classes, the class of all ordinals exists by Theorem 7.13, Chapter 4. We let On be the class of all ordinals.

Thus On is the class that is minimally superinductive under σ. Then the following is immediate.

Theorem 1.2. *The class of ordinals has the following properties.*

(1) *0 is an ordinal.*

(2) *If α is an ordinal, so is α^+.*

(3) *The union of any chain of ordinals is an ordinal.*

(4) *(Superinduction principle) Suppose A is a class satisfying the following three conditions:*

   (a) $0 \in A$.

   (b) *For any ordinal $\alpha \in A$, the ordinal $\alpha^+ \in A$.*

   (c) *The union of any chain of ordinals in A is in A.*

Then A contains all ordinals.

In the last chapter we proved (Theorem 3.3) that every g-tower is well ordered under inclusion. Hence we at once have the following basic result.

Theorem 1.3. *The class On of all ordinal numbers is well ordered under $\subseteq$.*

An ordinal is called a *successor ordinal*, *limit ordinal* if it is respectively a successor element, limit element in the well ordering of On under inclusion.

Corollary 1.4. *The union of any set of ordinals is an ordinal.*

Proof By Theorem 1.2 part 3 and the fact that any set of ordinals is a chain (by Theorem 1.3). ∎

Theorem 1.5. *Every natural number is an ordinal.*

Proof Since On is superinductive, it is inductive, and since ω is minimally inductive, then $\omega \subseteq On$. (Alternatively, by the parts 4a and 4b of Theorem 1.2, it follows by mathematical induction that every natural number is an ordinal.) ∎

Theorem 1.6. *Every element of an ordinal is an ordinal—i.e., the class On is transitive.*

Proof By superinduction. It is vacuously true that every element of 0 is an ordinal. Suppose that every element of α is an ordinal. Then every element of α^+ is either α, or an element of α, hence is an ordinal. Lastly, suppose A is a chain of ordinals with each member α having the property that every member of α is an ordinal; let $\beta = \cup A$. Every element of β is an element of some $\alpha \in A$, hence is an ordinal by assumption on A. Thus every element of β is an ordinal. This completes the proof. ∎

Theorem 1.7. *Every ordinal number is transitive.*

§1. ORDINAL NUMBERS 73

Proof By superinduction. Clearly 0 is transitive. If x is transitive, so is x^+ (we proved this in Chapter 3, when we showed that every natural number is transitive). Also the union of any class of transitive sets is transitive (Theorem 10.2 of Chapter 2), hence the union of any chain of transitive sets is transitive. This completes the proof. ∎

Theorem 1.8. *For α any ordinal, $\alpha \notin \alpha$.*

Proof In Chapter 3 we proved that no natural number is an element of itself, by showing that $0 \notin 0$ (obvious) and for any transitive set x, if $x \notin x$ then $x^+ \notin x^+$. To prove the present theorem for all ordinals α it merely remains to show that if A is a chain of ordinals, none of which is a member of itself, then $\cup A$ is not a member of itself. Equivalently, we show that if $\cup A \in \cup A$ (where A is a chain of ordinals) then for some $\alpha \in A$, $\alpha \in \alpha$.

So let A be a chain of ordinals, let $\alpha = \cup A$ (we know $\cup A$ is an ordinal) and suppose $\alpha \in \alpha$. Then $\alpha \in \cup A$, hence for some $\beta \in A$, $\alpha \in \beta$. But $\beta \in A$, so $\beta \subseteq \cup A$—i.e., $\beta \subseteq \alpha$. Also $\alpha \subseteq \beta$ (because $\alpha \in \beta$ and β is transitive), hence $\alpha = \beta$. Hence $\beta \in \beta$. Thus some element of A, namely β, is an element of itself. This concludes the proof. ∎

Corollary 1.9. $\alpha \subset \alpha^+$ *(α is a* proper *subset of α^+).*

Proof If $\alpha = \alpha^+$, we would have $\alpha \in \alpha$ (since $\alpha \in \alpha^+$), which is contrary to Theorem 1.8. Hence $\alpha \neq \alpha^+$. But $\alpha \subseteq \alpha^+$, so $\alpha \subset \alpha^+$. ∎

We thus see that the function σ is *strictly* progressing (as well as slowly progressing). Then by Theorem 2.9 of Chapter 4 we have the following.

Theorem 1.10. *The class On of ordinals is not a set.*

Next, by Theorem 3.3 of Chapter 4, and the fact that no α is the greatest element of *On*, we have the following.

Theorem 1.11. *On is well ordered under inclusion in such a way that the following three conditions hold:*

(1) the least ordinal is 0;

(2) the successor of any ordinal α is α^+;

(3) each limit ordinal is the union of the set of lesser ordinals.

We understand $\alpha < \beta$ to mean: α precedes β in the usual ordering of the ordinals—in other words, $\alpha \subset \beta$. Thus also $\alpha \leq \beta$ is tantamount to $\alpha \subseteq \beta$.

Theorem 1.12. $\alpha < \beta$ *iff* $\alpha \in \beta$.

Proof
(1) If $\alpha \in \beta$, then $\alpha \subseteq \beta$ (since β is transitive), but if $\alpha = \beta$, we would have $\alpha \in \alpha$, contrary to Theorem 1.8, hence $\alpha \in \beta$ implies $\alpha < \beta$.
(2) If $\alpha < \beta$ then $\alpha^+ \subseteq \beta$ (by Theorem 2.4 of Chapter 4), and since $\alpha \in \alpha^+$, then $\alpha \in \beta$. ∎

Corollary 1.13. *Every ordinal is the set of all lesser ordinals.*

We recall that in a well ordering, we are using $L_<(x)$ to mean the set of all elements less than x. Thus the corollary above says that for any ordinal α, $L_<(\alpha) = \alpha$.

Corollary 1.14. *For any ordinals α and β, either $\alpha \in \beta$ or $\beta \in \alpha$ or $\alpha = \beta$.*

Proof By Theorem 1.3, On is a nest, hence for any ordinals α and β, either $\alpha \subset \beta$ or $\beta \subset \alpha$ or $\alpha = \beta$. But $\alpha \subset \beta$ iff $\alpha \in \beta$ and $\beta \subset \alpha$ iff $\beta \in \alpha$, hence the result follows. ∎

Corollary 1.15. *For any two distinct elements x, y of α, either $x \in y$ or $y \in x$.*

Proof By Corollary 1.14 and the fact that every element of an ordinal is an ordinal (Theorem 1.6). ∎

Corollary 1.16. *For any ordinals α, β, γ: $(\alpha \in \beta \wedge \beta \in \gamma) \supset \alpha \in \gamma$.*

Proof By Theorem 1.12, this is equivalent to $(\alpha \subset \beta \wedge \beta \subset \gamma) \supset \alpha \subset \gamma$. ∎

Theorem 1.17. *If $\alpha^+ = \beta^+$ then $\alpha = \beta$.*

Proof In the well ordering of On under $\subseteq$, α^+ is the successor of α and β^+ is the successor of β, and no two distinct elements can have the same successor, so if $\alpha^+ = \beta^+$ then α must be the same ordinal as β. ∎

Theorem 1.18. *If S is a set of ordinals then there is an ordinal greater than every element of S. More specifically, if S contains a greatest ordinal α, then α^+ is greater than every element of S, and if S doesn't contain a greatest element, then $\cup S$ is an ordinal greater than every element of S.*

Proof Since α^+ is greater than α, then if α is the greatest element of S, α^+ is greater than all elements of S.

Now suppose that S has no greatest element. Then $\cup S \notin S$ (for if $\cup S$ were an element of S, it would be the greatest element), and so for every ordinal $\alpha \in S$, $\alpha \neq \cup S$, yet $\alpha \subseteq \cup S$, hence $\alpha \subset \cup S$. Also $\cup S$ is an ordinal (Corollary 1.4), so $\alpha < \cup S$. Thus $\cup S$ is an ordinal greater than any ordinal in S. ∎

Theorem 1.19. *If λ is a limit ordinal, then for any ordinal α, if $\alpha < \lambda$ then $\alpha^+ < \lambda$.*

Proof Suppose λ is a limit ordinal and $\alpha < \lambda$. Then $\alpha^+ \leq \lambda$ (since α^+ is the successor of α). But $\alpha^+ \neq \lambda$ (since λ is not a successor ordinal), so $\alpha^+ < \lambda$. ∎

Theorem 1.20.

(1) $\cup \alpha \subseteq \alpha$.

(2) $\cup \alpha^+ = \alpha$.

(3) For any limit ordinal λ, $\cup \lambda = \lambda$.

Proof

(1) We show $x \in \cup\alpha$ implies $x \in \alpha$. So suppose $x \in \cup\alpha$. Then for some β in α, $x \in \beta$. β is an ordinal, by Theorem 1.6. Thus $x \in \beta$ and $\beta \in \alpha$, so $x \in \alpha$ (Corollary 1.16).

(2) Suppose $x \in \cup\alpha^+$. Then for some ordinal β, $x \in \beta$ and $\beta \in \alpha^+$. Since $\beta < \alpha^+$, then $\beta \leq \alpha$, and since $x < \beta$, then $x < \alpha$, thus $x \in \alpha$. Thus $\cup\alpha^+ \subseteq \alpha$.

Now suppose $x \in \alpha$. Since also $\alpha \in \alpha^+$, then $x \in \cup\alpha^+$. Thus $\alpha \subseteq \cup\alpha^+$. Thus $\cup\alpha^+ \subseteq \alpha$ and $\alpha \subseteq \cup\alpha^+$, so $\cup\alpha^+ = \alpha$.

(3) By (1), $\cup\lambda \subseteq \lambda$. Now suppose $x \in \lambda$, where λ is a limit ordinal. Since λ is a limit ordinal, $x^+ < \lambda$ (Theorem 1.19), and so $x \in x^+$ and $x^+ \in \lambda$, hence $x \in \cup\lambda$. Thus $\lambda \subseteq \cup\lambda$, and so $\cup\lambda = \lambda$. ∎

§2 Ordinals and transitivity

We wish to show that every transitive set of ordinals is an ordinal and that the only transitive class of ordinals which is not a set is the class On of all ordinals. First the following.

Lemma 2.1. *If A is a transitive class of ordinals, then for any ordinal α not in A, $A \subseteq \alpha$.*

Proof Suppose that A is a transitive class of ordinals and that α is an ordinal outside A. Let β be any ordinal in A. We are to show that $\beta \in \alpha$ (and hence that $A \subseteq \alpha$). Since $\beta \in A$ and $\alpha \notin A$, then $\beta \neq \alpha$. Since $\beta \in A$ and A is transitive, then all elements of β are in A, but since $\alpha \notin A$, then $\alpha \notin \beta$. So $\alpha \neq \beta$ and $\alpha \notin \beta$, hence $\beta \in \alpha$ (Corollary 1.14). ∎

Theorem 2.2.

(1) Every transitive set of ordinals is an ordinal.

(2) The only transitive class of ordinals that is not a set is the class On of all ordinals.

Proof

(1) Suppose x is a transitive set of ordinals. By Theorem 1.18, not every ordinal is in x, so let α be the least ordinal not in x. Then every lesser ordinal is in x, so $\alpha \subseteq x$. Also $x \subseteq \alpha$ by Lemma 2.1. Therefore $x = \alpha$, so x is an ordinal.

In short, if x is a transitive set of ordinals, then x is the smallest ordinal not in x.

(2) Suppose A is a transitive class of ordinals. If A lacked an ordinal α, then A would be a subclass of α (by Lemma 2.1), which would make A a set. And so if A is not a set, A lacks no ordinal, and hence $A = On$. ∎

The following characterization of ordinals is particularly neat.

Theorem 2.3. *x is an ordinal if and only if every transitive proper subset of x is an element of x.*

Proof
(1) Consider any ordinal α. If y is a proper subset of α and y is transitive, then y is a transitive set of ordinals, hence y is an ordinal (Theorem 2.2), and since y is a proper subset of α, then $y < \alpha$, so $y \in \alpha$. Thus every transitive proper subset of α is an element of α.

(2) Conversely, suppose x is a set such that every transitive proper subset of x is an element of x. Let α be the least ordinal not in x. We will show that $x = \alpha$.

Since α is the least ordinal outside x, then every ordinal less than α is in x, hence $\alpha \subseteq x$. If α were a *proper* subset of x, then α would be an element of x (since all transitive proper subsets of x are elements of x, by hypothesis). Therefore $\alpha = x$, so x is an ordinal. ∎

Theorem 2.4. *If K is a class such that every transitive subset of K is an element of K, then K contains all ordinals.*

Proof Assume the hypothesis. Since On is well ordered under $\subseteq$ we can use transfinite induction on On—we will use the first form (Theorem 1.8 of Chapter 4)—and show that every ordinal is in K.

Suppose α is an ordinal such that every ordinal less than α is in K. Then $\alpha \subseteq K$ and α is transitive, so $\alpha \in K$ (by hypothesis). Thus if all ordinals less than α are in K, so is α. Therefore by Theorem 1.8 of Chapter 4, all ordinals are in K. ∎

EXERCISES

Exercise 2.1. Let us say that a class K is T-*closed* if every transitive subset of K is an element of K. Theorem 2.4 asserts that if x is an ordinal, then x belongs to all T-closed classes. Is it conversely true that if x belongs to all T-closed classes, then x must be an ordinal?

Exercise 2.2. State whether the following is true or false: The intersection of any non-empty set of ordinals is an ordinal.

Exercise 2.3. Prove that for any set x, x is an ordinal if and only if x can be well ordered in such a way that each element y of x is the set of all elements of x that precede y in the ordering. (This constitutes one standard definition of *ordinal*.)

§3 Some ordinals

For any ordinal α, by $\alpha + 1$ is meant α^+, by $\alpha + 2$ is meant α^{++}, etc.—in general for any natural number n, $\alpha + (n+1)$ is the ordinal $(\alpha + n)^+$.

The set ω of natural numbers is a transitive set of ordinals, hence is an ordinal. It is the smallest ordinal that follows all the natural numbers, and is a limit ordinal. Since ω is an ordinal, so are $\omega + 1$, $\omega + 2$, ..., $\omega + n$, ..., and so we have denumerably many infinite ordinals. The class of ordinals of the form $\omega + n$, together with the natural numbers, is a denumerable class and is denoted $\omega \cdot 2$ (read "ω times 2"). This class is indeed a transitive class of ordinals, but it is not possible from the Zermelo axioms to prove that $\omega \cdot 2$ is a set! However in the next chapter we will add the axiom of substitution (since we can't get any further in ordinal theory without it), a consequence of which is that any class that can be put into a 1-1 correspondence with a set is itself a set. With this axiom, $\omega \cdot 2$ then becomes a set, and hence an ordinal number. Having the ordinal $\omega \cdot 2$ we then have $\omega \cdot 2 + 1, \omega \cdot 2 + 2, \ldots, \omega \cdot 2 + n, \ldots,$ and after this sequence we have the ordinal $\omega \cdot 3$, and so on.

We have seen that (assuming the axiom of substitution), the sequence of ordinals begins $0, 1, 2, \ldots, \omega, \omega + 1, \omega + 2, \ldots, \omega \cdot 2, \omega \cdot 2 + 1, \omega \cdot 2 + 2, \ldots, \omega \cdot 3, \omega \cdot 3 + 1, \ldots$. At the end of the next chapter we will introduce ordinal arithmetic, and then it will be seen that $\omega \cdot 3$ really is the product of ω and 3, and so on.

CHAPTER 6

ORDER ISOMORPHISM AND TRANSFINITE RECURSION

In this chapter we first prove a basic theorem on well ordering and then introduce the axiom of substitution and prove two other fundamental theorems, the second of which is the basis for several of the chapters that follow. In this chapter we will begin to see the importance of ordinals.

§1 A few preliminaries

Definition 1.1. For any class A and function F, by $F''(A)$ is meant the class of all elements $F(x)$ where x is in A. Thus $y \in F''(A)$ if and only if $y = F(x)$ for some $x \in A$.

In general, $F(x) \neq F''(X)$. As an example, suppose that F is a function from natural numbers to natural numbers and for any natural number x, $F(x) = 5x$ (5 times x). Then $F(4) = 20$, but since 4 is the set $\{0, 1, 2, 3\}$, then $F''(4)$ is $\{F(0), F(1), F(2), F(3)\}$, which is the set $\{0, 5, 10, 15\}$.

Thus $F(x)$ is the result of *applying* F to x, whereas $F''(X)$ is the result of applying F to all the elements of x, and then collecting the results.

We recall the notation $F \upharpoonright A$, which is the class of all ordered pairs $\langle a, F(a) \rangle$, where $a \in A$. Thus $F''(A)$ is simply the range of $F \upharpoonright A$.

We note that $F \upharpoonright A$ is a subclass of the Cartesian product $A \times F''(A)$.

Proposition 1.2. *For any set x, $F''(x)$ is a set if and only if $F \upharpoonright x$ is a set.*

Proof If $F \upharpoonright x$ is a set, so is its range, which is $F''(x)$. Conversely, if $F''(x)$ is a set, then so is the Cartesian product $x \times F''(x)$, but $F \upharpoonright x$ is a subclass of this Cartesian product, and hence is a set. ∎

Inverses For any relation R, by its *inverse* R^{-1}, is meant the class of all ordered pairs $\langle y, x \rangle$ such that $\langle x, y \rangle \in R$. Thus xRy iff $yR^{-1}x$. For any *function* φ, φ^{-1} is a relation, but in general not a function. However, if φ is a 1-1 function, then φ^{-1} is also a function, and the statements $\varphi(x) = y$ and $\varphi^{-1}(y) = x$ are equivalent. Obviously the domain (range) of φ^{-1} is the range (domain) of φ. Also, for any element x in the domain of φ^{-1}, we have $\varphi(\varphi^{-1}(x)) = x$.

Composition For two functions f and g such that the range of g is included in the domain of f, by the function fg (sometimes written $f \circ g$, and called the *composite* of f with g) we mean the function that assigns to every element x in the domain of g the element $f(g(x))$. Thus $(fg)(x) = (f(g(x))$.

Identity functions We call f an *identity function* if for every x in the domain of f, $f(x) = x$. If φ_1 and φ_2 are functions from a class A onto a class B and if $\varphi_1 \varphi_2^{-1}$ is an identity function, then obviously $\varphi_1 = \varphi_2$.

Lower sections of well orderings Suppose that A is linearly ordered under $\leq$. A subclass B of A is called a *lower section* of A (with respect to $\leq$) if all elements inside B are less than all elements of A that are not in B. B is called a *proper* lower section of A if B is a lower section of A and $B \neq A$.

We recall the notation $L_<(x)$, meaning the class of all elements y such that $y < x$. In the literature, $L_<(x)$ is sometimes written "$seg(x)$," and called "the initial segment of x." Obviously $L_<(x)$ is a proper lower section, so every initial segment is also a proper lower section. If the ordering $\leq$ is a *well* ordering, then conversely every proper lower section B is also an initial segment—namely $L_<(x)$, where x is the *least* element of A not in B.

By a *proper* well ordering is meant a well ordering in which every proper lower section is a set. If $\leq$ is a proper well ordering of a class A, then we say that A is *properly* well ordered under $\leq$. Obviously every well ordering on a *set* is a proper well ordering, but it is easily possible to construct well orderings on classes that are not proper.

Proposition 1.3. *Every g-tower is* properly *well ordered under* $\subseteq$.

Proof Any proper lower section is an initial segment $L_<(x)$ for some x. Since each element of $L_<(x)$ is a subset of x, then $L(x) \subseteq \mathcal{P}(x)$, and $\mathcal{P}(x)$ is a set, hence $L_<(x)$ is a set. ∎

Nests of functions We shall frequently be interested in nests of *functions*—i.e., nests whose elements are functions. For two functions f, g, to say that $f \subseteq g$ is equivalent to the condition that for every x in the domain of f, $f(x) = g(x)$. If $f \subseteq g$, we sometimes refer to g as an *extension* of f; if $f \subset g$, then we say that g is a *proper* extension of f.

The basic simple fact about nests of functions is the following proposition, whose proof is a routine exercise.

Proposition 1.4. *Let N be a nest of functions. Then $\cup N$ is a function. Further, the domain (range) of $\cup N$ is the union of the class of domains (respectively ranges) of the elements of N. If every element of N is 1-1, so is $\cup N$.*

EXERCISES

Exercise 1.1. Prove Proposition 1.4.

Exercise 1.2. Give an example of a well ordering that is not proper.

§2 Isomorphisms of well orderings

In what follows, A and B shall be understood to be properly well ordered classes. By an *isomorphism* φ from A onto B we mean a 1-1 function which is order preserving, i.e., which is such that for any a_1, a_2 in A, a_1 precedes a_2 in A iff $\varphi(a_1)$ precedes $\varphi(a_2)$

§2. ISOMORPHISMS OF WELL ORDERINGS

in B. (Strictly speaking, we should define φ as being an isomorphism from A with respect to one ordering relation $\leq_1$ onto B with respect to another ordering relation $\leq_2$, i.e., φ is really an isomorphism from the *system* $\langle A, \leq_1 \rangle$ to the system $\langle B, \leq_2 \rangle$. But in practice, no ambiguity will result from omitting explicit mention of $\leq_1$ and $\leq_2$.)

In this section we study some basic properties of isomorphisms of well orderings.

Theorem 2.1. *Let φ be an isomorphism from a well ordered system $\langle A, \leq \rangle$ into itself. Under φ, no elementا moves backwards—i.e., for each $a \in A$, $a \leq \varphi(a)$.*

Proof Suppose $a > \varphi(a)$. Applying φ to both sides, $\varphi(a) > \varphi(\varphi(a))$. Thus if some element a moves backwards, then so does some lesser element, namely $\varphi(a)$. Hence there is no *first* element of A that moves backwards, so there can be no element of A at all that moves backwards. ∎

Corollary 2.2. *The only isomorphism from a well ordered class* onto *itself is the identity.*

Proof Let φ be an isomorphism from A onto itself. Then the inverse φ^{-1} is also an isomorphism from A onto A. Hence (by Theorem 2.1) we have both
(1) $a \leq \varphi(a)$
(2) $a \leq \varphi^{-1}(a)$.
From (2) we have (applying φ to both sides) $\varphi(a) \leq \varphi(\varphi^{-1}(a))$. Hence
(2') $\varphi(a) \leq a$.
By (1) and (2'), $a = \varphi(a)$. ∎

Corollary 2.3. *Two well ordered classes are isomorphic in at most one way.*

Proof Suppose φ_1 and φ_2 are both isomorphisms from A onto B. Then φ_2^{-1} is an isomorphism from B onto A. Hence $\varphi_1 \varphi_2^{-1}$ is an isomorphism from A onto A. By the corollary above, $\varphi_1 \varphi_2^{-1}$ must be the identity map, so $\varphi_1 = \varphi_2$. ∎

Corollary 2.4. *No well ordered class can be isomorphic to any of its initial segments.*

Proof Let φ be an isomorphism from A into itself. Take any $a \in A$. To show that $L_<(a)$ is not the range of φ, it suffices to show that at least one element $\varphi(x)$ is not in $L_<(a)$. But obviously $\varphi(a) \notin L_<(a)$, since $\varphi(a) \geq a$. ∎

Corollary 2.5. *No two distinct ordinals α, β are isomorphic.*

Corollary 2.6. *No two distinct lower sections of the same well ordered system are isomorphic to each other.*

Corollary 2.7. *A lower section of A can be isomorphic to at most one lower section of B and isomorphic to it in at most one way.*

The next corollary is particularly critical in the proof of the comparability theorem for proper well orderings, Theorem 2.9.

Corollary 2.8. *Let I_1 be an isomorphism from a lower section of A onto a lower section of B; let I_2 be another such isomorphism. Then either $I_1 \subseteq I_2$ or $I_2 \subseteq I_1$.*

Proof Let D_1, D_2 be the respective domains of I_1, I_2. Clearly D_1, D_2 are comparable (they are both lower sections of A); we suppose $D_1 \subseteq D_2$. Then $I_2 \upharpoonright D_1$ is an isomorphism from D_1 onto a lower section of B. By the previous corollary, $I_2 \upharpoonright D_1 = I_1$—i.e., $I_1 \subseteq I_2$. Similarly if $D_2 \subseteq D_1$ then $I_2 \subseteq I_1$. ∎

The comparability theorem The following theorem is sometimes called "the fundamental theorem of well ordering," sometimes "the comparability theorem for well orderings."

Theorem 2.9. *For any two properly well ordered classes A, B, either A is isomorphic to a lower section of B (perhaps all of B) or B is isomorphic to a lower section of A (perhaps all of A).*

Proof Let $\mathcal{N}$ be the class of all isomorphisms from lower sections of A which are *sets* onto lower sections of B which are sets (each such isomorphism is itself obviously a set). By our last corollary, $\mathcal{N}$ is a *nest*. It may or may not contain a maximal element, the important point now to observe is that no isomorphism from a *proper* lower section of A onto a *proper* lower section of B can be a maximal element of $\mathcal{N}$. For if J is an isomorphism from a proper lower section $L_<(a)$ of A onto a proper lower section $L_<(b)$ of B, then $J \cup \{\langle a,b\rangle\}$ is obviously an element of $\mathcal{N}$ which properly extends J.

Now let $I = \cup \mathcal{N}$. Using the fact that $\mathcal{N}$ is a nest, it is easily seen that I must be an isomorphism from a lower section of A (not necessarily proper) onto a lower section of B (not necessarily proper). What remains to be shown is that either the domain of I is *all* of A or the range of I is *all* of B. Equivalently, we must show that I cannot be from a *proper* lower section of A onto a *proper* lower section of B.

Well, if it were, then it would surely be a maximal member of $\mathcal{N}$ (since $I = \cup \mathcal{N}$). But we have shown that no isomorphism from a proper lower section of A onto a proper lower section of B can be a maximal member of $\mathcal{N}$. Therefore I cannot be an isomorphism from a proper lower section of A onto a proper lower section of B. This concludes the proof. ∎

§3 The axiom of substitution

We have defined V to be a Zermelo universe if it satisfies axioms A_1–A_7. At this point we need Fraenkel's axiom of substitution (also called the axiom of *replacement*).

A_8 For any function F and any *set* x, $F''(x)$ is a set.

The intuitive idea is that if we take any set x and replace each element y of x by the set $F(y)$, the resulting class (which is $F''(x)$) is a set.

Definition 3.1. We now say V is a *Zermelo-Fraenkel* universe if it satisfies the axioms A_1–A_8.

Until further notice, V will be assumed to be a Zermelo-Fraenkel universe.

Discussion In the presence of A_8, axiom A_2 (which is that every subclass of a set is a set) becomes superfluous, by the following argument.

Suppose that x is a set and A is a class such that $A \subseteq x$. Using A_8, we wish to show that A is a set. If A is empty, then of course A is a set, so we assume that A is non-empty. Let c be any element of A. Now let F be the class of all ordered pairs $\langle a, a \rangle$ where $a \in A$, together with all ordered pairs $\langle y, c \rangle$, where $y \in x - A$. Thus F is a function whose domain is x and for any y in x, if $y \in A$, then $F(y) = y$, but if $y \notin A$, then $F(y) = c$. Thus for every y in x, $F(y) \in A$, and in fact $F''(x) = A$. Since x is a set, then so is A, by the axiom of substitution.

EXERCISES

Exercise 3.1.

(a) Show that the axiom of substitution is equivalent to the following axiom A_8': For any function F and any set x, the function $F \upharpoonright x$ is a set.

(b) Show that A_8 is also equivalent to the following axiom A_8'': Any function whose domain is a set is a set.

Exercise 3.2. The axiom of substitution obviously implies that any class that can be put into a 1-1 correspondence with a set is a set. This ostensibly weaker condition, however, is actually equivalent to the axiom of substitution if the axiom of choice is assumed. Why is this?

§4 The counting theorem

Now we turn to the first major application of the axiom of substitution.

Theorem 4.1 (The counting theorem). *Every well ordered set is isomorphic to exactly one ordinal number α. Every properly well ordered class which is not a set is isomorphic to the class On of all ordinals (as ordered in the natural manner, under $\subseteq$).*

Proof The proof is easy now that we have the comparability theorem! Let A be properly well ordered. By the comparability theorem 2.9, either A is isomorphic to a lower section of On, or On is isomorphic to a lower section of A. If A is a set, then every lower section of A is a set, hence by the axiom of substitution, it is not possible for On to be isomorphic to a lower section of A (since On would then be in a 1-1 correspondence with it), so it must be that A is isomorphic to a lower section of On, which must be a *proper* lower section of On (since On is not a set), and so A is then isomorphic to an ordinal number (a unique one, by Corollary 2.5).

Suppose now that A is not a set. Then by the axiom of substitution, it cannot be that either A or On is isomorphic to a *proper* lower section of the other, hence it must be that A is isomorphic to On. ∎

Discussion The ordinal numbers would fail in their major role, if it were not for the counting theorem. This theorem is really to the effect that for any properly well ordered system A, we can assign ordinal numbers as "indices" of the elements, and thus speak of the first, second, ..., α^{th}, ... elements of the system; moreover we can assign these indices in an order-preserving manner—i.e., for any α and β, $\alpha < \beta$ iff the αth element of the system precedes the βth element.

In the next section we do this for *minimally superinductive classes* and obtain a neat proof of some basic results known as *transfinite recursion theorems*.

§5 Transfinite recursion theorems

Let g be a strictly progressing function (strictly progressing in the sense that it is never the case that $g(x) = x$), and let M be the class that is minimally superinductive under g. We know by Theorems 2.9 and 3.3 of Chapter 4 that M is not a set, and that M is well ordered under $\subseteq$ in such a way that:

(1) the least element of M is $\emptyset$;

(2) the successor of any element x (in the well ordering) is $g(x)$;

(3) each limit element is the union of its set of predecessors.

We know by the counting theorem that there exists a unique isomorphism from On onto M. We let M_α be the element of M corresponding to α under this isomorphism (we read "M_α" as "the αth element of M"). It is immediate from the notion of "isomorphism" that M_0 is the least element of M under the well ordering induced by $\subseteq$, so $M_0 = \emptyset$. It is also obvious that M_{α^+} is the successor of M_α, i.e., $M_{\alpha^+} = g(M_\alpha)$. Also if λ is a limit ordinal, then M_λ must be a limit element of M, hence M_λ is the *union* of the set of all M_α, where $\alpha < \lambda$. We write this union as $\cup_{\alpha<\lambda} M_\alpha$. We then have the following.

Theorem 5.1. *Let M be minimally superinductive under g, where g is strictly progressing. For an ordinal α, let M_α be the αth element of M in the well ordered system $\langle M, \subseteq \rangle$. Then for every ordinal α and every limit ordinal λ we have:*

(1) $M_0 = \emptyset$;

(2) $M_{\alpha^+} = g(M_\alpha)$;

(3) $M_\lambda = \cup_{\alpha<\lambda} M_\alpha$.

This theorem is one form of the transfinite recursion theorem. From it, we now proceed to derive several other forms.

By an *ordinal sequence* we shall understand a function θ whose domain is an ordinal number α (i.e., the domain of θ is the set of all $\gamma < \alpha$); we also shall refer to such a sequence as an α-sequence, or an ordinal sequence of *length* α. By an On-sequence we mean a function whose domain is On.

By an *extending operation* E we shall mean a function defined on all ordinal sequences and which, when applied to any ordinal sequence of length α, yields an ordinal sequence of length α^+. (We paraphrase this last condition by saying that E *extends* every ordinal sequence of length α to an ordinal sequence of length α^+.)

§5. TRANSFINITE RECURSION THEOREMS

Looking at an α-sequence θ as a set of ordered pairs, $E(\theta)$ consists of the set θ with one more ordered pair $\langle \alpha, x \rangle$ thrown in—thus E is a slowly but strictly progressing function.

Before proceeding further, the reader should verify the following simple lemma.

Lemma 5.2.

(1) The union of any chain C of ordinal sequences is an ordinal sequence. If λ is a limit ordinal such that all elements of C are of length less than λ, and if for every $\alpha < \lambda$ there is an element of C of length α, then $\cup C$ is of length λ.

(2) If N is a nest of ordinal sequences, and N is not a set, then $\cup N$ is a function whose domain is On.

In what follows, we shall always use "F" to denote a function whose domain is On. We use "E" to denote any extending operation.

Theorem 5.3. *For any extending operation E there exists an F on On such that for every α:*

$$F \upharpoonright \alpha^+ = E(F \upharpoonright \alpha).$$

Proof Let M be the class that is minimally superinductive under E. We use Theorem 5.1 (reading "E" for "g"). We know $M_0 = 0$, so M_0 is a 0-sequence. If M_α is an α-sequence then M_{α^+} is an α^+-sequence (because $M_{\alpha^+} = E(M_\alpha)$). If λ is a limit ordinal, and if for each $\alpha < \lambda$, M_α is an α-sequence, then M_λ is a λ-sequence (by Lemma 5.2). By transfinite induction it then follows that for every α, M_α is an ordinal sequence of length α. Let $F = \cup M$. Then F is a function on On (by Lemma 5.2 again). Since each $M_\alpha \subseteq F$, and $\text{Dom}(M_\alpha) = \alpha$, then $M_\alpha = F \upharpoonright \alpha$. Thus condition (2) of Theorem 5.1 yields $F \upharpoonright \alpha^+ = E(F \upharpoonright \alpha)$. ∎

Remark It may be surprising that $F \upharpoonright \alpha^+ = E(F \upharpoonright \alpha)$ determines a unique function F. The fact, however, is that *every* function F on On obeys the following two conditions:

$F \upharpoonright 0 = 0$

$F \upharpoonright \lambda = \cup_{\alpha < \lambda} F \upharpoonright \alpha$, for λ a limit ordinal (for $F \upharpoonright \lambda$ is an extension of $F \upharpoonright \alpha$ for every $\alpha < \lambda$).

Thus the two conditions above need not be explicitly stated.

We now wish to show that for any function g, whose domain includes at least all ordinal sequences, there is a unique F on On such that for every ordinal α, $F(\alpha) = g(F \upharpoonright \alpha)$. We first prove the following elementary lemma (which does not require the axiom of substitution).

Lemma 5.4. *For any function g, whose domain includes all ordinal sequences, there is an extending operation E such that for every function F on On and every ordinal α, $(E(F \upharpoonright \alpha))(\alpha) = g(F \upharpoonright \alpha)$.*

Proof For each ordinal sequence θ_α of length α, define $E(\theta_\alpha)$ to be $\theta_\alpha \cup \{\langle \alpha, g(\theta_\alpha) \rangle\}$. Thus $E(\theta_\alpha)$ is an ordinal sequence of length α^+ and, for any $\beta < \alpha$, $(E(\theta_\alpha))(\beta) = \theta_\alpha(\beta)$, but $(E(\theta_\alpha))(\alpha) = g(\theta_\alpha)$. Then for any function F on On, $F \upharpoonright \alpha$ is an ordinal sequence of length α, hence $(E(F \upharpoonright \alpha))(\alpha) = g(F \upharpoonright \alpha)$. ∎

Now we have the following, which is a more usual form of the transfinite recursion theorem.

Theorem 5.5. *For any function g whose domain includes all ordinal sequences there is a (unique) function F on On such that for every ordinal α:*

$$F(\alpha) = g(F \upharpoonright \alpha).$$

Proof By Lemma 5.4 there is an extending operation E such that for every function F on On and every ordinal α, $(E(F \upharpoonright \alpha))(\alpha) = g(F \upharpoonright \alpha)$. By Theorem 5.3 there is a function F on On such that for every α: $F \upharpoonright \alpha^+ = E(F \upharpoonright \alpha)$. Then for every α,

$$F(\alpha) = (F \upharpoonright \alpha^+)(\alpha) = (E(F \upharpoonright \alpha))(\alpha) = g(F \upharpoonright \alpha).$$

Uniqueness of F is an easy exercise. ∎

From Theorem 5.5 we in turn get the following form of the transfinite recursion theorem.

Theorem 5.6. *For any functions h and f, defined on all sets, and for any set c, there is a unique function F on On such that:*

(1) $F(0) = c$;

(2) $F(\alpha^+) = h(F(\alpha))$ *(where α is any ordinal);*

(3) $F(\lambda) = f(F''(\lambda))$ *(where λ is any limit ordinal).*

Proof For any ordinal sequence θ, define $g(\theta)$ by the following explicit conditions.

(1) If length of θ is 0, take $g(\theta) = c$.

(2) If length of θ is α^+, take $g(\theta) = h(\theta(\alpha))$.

(3) If length of θ is λ, where λ is a limit ordinal, take $g(\theta) = f(\text{Ran}(\theta))$.

By Theorem 5.5 there is a function F on On such that for all ordinals α, $F(\alpha) = g(F \upharpoonright \alpha)$. Hence we have:

(1) $F(0) = g(F \upharpoonright 0) = c$ (since $F \upharpoonright 0$ is of length 0);

(2) $F(\alpha^+) = g(F \upharpoonright \alpha^+) = h((F \upharpoonright \alpha^+)(\alpha))$ (since $F \upharpoonright \alpha^+$ is of length α^+);

(3) for a limit ordinal λ, $F(\lambda) = g(F \upharpoonright \lambda) = f(\text{Ran}(F \upharpoonright \lambda))$ (since $F \upharpoonright \lambda$ is of length λ) $= f(F''(\lambda))$. ∎

§5. TRANSFINITE RECURSION THEOREMS

Next we consider the special case of Theorem 5.6 in which $f(x) = \cup x$, so $f(F''(\lambda))$ is then $\cup(F''(\lambda))$, which is $\cup_{\alpha<\lambda} F(\alpha)$. We let S_α be the set $F(\alpha)$, and so we have the following form of the transfinite recursion theorem, which is the form we will be using most often in subsequent chapters.

Theorem 5.7. *For every function g defined on all sets, and any element c, there is an On-sequence $S_0, S_1, \ldots, S_\alpha, \ldots$ such that:*

(1) $S_0 = c$;

(2) for each α, $S_{\alpha+1} = g(S_\alpha)$;

(3) for each limit ordinal λ, $S_\lambda = \cup_{\alpha<\lambda} S_\alpha$.

Note The On-sequence $S_0, S_1, \ldots, S_\alpha, S_{\alpha^+}, \ldots$, is unique (as is seen by an obvious transfinite induction argument) and is said to be the On-sequence *defined* from g and c by *transfinite recursion*.

Theorem 5.1 is really but a special case of Theorem 5.7, taking c to be 0, and specializing g to be a strictly progressing function. However, Theorem 5.7 is derivable from Theorem 5.1 without further use of the axiom of substitution. In fact, in the absence of the axiom of substitution, Theorems 5.1, 5.3, 5.5, 5.6 and 5.7 are all equivalent. Also, the full force of the axiom of substitution is apparently not required to prove these transfinite recursion theorems; the counting theorem (which we derived from the axiom of substitution, and which appears to be weaker) is enough to yield the transfinite recursion theorems. Also, in the absence of the axiom of substitution, any of the transfinite recursion theorems, if taken as an axiom, yields the counting theorem (Exercise 5.1 below).

The following is another form of the transfinite recursion theorem that will be seen to have an interesting generalization in Chapter 10.

Theorem 5.8. *For any function h there is a unique function F on On such that for every ordinal α,*

$$F(\alpha) = h(F''(\alpha)).$$

Proof By Theorem 5.5, taking $g(x) = h(\mathsf{Ran}(x))$, if x is a function (otherwise take $g(x) = x$). ∎

EXERCISES

Exercise 5.1. Prove that if we delete the axiom of substitution and add Theorem 5.8 as an axiom, we can then prove the counting theorem.

Exercise 5.2. Using any of the transfinite recursion theorems, prove that for any function g there is a class M that is minimally superinductive under g.

Exercise 5.3. Here is the outline of another proof of Zermelo's well ordering theorem. This proof uses the transfinite recursion theorem (and hence is available, if we assume the axiom of substitution).

Consider a non-empty set S and a choice function C for S. For any set A, define $h(A)$ as follows: If A is a *proper* subset of S, take $h(A) = C(S - A)$. If $A = S$, or if A is not a subset of S, take $h(A)$ to be some element outside S (which one makes no difference).

By Theorem 5.8 there is a function F on On such that for every ordinal α, $F(\alpha) = h(F''(\alpha))$. Now show the following.

(a) There is an ordinal δ such that $F(\delta) \notin S$.

(b) Let β be the least ordinal such that $F(\beta) \notin S$. Show $F''(\beta) = S$ and $F \upharpoonright \beta$ is 1-1 (hence β can be put into a 1-1 correspondence with S, and so S can be well ordered).

Exercise 5.4. Suppose that F is a function defined on On that satisfies the following two conditions:

(a) for every ordinal α, $F(\alpha) \subseteq F(\alpha^+)$;

(b) for every limit ordinal λ, and any ordinal α, $\alpha < \lambda$ implies $F(\alpha) \subseteq F(\lambda)$.

Prove that for any ordinals α and β, if $\alpha \leq \beta$ then $F(\alpha) \subseteq F(\beta)$.

§6 Ordinal arithmetic

Ordinal numbers generalize the counting aspect of natural numbers—their *first, second, third,* ... role. Later we will also examine infinite cardinal numbers, which generalize the size designating role of natural numbers. These two generalizations are quite different, though they happen to coincide in the finite case. We are all familiar with elementary arithmetic operations on the natural numbers: addition, multiplication, and so on. These extend to both infinite ordinal and infinite cardinal numbers in ways that are quite intuitive, but they do so in markedly different ways. In this section we briefly consider ordinal arithmetic; cardinal arithmetic will be taken up in Chapter 9.

The first approach to ordinal addition, historically, is still the easiest to understand intuitively. Suppose we want to evaluate $\alpha + \beta$, where α and β are ordinals. Find two well ordered sets a and b such that a is isomorphic to α, b is isomorphic to β, and a and b are disjoint. (This can always be done—take a to be $\alpha \times \{0\}$ and b to be $\beta \times \{1\}$, where a and b are ordered according to first components.) Now "follow" a with b. More precisely, define an ordering $<$ on $a \cup b$ as follows: if $x, y \in a$, $x < y$ iff $x < y$ in the ordering on a; if $x, y \in b$, $x < y$ iff $x < y$ in the ordering on b; if $x \in a$ and $y \in b$, $x < y$ is true; and if $x \in b$ and $y \in a$, $x < y$ is false. It is not hard to show this is a well ordering of $a \cup b$, and so there is some ordinal γ to which $a \cup b$ is isomporphic. This ordinal is taken to be $\alpha + \beta$.

Using this characterization, let us evaluate $1 + \omega$. The set $\{1, 2, 3, \ldots\}$, using the usual ordering, is isomorphic to ω, while 1 *is* $\{0\}$. Combining these as directed above, we get $\{0, 1, 2, \ldots\}$, with the usual ordering—in other words, we get ω. Thus $1+\omega = \omega$! On the other hand, $\omega + 1$ is the successor of ω, as discussed in §3 of Chapter 5. So ordinal addition is not commutative.

§6. ORDINAL ARITHMETIC

While this characterization of ordinal addition is easy to understand, it is somewhat complicated to use. Consequently we take a different route, extending the Peano characterization of the basic arithmetic operations for the natural numbers.

Theorem 5.6 allows us to define a function F with one argument. Addition, however, has two arguments, so we need an analog of the theorem. The following can be proved in essentially the same way—we omit the proof. (Part 3 is stated slightly differently, because the F'' notation does not work well for two-place functions.)

Theorem 6.1. *For any functions h, f, and g, on ordinals, there is a unique two-place function F, defined on all ordinals, such that:*

(1) $F(\alpha, 0) = g(\alpha)$;

(2) $F(\alpha, \beta^+) = h(F(\alpha, \beta), \alpha, \beta)$;

(3) for a limit ordinal λ, $F(\alpha, \lambda) = f(\{F(\alpha, \beta) \mid \beta < \lambda\})$.

Now we use this theorem, taking h to be the successor function on its first argument, f to be $\cup$, and g to be the identity function. Here is the resulting definition of ordinal addition, which we write as $\alpha + \beta$, rather than the less readable $F(\alpha, \beta)$.

(1) $\alpha + 0 = \alpha$;

(2) $\alpha + \beta^+ = (\alpha + \beta)^+$;

(3) $\alpha + \lambda = \cup_{\beta < \lambda}(\alpha + \beta)$, for a limit ordinal λ.

If we ignore the third item, this is the usual Peano definition of addition for the natural numbers. Consequently, if α and β are natural numbers, $\alpha + \beta$, as defined above, is just the usual natural number sum.

According to this definition, $\omega + 1 = (\omega + 0)^+ = \omega^+$. That is, $\omega + 1$ is the ordinal immediately following ω—as we already observed in §3 of Chapter 5. On the other hand, since ω is a limit ordinal, $1 + \omega = \cup_{n<\omega}(1 + n) = \cup_{n\in\omega} n^+ = \omega$. Thus, as we saw using our earlier characterization, $1 + \omega \neq \omega + 1$. On the other hand, various other properties of addition for natural numbers do carry over to ordinals. Exercise 6.1 lists some of these.

Ordinal multiplication and exponentiation have definitions that also extend the usual Peano characterizations for the natural numbers.

(1) $\alpha \cdot 0 = 0$;

(2) $\alpha \cdot \beta^+ = (\alpha \cdot \beta) + \alpha$;

(3) $\alpha \cdot \lambda = \cup_{\beta<\lambda}(\alpha \cdot \beta)$, for a limit ordinal λ.

For $\alpha \neq 0$:

(1) $\alpha^0 = 1$;

(2) $\alpha^{\beta^+} = \alpha^\beta \cdot \alpha$;

(3) $\alpha^\lambda = \cup_{\beta<\lambda}(\alpha^\beta)$, for a limit ordinal λ.

The exercises give a few of the properties of these operations.

EXERCISES

Exercise 6.1. Prove the following.

(a) $0 + \alpha = \alpha$.

(b) $\alpha \leq \alpha + \beta$.

(c) $\alpha + (\beta + \gamma) = (\alpha + \beta) + \gamma$. Hint: use transfinite induction on γ.

(d) If $\alpha \leq \beta$ then $\alpha + \gamma \leq \beta + \gamma$. (Can $\leq$ be replaced by $<$ here?)

Exercise 6.2. Prove the following.

(a) $\omega \cdot 2 \neq 2 \cdot \omega$.

(b) $1 \cdot \alpha = \alpha \cdot 1 = \alpha$.

(c) $\alpha \cdot (\beta \cdot \gamma) = (\alpha \cdot \beta) \cdot \gamma$.

Exercise 6.3. Prove the following.

(a) $\alpha^1 = \alpha$.

(b) $1^\alpha = 1$.

(c) $\alpha^{(\beta+\gamma)} = (\alpha^\beta) \cdot (\alpha^\gamma)$.

CHAPTER 7

RANK

§1 The notion of rank

For each ordinal α we define the set R_α by the following transfinite recursion (Theorem 5.7 Chapter 6):

$R_0 = \emptyset$;
$R_{\alpha+1} = \mathcal{P}(R_\alpha)$ (the power set of R_α);
$R_\lambda = \cup_{\alpha<\lambda} R_\alpha$ (for any limit ordinal λ).

We say a set has *rank* if it is a member of some R_α, and if a set x does have rank, we define its rank to be the first ordinal α such that $x \in R_{\alpha+1}$ (or what is the same thing, the first α such that $x \subseteq R_\alpha$).

We define R_Ω to be the union of all the R_α's—thus R_Ω is the class of all sets with rank.

Remarks Clearly $R_0 = 0$, $R_1 = 1$, $R_2 = 2$, but $R_3 \neq 3$; R_3 is the set of all subsets of R_2, hence has four elements—specifically $R_3 = \{0, \{0\}, \{1\}, \{0, 1\}\}$. For any finite n, R_{n+1} has $2^{\text{card}(R_n)}$ elements, so an easy induction shows that R_{n+1} has 2^n elements.

We will show that R_Ω is itself a Zermelo-Fraenkel universe (assuming that V is) and that $R_{\omega \cdot 2}$ is already a Zermelo universe (but not a Zermelo-Fraenkel universe), and that R_ω satisfies all the conditions of a Zermelo universe except for the axiom of infinity (R_ω is a basic universe). Also (as we will see), for any limit ordinal $\lambda > \omega$, the set R_λ is a Zermelo universe.

R_α is also denoted V_α by some authors.

Now we show some basic properties of the R_α's.

Proposition 1.1. *Each R_α is transitive.*

Proof By transfinite induction. R_0, being empty, is vacuously transitive. If R_α is transitive, so is R_{α^+} (because the power set of a transitive set is transitive). For any limit ordinal λ, if R_α is transitive for each $\alpha < \lambda$, then R_λ is transitive, because the union of any set of transitive sets is transitive. This completes the induction. ∎

Proposition 1.2. *For each α, $R_\alpha \in R_{\alpha+1}$ and $R_\alpha \subseteq R_{\alpha+1}$.*

Proof $R_\alpha \subseteq R_\alpha$, and hence $R_\alpha \in \mathcal{P}(R_\alpha) = R_{\alpha+1}$. Then $R_\alpha \subseteq R_{\alpha+1}$ follows by the transitivity of $R_{\alpha+1}$. ∎

Proposition 1.3. *For each α, $R_\alpha \subseteq R_{\alpha+1}$, but $R_\alpha \neq R_{\alpha+1}$.*

Proof Since $R_\alpha \in R_{\alpha+1}$ and $R_{\alpha+1}$ is transitive, $R_\alpha \subseteq R_{\alpha+1}$. Further, by Cantor's theorem, $\mathcal{P}(R_\alpha)$ is larger than R_α, so $\mathcal{P}(R_\alpha) \neq R_\alpha$. Thus R_α is a *proper* subset of $R_{\alpha+1}$. ∎

Before establishing more facts about the R_α's, we pause to turn to a more general situation that will also cover some facts crucially needed for later chapters.

§2 Ordinal hierarchies

Let $\mathcal{O}$ be an *On*-sequence $\mathcal{O}_0, \mathcal{O}_1, \ldots, \mathcal{O}_\alpha, \ldots$ of subsets of a class A. We shall call $\mathcal{O}$ an *ordinal hierarchy on A* if the following three conditions hold.

(1) $\mathcal{O}_0 = \emptyset$.

(2) For each ordinal α, $\mathcal{O}_\alpha \subseteq \mathcal{O}_{\alpha+1}$.

(3) For each limit ordinal λ, $\mathcal{O}_\lambda = \cup_{\alpha<\lambda}\mathcal{O}_\alpha$.

We have shown that $R_\alpha \subseteq R_{\alpha+1}$, and since $R_0 = \emptyset$ and $R_\lambda = \cup_{\alpha<\lambda}R_\alpha$ for λ a limit ordinal, we see that the R_α sequence (i.e., the *On* sequence $R_0, R_1, \ldots, R_\alpha, \ldots$) is an ordinal hierarchy, and so what we shall prove about ordinal hierarchies in general will have applications to the R_α's in particular. (Three more ordinal hierarchies of major importance will come up in the course of this volume.)

Suppose now that $\mathcal{O}$ is an ordinal hierarchy on A. We shall say that an element x of A has *$\mathcal{O}$-rank* if it is a member of some $\mathcal{O}_\alpha$, in which case we define its $\mathcal{O}$-rank to be the first ordinal α such that $x \in \mathcal{O}_{\alpha+1}$ (as we did for the R_α sequence).

Theorem 2.1. *For any ordinal hierarchy $\mathcal{O}$ on A the following conditions hold:*

O$_1$ $\alpha \leq \beta$ implies $\mathcal{O}_\alpha \subseteq \mathcal{O}_\beta$.

O$_2$ *If for each α, $\mathcal{O}_\alpha$ is a proper subset of $\mathcal{O}_{\alpha+1}$, then for all ordinals α and β, $\alpha < \beta$ iff $\mathcal{O}_\alpha \subset \mathcal{O}_\beta$.*

O$_3$ *For any $x \in A$ and any α, $x \in \mathcal{O}_\alpha$ iff $x \in \mathcal{O}_{\beta+1}$, for some $\beta < \alpha$.*

O$_4$ $x \in \mathcal{O}_\alpha$ *iff x has $\mathcal{O}$-rank $< \alpha$.*

O$_5$ *x has $\mathcal{O}$-rank α iff $x \in \mathcal{O}_{\alpha+1} - \mathcal{O}_\alpha$.*

O$_6$ *For any set S of elements with $\mathcal{O}$-rank, there is an ordinal α such that $S \subseteq \mathcal{O}_\alpha$.*

Proof

O$_1$ Since for each α, $\mathcal{O}_\alpha \subseteq \mathcal{O}_{\alpha+1}$, it follows by an obvious transfinite induction on the class of ordinals $> \alpha$ that for any $\beta > \alpha$, $\mathcal{O}_\alpha \subseteq \mathcal{O}_\beta$.

O$_2$ Suppose that for each α, $\mathcal{O}_\alpha$ is a *proper* subset of $\mathcal{O}_{\alpha+1}$.

§3. APPLICATIONS TO THE R_α SEQUENCE

(1) Again it follows by an obvious transfinite induction on the class of ordinals $> \alpha$ that $\alpha < \beta$ implies $\mathcal{O}_\alpha \subset \mathcal{O}_\beta$.

(2) Conversely, suppose $\mathcal{O}_\alpha \subset \mathcal{O}_\beta$. Then $\mathcal{O}_\beta \subseteq \mathcal{O}_\alpha$ doesn't hold, hence $\beta \leq \alpha$ doesn't hold (by $\mathbf{O}_1$), hence $\alpha < \beta$.

$\mathbf{O}_3$

(1) Suppose $x \in \mathcal{O}_\alpha$. Then of course $\alpha \neq 0$ (since $\mathcal{O}_0 = \emptyset$). If α is a successor ordinal, then $x \in \mathcal{O}_{\beta+1}$, where β is the predecessor of α. Now suppose α is a limit ordinal. Then for some $\beta < \alpha$, $x \in \mathcal{O}_\beta$, hence also $x \in \mathcal{O}_{\beta+1}$ (since $\mathcal{O}_\beta \subseteq \mathcal{O}_{\beta+1}$).

(2) Conversely, suppose $x \in \mathcal{O}_{\beta+1}$ and $\beta < \alpha$. Then $\beta+1 \leq \alpha$, hence $\mathcal{O}_{\beta+1} \subseteq \mathcal{O}_\alpha$ (by $\mathbf{O}_1$), hence $x \in \mathcal{O}_\alpha$.

$\mathbf{O}_4$

(1) Suppose $x \in \mathcal{O}_\alpha$. By $\mathbf{O}_3$ there is an ordinal $\beta < \alpha$ such that $x \in \mathcal{O}_{\beta+1}$; let β be the smallest ordinal such that $x \in \mathcal{O}_{\beta+1}$. Then β must be less than α, and β is the $\mathcal{O}$-rank of x. Thus x has $\mathcal{O}$-rank β and $\beta < \alpha$.

(2) Suppose x has $\mathcal{O}$-rank β and $\beta < \alpha$. Then $x \in \mathcal{O}_{\beta+1}$ and $\beta + 1 \leq \alpha$ (since $\beta < \alpha$), hence $\mathcal{O}_{\beta+1} \subseteq \mathcal{O}_\alpha$ (by $\mathbf{O}_1$), hence $x \in \mathcal{O}_\alpha$.

$\mathbf{O}_5$

(1) Suppose x has $\mathcal{O}$-rank α. Then of course $x \in \mathcal{O}_{\alpha+1}$. If x were in $\mathcal{O}_\alpha$ then x would have $\mathcal{O}$-rank $< \alpha$ (by $\mathbf{O}_4$), hence $x \notin \mathcal{O}_\alpha$. Thus $x \in \mathcal{O}_{\alpha+1} - \mathcal{O}_\alpha$.

(2) Conversely, suppose $x \in \mathcal{O}_{\alpha+1} - \mathcal{O}_\alpha$. Since $x \notin \mathcal{O}_\alpha$ then for no $\beta \leq \alpha$ is it the case that $x \in \mathcal{O}_\beta$ (because if $\beta \leq \alpha$ then $\mathcal{O}_\beta \subseteq \mathcal{O}_\alpha$ by $\mathbf{O}_1$, hence $x \in \mathcal{O}_\beta \supset x \in \mathcal{O}_\alpha$). Thus for no $\beta < \alpha + 1$ is $x \in \mathcal{O}_\beta$, hence the smallest ordinal β such that $x \in \mathcal{O}_\beta$ is $\alpha + 1$, and so α is the $\mathcal{O}$-rank of x.

$\mathbf{O}_6$ For each $x \in S$, let $\varphi(x)$ be the $\mathcal{O}$-rank of x. Since x is a set, so is $\varphi''(S)$ (by the axiom of substitution), hence there is an ordinal α greater than all the elements of $\varphi''(S)$. Thus every element of S has $\mathcal{O}$-rank $< \alpha$, hence by $\mathbf{O}_4$, $S \subseteq \mathcal{O}_\alpha$. ∎

§3 Applications to the R_α sequence

Now we establish some more properties of the R_α-sequence. By $\mathbf{O}_1$ of Theorem 2.1 we have:

$\mathbf{P}_1$ $\alpha \leq \beta$ implies $R_\alpha \subseteq R_\beta$.

Next, for each α, R_α is a *proper* subset of R_β (by Proposition 1.3), so by $\mathbf{O}_2$ of Theorem 2.1 we have:

$\mathbf{P}_2$ $\alpha < \beta$ iff $R_\alpha \subset R_\beta$.

Next, by **O₃**, **O₄**, and **O₅** of Theorem 2.1 we have:

P₃ $x \in R_\alpha$ iff $x \in R_{\beta+1}$ for some $\beta < \alpha$.

P₄ $x \in R_\alpha$ iff x has rank $< \alpha$.

P₅ x has rank α iff $x \in R_{\alpha+1} - R_\alpha$.

Now we establish some further properties of the R_α's that do not necessarily hold for ordinal hierarchies in general, but are particular for the R_α-sequence.

P₆ Every element of a set with rank has lower rank.

Proof Suppose y has rank α and $x \in y$. Since y has rank α then $y \in R_{\alpha+1}$, hence $y \subseteq R_\alpha$. Hence $x \in R_\alpha$, and so x has rank $< \alpha$ by **P₄**. ∎

P₇ Each ordinal α has rank α.

Proof The proof has two parts, since we must show $\alpha \in R_{\alpha+1}$ but also $\alpha \notin R_\alpha$.
(a) We first show by transfinite induction that $\alpha \subseteq R_\alpha$.
Suppose that for each $\beta < \alpha, \beta \subseteq R_\beta$. Then for each $\beta < \alpha, \beta \in R_{\beta+1}$, hence $\beta \in R_\alpha$ (because $\beta + 1 \leq \alpha$, so $R_{\beta+1} \subseteq R_\alpha$). Therefore, if for each $\beta < \alpha, \beta \subseteq R_\beta$, then for each $\beta < \alpha, \beta \in R_\alpha$. But, to say that each $\beta < \alpha$ is in R_α is to say that $\alpha \subseteq R_\alpha$. And so if, for each $\beta < \alpha, \beta \subseteq R_\beta$, then $\alpha \subseteq R_\alpha$. Hence by transfinite induction, it follows that for every $\alpha, \alpha \subseteq R_\alpha$.
(b) Next we show that if $\alpha \in R_\alpha$, then for some $\beta < \alpha, \beta \in R_\beta$ (and hence by transfinite induction, $\alpha \in R_\alpha$ is always false).
Well, suppose $\alpha \in R_\alpha$. Then by **P₄**, for some $\beta < \alpha, \alpha \in R_{\beta+1}$—and hence $\alpha \subseteq R_\beta$. But since $\beta < \alpha$, then $\beta \in \alpha$, and so $\beta \in \alpha \subseteq R_\beta$, which implies that $\beta \in R_\beta$.

By (a) and (b) we see that for each ordinal $\alpha, \alpha \in R_{\alpha+1} - R_\alpha$, and hence α has rank α by **P₅**. ∎

P₈ Each R_α is supercomplete (i.e., R_α is transitive and contains with each element x all subsets of x as well).

Proof R_α is transitive by Proposition 1.1. Now suppose that $x \in R_\alpha$ and $y \subseteq x$. We are to show that $y \in R_\alpha$.
Let β be the rank of x. Then $\beta < \alpha$ (by **P₄**) hence $\beta + 1 \leq \alpha$, hence $R_{\beta+1} \subseteq R_\alpha$ (by **P₁**). Since x has rank β, then $x \subseteq R_\beta$, hence $y \subseteq R_\beta$ (since $y \subseteq x$), hence $y \in R_{\beta+1}$, hence $y \in R_\alpha$ (since $R_{\beta+1} \subseteq R_\alpha$). ∎

And as a corollary we have the following.

P₉ The class R_Ω of all sets with rank is supercomplete.

Proof Suppose x has rank. Then $x \in R_\alpha$ for some α. Then all elements of x and all subsets of x are in R_α (since R_α is supercomplete), hence they are all in R_Ω. ∎

§3. APPLICATIONS TO THE R_α SEQUENCE

Next we shall consider some properties of rank that will be directly used in the proofs of our main theorems.

Q_1 [Unordered pairs]

(1) If $x \in R_\alpha$ and $y \in R_\alpha$ then $\{x,y\} \in R_{\alpha+1}$.

(2) For any limit ordinal λ, if $x \in R_\lambda$ and $y \in R_\lambda$ then $\{x,y\} \in R_\lambda$.

(3) If x and y both have rank, so does $\{x,y\}$.

Proof

(1) Suppose $x \in R_\alpha$ and $y \in R_\alpha$. Then $\{x,y\} \subseteq R_\alpha$, so $\{x,y\} \in R_{\alpha+1}$.

(2) For a limit ordinal λ, if $x \in R_\lambda$ and $y \in R_\lambda$, then for some $\alpha_1 < \lambda$ and $\alpha_2 < \lambda$, $x \in R_{\alpha_1}$ and $y \in R_{\alpha_2}$. Let α be the maximum of α_1, α_2. Then x, y are both in R_α, hence by the previous part, $\{x,y\} \in R_{\alpha+1}$. Therefore $\{x,y\} \in R_\lambda$ (since $\alpha + 1 \leq \lambda$, and hence $R_{\alpha+1} \subseteq R_\lambda$).

(3) This simply says that if x, y are both in R_Ω, so is $\{x,y\}$. The proof is the same as the previous part, using "Ω" in place of "λ."

∎

Q_2 [Unions]

(1) $\cup R_{\alpha+1} = R_\alpha$.

(2) For any limit ordinal λ, $\cup R_\lambda = R_\lambda$.

(3) For any successor ordinal α, if $x \in R_{\alpha+1}$ then $\cup x \in R_\alpha$.

(4) For any limit ordinal λ, if $x \in R_\lambda$ then $\cup x \in R_\lambda$.

(5) If x has rank, so does $\cup x$.

Proof We first recall the following facts:

$\cup(\mathcal{P}(x)) = x$;
$x \subseteq y$ implies $\cup x \subseteq \cup y$.

Now using these:

(1) $\cup R_{\alpha+1} = \cup(\mathcal{P}(R_\alpha)) = R_\alpha$.

(2) For any ordinal α, $\cup R_\alpha \subseteq R_\alpha$, since R_α is transitive, so we must show that for any limit ordinal λ, $R_\lambda \subseteq \cup R_\lambda$. Well, suppose $x \in R_\lambda$. Then for some $\alpha < \lambda$, $x \in R_\alpha$. Also $R_\alpha \in R_{\alpha+1}$, hence $R_\alpha \in R_\lambda$ (since $R_{\alpha+1} \subseteq R_\lambda$). Since $x \in R_\alpha$ and $R_\alpha \in R_\lambda$, then $x \in \cup R_\lambda$. Thus $x \in R_\lambda$ implies $x \in \cup R_\lambda$, so $R_\lambda \subseteq \cup R_\lambda$.

(3) Suppose $\alpha = \beta + 1$ and $x \in R_{\alpha+1}$. Then $x \subseteq R_\alpha$, so $x \subseteq R_{\beta+1}$, hence $\cup x \subseteq \cup R_{\beta+1}$, so $\cup x \subseteq R_\beta$, hence $\cup x \in R_{\beta+1} = R_\alpha$. Thus $\cup x \in R_\alpha$.

(4) Suppose $x \in R_\lambda$ and λ is a limit ordinal. Then $x \in R_\beta$ for some $\beta < \lambda$. Let $\alpha = \beta^+$, so α is a successor ordinal $< \lambda$ and $x \in R_{\alpha^+}$ (since $\beta < \alpha^+$), so $\cup x \in R_\alpha$ (by the previous part), hence $\cup x \in R_\lambda$.

(5) The proof is like the previous part, using On in place of λ. ∎

Q₃ [Power]

(1) $x \in R_{\alpha+1}$ implies $\mathcal{P}(x) \in R_{\alpha+2}$.

(2) $x \in R_\lambda$ implies $\mathcal{P}(x) \in R_\lambda$ (for λ a limit ordinal).

(3) If x has rank, so does $\mathcal{P}(x)$.

Proof We first note that $x \subseteq y$ implies $\mathcal{P}(x) \subseteq \mathcal{P}(y)$. Now, for the first part, suppose $x \in R_{\alpha+1}$. Then $x \subseteq R_\alpha$, hence $\mathcal{P}(x) \subseteq \mathcal{P}(R_\alpha)$, thus $\mathcal{P}(x) \subseteq R_{\alpha+1}$, and hence $\mathcal{P}(x) \in R_{\alpha+2}$. The other two parts are left to the reader. ∎

§4 Zermelo universes

We recall that $\omega \cdot 2$ is the first limit ordinal $> \omega$.

Theorem 4.1.

(1) For any limit ordinal λ, R_λ is a basic universe—i.e., it satisfies all the conditions of being a Zermelo universe except possibly for the axiom of infinity.

(2) If also $\lambda \geq \omega \cdot 2$, then R_λ is a Zermelo universe. In particular, $R_{\omega \cdot 2}$ is a Zermelo universe.

Proof

(1) Let λ be any limit ordinal. We have already shown that R_λ is supercomplete ($\mathbf{P_8}$) and that the pairing, union and power axioms hold for R_λ (by $\mathbf{Q_1}, \mathbf{Q_2}, \mathbf{Q_3}$). Thus all conditions hold except possibly for the axiom of infinity.

(2) If furthermore $\lambda > \omega$, then $\omega \in R_\lambda$, and so the axiom of infinity also holds for R_λ. Thus for any limit ordinal $\lambda > \omega$ (or equivalently, $\lambda \geq \omega \cdot 2$), R_λ is a Zermelo universe. In particular, $R_{\omega \cdot 2}$ is a Zermelo universe.

∎

Discussion Theorem 4.1 has an interesting metamathematical corollary: without the axiom of substitution, one cannot prove that $\omega \cdot 2$ is a set! The reason is: $R_{\omega \cdot 2}$ is a Zermelo universe in which $\omega \cdot 2$ is *not* an element.

Theorem 4.2. *If R_α is a Zermelo universe, then α must be a limit ordinal $\geq \omega \cdot 2$.*

§4. ZERMELO UNIVERSES

Proof For any successor ordinal $\beta + 1$, $R_{\beta+1}$ fails to be a Zermelo universe on several counts: for one thing $\beta \in R_{\beta+1}$, but $\beta + 1 \notin R_{\beta+1}$ (since $\beta + 1$ has rank $\beta + 1$), but in a Zermelo universe, each ordinal of the universe must have a successor in the universe. So $R_{\beta+1}$ cannot be a Zermelo universe. Also R_0 is obviously not a Zermelo universe. So if R_α is a Zermelo universe, then α must be a limit ordinal.

Now suppose R_α is a Zermelo universe. We have just seen that α must be a limit ordinal. Also R_α satisfies the axiom of infinity, and so $\omega \in R_\alpha$, hence $\omega < \alpha$ (because $\omega \in R_\alpha$ implies the rank of ω is less than α, but the rank of ω is ω). Thus α is a limit ordinal $> \omega$, and so $\alpha \geq \omega \cdot 2$. ∎

Zermelo-Fraenkel Universes

We now use the axiom of substitution for V to obtain the axiom of substitution for R_Ω. We first need the following property.

Q$_4$ If $x \subseteq R_\Omega$ and x is a *set*, then $x \in R_\Omega$. Stated otherwise, every set x of sets with rank itself has a rank.

Proof Suppose that x is a set and that every element of x has rank. Then by **O$_6$**, $x \subseteq R_\alpha$ for some α, hence $x \in R_{\alpha+1}$, and so by **P$_4$**, x has rank $< \alpha$. ∎

Theorem 4.3. R_Ω *is a Zermelo-Fraenkel universe (assuming, of course, that V is).*

Proof That R_Ω is a Zermelo universe can be proved in an analogous manner to the proof that R_λ is a Zermelo universe for any limit ordinal $\lambda \geq \omega \cdot 2$ (Theorem 4.1).

It remains to show that R_Ω satisfies the axiom of substitution. Well, let F be any function from sets with rank to sets with rank (F is from R_Ω to R_Ω). Take any $x \in R_\Omega$. Since $x \in R_\Omega$, then of course x is a set, and so by the axiom of substitution (applied in V) $F''(x)$ is a set. Thus $F''(x)$ is a set all of whose members have rank, and so by **Q$_4$**, $F''(x)$ has rank—i.e., $F''(x) \in R_\Omega$. This completes the proof. ∎

EXERCISES

Exercise 4.1. Show that R_Ω is the intersection of all subclasses C of V such that C is a Zermelo-Fraenkel universe containing all ordinals of V.

Exercise 4.2. The objects that mathematics customarily deals with can all be developed within set theory. In particular, the set of positive real numbers can be created as follows. A fraction is an ordered pair of non-zero integers; *mathbbF* is the set of fractions. A rational is a set of fractions—loosely, the set of all fractions that informally name the same thing. $\mathbb{Q}$ is the set of rationals. A real is a Dedekind cut of rationals—details don't matter here, but a real is a special kind of set of rationals. $\mathbb{R}$ is the set of reals.

Show the set $\mathbb{R}$ of reals is a set with rank, and determine what the rank is.

Exercise 4.3. (See previous exercise). The real numbers can equally well be defined using Cauchy sequences. A real-representative is a Cauchy convergent sequence r_0,

$r_1, r_2, \ldots$, of rationals. As before, details don't matter. What counts is that a real-representative is a special kind of sequence of rationals. Then a real is an equivalence class of real-representatives—loosely, all Cauchy sequences that converge to the same limit. Thus a real, in this approach, is a set of real-representatives. Let $\mathbb{R}$ be the set of reals.

Show $\mathbb{R}$, defined in this way, is a set with rank, and determine the rank.

CHAPTER 8

FOUNDATION, $\in$-INDUCTION, AND RANK

§1 The notion of well-foundedness

Definition 1.1. An element x of a class A is called an *initial element* of A if all members of x lie outside A, i.e., if $x \cap A = \emptyset$.

Definition 1.2. A class A is called *well founded* if every non-empty subclass of A contains an initial element.

It is obvious that if A is well founded, so is every subclass of A.

Example Let B be any non-empty class of ordinals. Then the smallest ordinal α of B is clearly an initial element of B (why?). Thus every non-empty class of ordinals contains an initial element, i.e., every subclass of On contains an initial element, so On is well founded.

Some systems of set theory use the following—taking its name from (Gödel 1940).

Axiom D V is well founded, i.e., every non-empty class (subcollection of V) contains an initial element.

We shall not be primarily interested in "using" Axiom D, so much as "discussing" it. That is to say, our theorems will be more of the form "in every well founded Zermelo (or Zermelo Fraenkel) universe V, such and such must hold." We shall also be interested in other properties of V which are equivalent to Axiom D. For example, we will see that Axiom D is equivalent to the statement that every element of V has rank! This is curious in that the notion of *rank* presupposes a lot of antecedent machinery—results on ordinals, transfinite recursion, etc., whereas the definition of well foundedness can be given at the very beginning of the development of set theory, yet the notions are equivalent!

Early in the history of set theory, sets that violated Axiom D were sometimes explicitly considered, but as time went on, assuming all sets were well founded became the standard. More recently, Aczel (1988) revived interest in non-well founded sets, and it turns out they have applications to linguistics and to computer science. It is not enough, of course, to simply drop Axiom D. One needs a more positive approach, and so-called *anti-well foundedness axioms* have been proposed. We do not discuss them here—we recommend (Aczel 1988) for further reading.

EXERCISES

Exercise 1.1. Show that each R_α is well founded. Show that R_Ω is well founded. (Hint: If A is a non-empty subclass of R_Ω show that any element of A of lowest rank must be an initial element of A.)

§2 Descending ∈-chains

Axiom D rules out sets which are members of themselves, or pairs x, y such that $x \in y \in x$. More generally it rules out any of the so-called "∈-cycles," i.e., sets $x_1, x_2, \ldots, x_n$ such that $x_1 \in x_2 \in \ldots \in x_n \in x_1$.

Let us see how this works. Suppose V does obey Axiom D, i.e., suppose that V is well founded. We wish to show that $x \notin x$. Consider the unit set $\{x\}$. It is clearly non-empty (since it contains x), hence it must contain an initial element. This element can only be x itself. Thus x is an initial element of $\{x\}$, i.e. $x \cap \{x\} = \emptyset$. This means that $\{x\}$ has no member in common with x, i.e., no member of $\{x\}$ is a member of x, which means that x is not a member of x.

Consider now two sets x, y. Then $\{x, y\}$ contains an initial element; it must be either x or y (possibly both). Suppose it is x. Then x has no members in common with $\{x, y\}$, so $y \notin x$. If it is y that is an initial element of $\{x, y\}$, then $x \notin y$. Hence $x \notin y$ or $y \notin x$.

More generally, consider n sets $x_1, \ldots, x_n$. Then the set $\{x_1, \ldots, x_n\}$ contains some initial element x_i. Then no $x_j \in x_i$ for any j from i to n. Thus if $i \neq 1$, $x_{i-1} \in x_i$ is false. If $i = 1$ then $x_n \in x_i$ is false. Hence at least one of the statements $x_1 \in x_2$, $x_2 \in x_3$, $\ldots$, $x_{n-1} \in x_n$, $x_n \in x_1$ is false. We thus have the following.

Theorem 2.1. *A well founded Zermelo universe contains no ∈-cycles.*

Remarks In Chapter 5 we proved that no ordinal can be a member of itself, and that for two ordinals α, β, $\alpha \in \beta$ and $\beta \in \alpha$ are not jointly possible. Of course these facts would be immediate from Theorem 2.1, had we used Axiom D, but it is of significant interest that ∈-cycles for ordinals can be ruled out without using Axiom D.

Descending ∈-chain condition We write $x \ni y$ to mean $y \in x$. By a descending ∈-chain is meant a finite sequence $x_1, x_2, \ldots, x_n$ such that $x_1 \ni x_2 \ni \ldots \ni x_n$, or a denumerable sequence $x_1, x_2, \ldots, x_n, \ldots$ such that $x_1 \ni x_2 \ni \ldots \ni x_n \ni \ldots$.

A class A is said to satisfy the *descending ∈-chain condition* if it contains no infinite descending ∈-chains, i.e., if there is no infinite descending ∈-chain whose elements belong to A.

Theorem 2.2. *If A is well founded, then A satisfies the descending ∈-chain condition.*

Proof Suppose there were an infinite descending ∈-chain $a_1 \ni a_2 \ni a_3 \ni \ldots$ (where each $a_i \in A$). Then the class $B = \{a_1, a_2, \ldots, a_n, \ldots\}$ obviously contains no initial element (because each a_i contains an element of B, namely a_{i+1}). Hence A is not well founded. Thus if A is well founded, then A must satisfy the descending ∈-chain condition. ∎

Remark Theorem 2.2 affords an alternate proof of Theorem 2.1, for the existence of an $\in$-cycle $x_1 \ni x_2 \ni \ldots \ni x_n \ni x_1$ would imply the existence of the infinite descending $\in$-chain $x_1 \ni x_2 \ni \ldots \ni x_n \ni x_1 \ni \ldots$. (In particular, e.g., if $x \in x$ we would have an infinite descending $\in$-chain $x \ni x \ni x \ni \ldots$.) Our definition of a descending $\in$-chain $x_1 \ni x_2 \ni \ldots \ni x_i \ni \ldots$ does not require that the terms of the sequence $x_1, x_2, \ldots, x_i, \ldots$ be distinct.

Theorem 2.3. *If A satisfies the descending $\in$-chain condition, and if A can be well ordered, then A is well founded.*

Proof (Informal sketch)

Assume the hypothesis. Let B be any non-empty subclass of A. We must show that B contains an initial element.

Since B is non-empty, it contains an element b_1. If b_1 is an initial element of B, then we are done. Otherwise, b_1 contains at least one element of B. Let b_2 be the first such element (in the well-ordering of A). If b_2 is an initial element of B then we are done. If not, we can again pick some element b_3 of B which is in b_2. Continuing in this manner, we either stop when we come to some initial element of B or we obtain an infinite descending $\in$-chain. The latter is incompatible with our hypothesis, hence we must, after a finite number of steps, hit an initial element b_n. ∎

§3 $\in$-Induction and rank

Theorem 3.1. *Suppose A is well founded and B is a subclass of A. Suppose also that for every element x of A, if B contains all elements of $x \cap A$, then B contains x. Then B contains all elements of A.*

Proof Assume the hypothesis. We are to show that $A - B$ is empty. If $A - B$ were not empty, then $A - B$ must contain an initial element x of A (since A is well founded). Then no element of x lies in $A - B$, hence no element of $x \cap A$ lies in $A - B$. But all elements of $x \cap A$ lie in A, so all elements of $x \cap A$ lie in B, i.e., $x \cap A \subseteq B$. But this implies that $x \in B$ (by hypothesis) contrary to the fact that $x \in A - B$. Thus the assumption that $A - B$ is non-empty leads to a contradiction. Hence $A - B$ is empty, which means that $B = A$. ∎

As a corollary we have the following.

Theorem 3.2. *If A is transitive and well founded then it obeys the following induction principle: If a property P is such that for every element x of A, P holds for x, provided P holds for every element of x, then P holds for all elements of A.*

Proof Suppose A is transitive and that P is a property such that for every element x of A, if P holds for all elements of x, then P holds for x. Let B be the class of all elements of A for which P holds. We are to show that $B = A$. Since by hypothesis, A is well founded, then by Theorem 3.1, it suffices to show that for any element x of A, if B contains all elements of $x \cap A$, then B contains x. So suppose that $x \in A$ and B contains all elements of $x \cap A$. Then B contains all elements of x (because $x \cap A = x$, since A is transitive). Thus all elements of x have property P, hence x has property P (by our assumption concerning P), and since $x \in A$, then $x \in B$. This concludes the proof. ∎

Theorem 3.3. *If A is transitive and well founded, then every element of A has a rank.*

Proof Suppose A is transitive and well founded. Let $P(x)$ be the statement "x has a rank." We showed in Chapter 7 (Property Q_4 in §4), using the axiom of substitution, that if every element of a set x has a rank, then x has a rank. Thus if P holds for all elements of x, then P holds for x. Then by Theorem 3.2, P holds for every element of A, i.e., every element of A has rank. ∎

Theorem 3.4. *If every element of A has a rank, then A is well founded.*

Proof We stated as a previous exercise that R_Ω is well founded. Let us now prove this. Suppose A is a non-empty subclass of R_Ω. Let x be any element of A of lowest rank. Then all elements of x, being of lower rank, lie outside A. Thus x is an initial element of A, and so R_Ω is well founded.

If now every element of A has a rank, then $A \subseteq R_\Omega$ hence A, being a subclass of a well founded class, must be well founded. ∎

By virtue of Theorems 3.3 and 3.4, we see that for any transitive class A, A is well founded iff every element of A has a rank. Taking V for A, we thus have the following.

Theorem 3.5. *In the presence of the axiom of substitution, Axiom D is equivalent to the assertion that every set has a rank, i.e., that $V = R_\Omega$.*

Remarks Theorem 3.5 affords yet another proof of Theorem 2.1. For suppose that V is well founded. Then every element of V has a rank. Now take any set x. Since every element of x has a lower rank than x, $x \in x$ is impossible. Likewise $x_1 \ni x_2 \ni \ldots \ni x_n \ni x_1$ would imply the absurdity that x_1 has lower rank than x_1. And an infinite descending chain $x_1 \ni x_2 \ni \ldots \ni x_n \ni \ldots$ would imply the existence of an infinite descending chain of ordinals (viz. the ranks of the x_i) which is impossible. (Why?)

Theorem 3.6. *Let V be a Zermelo-Fraenkel universe. Then every well founded subclass of V which is a Zermelo-Fraenkel universe is either some R_α or the class R_Ω.*

Proof Suppose V_0 is a well founded Zermelo-Fraenkel universe and that $V_0 \subseteq V$. We let Ω be the set of all ordinals of our larger universe V. Since V_0 is transitive and well founded, then $V_0 \subseteq R_\Omega$ (by Theorem 3.3). Either V_0 contains all ordinals of V or it does not. Suppose it does. Then $R_\Omega \subseteq V_0$, hence $V_0 = R_\Omega$. If, on the other hand, V_0 does not contain all ordinals of V, then we let α be the first ordinal not in V_0. Then every ordinal less than α belongs to V_0, so $R_\alpha \subseteq V_0$. Also the rank of every element of V_0 must be in V_0, hence must be less than α, i.e., $V_0 \subseteq R_\alpha$. Hence $V_0 = R_\alpha$. ∎

EXERCISES

Exercise 3.1. Show that for any Zermelo-Fraenkel universe V, the intersection of all Zermelo sub-universes of V, each containing all the ordinals of V, is well founded.

Exercise 3.2. Suppose A is well founded, transitive, and G is a function from A into A. Using Theorem 3.2, prove that there exists a unique function F from A into A such that for every $x \in A$,
$$F(x) = G(F''(x \cap A)).$$

(This is an important generalization of the principle of definition by transfinite recursion.)

§4 Axiom E and Von Neumann's principle

As we formulated it, the axiom of choice postulates that each set of non-empty sets has a choice function. This can be strengthened by assuming there is a *single* choice function that picks a member from every non-empty set—in effect, a choice function for V. Some systems of set theory postulate this strengthened version. We do not, but we do investigate some of its consequences.

Axiom E There is a function F that simultaneously assigns to every non-empty set x an element of x.

Theorem 4.1. *Axiom E implies that R_Ω can be properly well ordered.*

Proof (Sketch)

For each α, let $S_\alpha = R_{\alpha+1} - R_\alpha$, i.e., S_α is the set of all sets of rank α. Every element of R_Ω is in one and only one of the sets S_α. Thus the class Σ of all sets S_α constitutes a partition of R_Ω into disjoint sets. By Axiom E (or even the weaker axiom of choice) each of the sets S_α can be well ordered; let $\mathcal{W}_\alpha$ be the class of all well orderings of S_α. Then each $\mathcal{W}_\alpha$ is non-empty. Also each $\mathcal{W}_\alpha$ is a set. (Reason: each element of $\mathcal{W}_\alpha$ is a binary relation on S_α, hence is an element of $S_\alpha \times S_\alpha$. Hence $\mathcal{W}_\alpha \subseteq \mathcal{P}(S_\alpha \times S_\alpha)$.) Now we need the axiom of choice in the strong form given by Axiom E in order to simultaneously pick one element from each $\mathcal{W}_\alpha$. More precisely, define $W(\alpha) = F(\mathcal{W}_\alpha)$, where F is a function satisfying Axiom E. Thus for each α, $W(\alpha)$ is a well ordering of the set of all elements of rank α.

Now for every x, y in R_Ω, we define $x < y$ iff either x is of lower rank than y or x, y have the same rank α but x precedes y in the well ordering $W(\alpha)$. It is a routine matter to check that the relation $x \leq y$ is a proper well ordering of the class R_Ω. ∎

Corollary 4.2. *Axioms D and E jointly imply that V (and hence that every class) can be properly well ordered.*

Von Neumann's principle Sometimes called Axiom F, Von Neumann's principle is the assertion: Every subclass of V is either a set, or can be put into a 1-1 correspondence with the class V of all sets.

Suppose we assume Axioms D and E and the Axiom of substitution. Let A be any class which is not a set. Then A can be properly well ordered (by Axioms D, E and Corollary 4.2). Then by the counting theorem (Chapter 6, Theorem 4.1) A can be put into a 1-1 correspondence with On. Hence any two proper classes A and B can be put into a 1-1 correspondence with each other. In particular, every proper class A can be put into a 1-1 correspondence with the proper class V. We thus have the following.

Theorem 4.3. *Axioms D, E, and the axiom of substitution jointly yield Von Neumann's principle.*

We next wish to prove that Von Neumann's principle yields both Axiom E and the axiom of substitution.

Lemma 4.4. *The axiom of choice (even in the weaker form) implies that for any sets x, y, and any function F, if $y = F''x$ then y can be put into a 1-1 correspondence with a subset of x.*

Proof For each element a of y, let S_a be the set of all elements of x which map to a under f. Each S_a is non-empty and is a subset of x. Also $a \neq b$ implies that S_a is disjoint from S_b. Using the axiom of choice, we can pick out one element from each set S_a—let x_0 be a set containing one element (exactly one) from each S_a. Then $F \restriction x_0$ is a 1-1 mapping from x_0 onto y. ∎

Lemma 4.5. *No set x can be put into a 1-1 correspondence with V. (This now, without necessitating the use of the axiom of substitution!)*

Proof By Cantor's theorem (Theorem 4.1, Chapter 1) for any set x, the set $\mathcal{P}(x)$ is larger than x. Therefore, if V were the same size as x, then $\mathcal{P}(x)$ would be larger than V, which is impossible, since $\mathcal{P}(x) \subseteq V$. ∎

Theorem 4.6. *Von Neumann's principle implies both Axiom E and the axiom of substitution.*

Proof Assume Von Neumann's principle. Then V can be put into a 1-1 correspondence with On, hence V can be properly well ordered. This of course yields Axiom E (take for $F(x)$ the first element of x in our well ordering of V).

Now for the axiom of substitution. Suppose x is a set and g is any function. Let $y = g(x)$. By Lemma 4.4, y can be put into a 1-1 correspondence with a subset x_0 of x. Therefore, if y could be put into a 1-1 correspondence with V, so could x_0, contrary to Lemma 4.5. And so y cannot be put into a 1-1 correspondence with V, hence by Von Neumann's principle, y must be a set. ∎

§5 Some other characterizations of ordinals

Axiom D, the axiom of well foundedness, leads to several alternative characterizations of ordinals.

Theorem 5.1. *A sufficient condition for a set a to be an ordinal number is that a is transitive, well founded, and every element of a is transitive.*

§5. SOME OTHER CHARACTERIZATIONS OF ORDINALS

Proof Suppose that a is transitive, well founded, and that every element of a is transitive. We first show that every element of a is an ordinal.

Since a is well founded and transitive, we can use the $\in$-induction principle of Theorem 3.2. So suppose that x is an element of a and that every element of x is an ordinal. Then x is (by hypothesis) a transitive set of ordinals, hence x is an ordinal. Thus if every element of x is an ordinal, then x is an ordinal. Hence by Theorem 3.2 every element of a is an ordinal.

Thus a is a set of ordinals. But a is transitive, hence a is an ordinal (Theorem 2.2 of Chapter 5). ∎

Remark Of course the condition of Theorem 5.1 is necessary as well as sufficient. Thus if we assume Axiom D, then a set a is an ordinal if and only if a is transitive and every element of a is transitive.

Theorem 5.2. *Suppose that a is transitive, well founded, and that for all x and y in a, if x is of lower rank than y, then $x \in y$. Then a is an ordinal.*

Proof Assume the hypothesis. By virtue of Theorem 5.1, it suffices to show that every element x of a is transitive. So suppose $x \in a$ and that $y \in x$ and $z \in y$; we must show that $z \in x$. Since a is well founded, then all elements of a have rank. Since $y \in x$, then y is of lower rank than x. Since $z \in y$, then z has lower rank than y, hence z has lower rank than x, hence $z \in x$ by hypothesis. This concludes the proof. ∎

Raphael Robinson's characterization A class A is called $\in$-connected if for any two distinct elements x, y of A, either $x \in y$ or $y \in x$. We know from Chapter 5 that every ordinal and also the class On is $\in$-connected. Also every ordinal is well founded and transitive. Thus every ordinal is transitive, well founded, and $\in$-connected. Robinson proved the converse, as follows.

Theorem 5.3. *If a set a is well founded, transitive, and $\in$-connected, then a is an ordinal.*

Proof Suppose a is well founded, transitive, and $\in$-connected. We will show that the hypothesis of Theorem 5.2 holds.

Suppose that x and y are elements of a and that x is of lower rank than y. Then $y \in x$ is not possible (because $y \in x$ implies that y is of lower rank than x), hence $x \in y$, since a is $\in$-connected. Then by Theorem 5.2, a is an ordinal. ∎

Discussion Several treatments of set theory, such as (Kelley 1975), assume Axiom D, and take as the definition of an ordinal α that α be transitive and $\in$-connected (well foundedness is automatic by Axiom D). From this definition one can, of course, derive all the usual properties of the ordinals. However this line of development seems to me far less intuitive than the course we have followed.

Theorem 5.4. *If A is a well founded, transitive, and $\in$-connected class that is not a set, then $A = On$.*

Proof Exercise 5.2 below. ∎

EXERCISES

Exercise 5.1. Suppose that A is well founded, transitive, and a proper class (i.e., not a set).

(a) Show that if every element of A is transitive then $A = On$.

(b) Suppose that for all x and y in A, if x is of lower rank than y, then $x \in y$. Prove that $A = On$.

Exercise 5.2. Now prove Theorem 5.4.

§6 More on the axiom of substitution

We will show that if V is well founded, then the axiom of substitution ($\mathbf{A}_8$) implies the following condition, often called the axiom of collection:

$\mathbf{A}_8^*$ For any relation R and set A, if for every $a \in A$ there is some x such that $R(a, x)$, then there is a set B such that for every $a \in A$ there is some $b \in B$ such that $R(a, b)$.

It is relatively trivial to show that $\mathbf{A}_8^*$ implies the axiom of substitution (even without the assumption that V is well founded), for suppose $\mathbf{A}_8^*$ holds. For any set A and function F, let A_0 be the set of all elements of A that are in the domain of F. Then $F''(A_0) = F''(A)$, and for every $a \in A_0$ there is at least one x (in fact exactly one x) such that $F(a) = x$. Then by $\mathbf{A}_8^*$ there is a set B such that for every $a \in A_0$ there is some $b \in B$ such that $F(a) = b$—in other words, for every $a \in A_0$, $F(a) \in B$, and so $F''(A_0) \subseteq B$, hence $F''(A_0)$ is a set. Therefore $F''(A)$ is a set.

This proves that $\mathbf{A}_8^*$ implies the axiom of substitution. And now we have the following.

Theorem 6.1. *If V is well founded, then the axiom of substitution implies $\mathbf{A}_8^*$.*

Proof Suppose that V is well founded and satisfies the axiom of substitution. Then every set has rank (Theorem 3.5). Now suppose that R is a relation and that A is a subset of the domain of R. For each $a \in A$, let $F(a)$ be the lowest rank of any x such that $R(a, x)$. By the axiom of substitution, $F''(A)$ is a set (of ordinals); let α be an ordinal greater than all the ordinals in $F''(A)$. Then for every $a \in A$ there is an x of rank $< \alpha$ such that $R(a, x)$, and so for every $a \in A$ there is some x in R_α such that $R(a, x)$. We thus take B to be R_α. ∎

CHAPTER 9

CARDINALS

Part I—Sizes of Sets

We have seen one infinitary generalization of the natural numbers: ordinal numbers. Using them, the notion of counting is extended. Now we come to the second generalization of the natural numbers: cardinal numbers. These extend the role of the natural numbers for specifying the size of a set.

Given two sets A and B, we are using the notations $A \cong B$ to mean that A can be put into a 1-1 correspondence with B; $A \leq B$ to mean that A can be put into a 1-1 correspondence with a subset of B (possibly all of B); and $A \prec B$ to mean that A can be put into a 1-1 correspondence with a proper subset of B, but not with the whole of B. Thus $A \prec B$ means that $A \leq B$ holds and $A \cong B$ doesn't hold.

If $A \cong B$, then we say that A and B are *similar* or *equinumerous*, or that A and B are of the *same size*, or that they have the *same cardinality*. We say that B has *higher cardinality* than A, or that A has *lower cardinality* than B if $A \prec B$, in which case we also say that A is *smaller than B*, and that B is *larger than A* (as we stated in Chapter 1).

For each finite set x, there is exactly one member of ω—one natural number—that is the same size as x. Thus natural numbers can be used to specify how big finite sets are. We need something like this to play a corresponding role for infinite sets. We cannot use the ordinals since, for example, both ω and $\omega + 1$ are the same size. Instead, we single out special ordinals for this purpose.

By a *cardinal number*—more briefly a *cardinal*—is meant an ordinal that cannot be put into a 1-1 correspondence with any lesser ordinal. For example, each natural number n is a cardinal and ω is a cardinal, but $\omega + 1$ is not a cardinal (since $\omega + 1 \cong \omega$). If we use the axiom of choice, then every set can be well ordered, hence every set can be put into a 1-1 correspondence with an ordinal. Then for any set x, by the *cardinality* of x, or the *cardinal* of x, is meant the least ordinal that is of the same size as x. Thus two sets x and y are of the same size if and only if they have the same cardinality.

In fact, we will not be using the axiom of choice for a while.

§1 Some simple facts

Proposition 1.1. *For any set A there are sets A_1 and A_2 of the same size as A such that A_1 is disjoint from A_2.*

Proof Take $A_1 = A \times \{1\}$ and $A_2 = A \times \{2\}$. ∎

Proposition 1.2. *If $A \le A'$ and $B \le B'$ and A' is disjoint from B' then $A \cup B \le A' \cup B'$.*

Proof Assume the hypothesis. Let φ_1 be a 1-1 mapping from A into A'. Since $B \le B'$ then of course $B - A \le B'$; let φ_2 be a 1-1 mapping from $B - A$ into B'. Since A is disjoint from $B-A$ and A' is disjoint from B' then φ_1 is disjoint from φ_2, hence $\varphi_1 \cup \varphi_2$ is a 1-1 mapping, and it maps $A \cup (B - A)$ into $A' \cup B'$, hence $A \cup (B - A) \le A' \cup B'$. But $A \cup (B - A) = A \cup B$, and so $A \cup B \le A' \cup B'$. ∎

We note that Proposition 1.2 of course implies that if $A \cong A'$ and $B \cong B'$, A is disjoint from B, and A' is disjoint from B', then $A \cup B \cong A' \cup B'$.

§2   The Bernstein-Schröder theorem

We say that two sets A and B are *comparable with respect to size* if either $A \le B$ or $B \le A$. In this chapter *comparable* will mean comparable with respect to size (not comparable with respect to the inclusion relation, as we used the term in Chapters 3 and 4).

The axiom of choice implies that any two sets A, B are comparable, because from this axiom it follows that A and B can both be well ordered, hence A can then be put into a 1-1 correspondence with an ordinal α and B can be put into a 1-1 correspondence with an ordinal β (using the axiom of substitution) and of course α and β are comparable with respect to size (since one of them is a subset of the other).

We will later show that the statement that any two sets are comparable (with respect to size) is not only implied by the axiom of choice, but is actually equivalent to it! But now *without the axiom of choice* we have the following.

Theorem 2.1 (Bernstein-Schröder).   *If $A \le B$ and $B \le A$, then $A \cong B$.*

Theorem 2.1 says that if there is some 1-1 mapping f from A *into* B (i.e., *onto* a subset of B) and if there is a 1-1 mapping g from B *into* A, then there is a 1-1 mapping h from A *onto* B.

We will give two proofs of this basic result. (The first is particularly well known and the second uses as a lemma a fixed-point theorem that is quite lovely in its own right!)

Actually, both proofs yield something a little stronger than that explicitly stated in the theorem. We like to present this theorem in the following form of a popular puzzle.

Suppose we have an infinite set M of men and an infinite set W of women. We are given that each of the men loves one and only one of the women and no two men love the same woman. However, some of the women may be left out—i.e., not loved by any man. Well, to symmetrize the situation, we are also given that each of the women loves one and only one of the men and no two women love the same man. (And again, some of the men may not be loved by any of the women.) The problem is to prove that not only can we marry all of the men to all of the women in a monogamous fashion (which merely says that the sets M and W are of the same size), but that we can marry them in such a manner that in each of the married couples, at least one of the couple is happy (i.e., either the husband loves his wife, or the wife loves her husband. In general, one can't guarantee that *both* partners will be happy).

§2. THE BERNSTEIN-SCHRÖDER THEOREM

How can this be done? The following shows how, and also constitutes a proof of the Bernstein-Schröder Theorem.

Solution For each man x in M, let $f(x)$ be the woman he loves. For each woman y in W, let $g(y)$ be the man she loves (not the man who loves her—there might not be any—but the man she loves).

Now, each person x_1 (man or woman) can be placed into one of three groups according to the following scheme: Take a person x_2 (if there is one) who loves x_1; then take a person x_3 (if there is one) who loves x_2; and keep going as long as possible. There are then three possibilities. (1) The process will terminate in some unloved person x_n in M, in which case we say that x_1 belongs to the *first* group. (2) The process will terminate in some unloved person x_n in W, in which case we say that x_1 belongs to the *second* group. (3) The process never terminates, in which case we say that x_1 belongs to the *third* group.

We let M_1, M_2, M_3 be the sets of men in the first, second, third groups respectively, and we let W_1, W_2, W_3 be the sets of women in the first, second, third groups respectively.

It is easy to verify the following three facts.

(1) f maps the set M_1 *onto* the set W_1.

(2) g maps the set W_2 *onto* the set M_2.

(3) f maps M_3 onto W_3 *and also* g maps W_3 onto M_3.

And so we marry each of the men in M_1 to the woman he loves. (This takes care of *all* the women in W_1.) Then we marry all the women in W_2 to the men they love. (This takes care of M_2.) As for M_3 and W_3, we have our choice of either marrying all the men of M_3 to the women they love, or of marrying all the women of W_3 to the men they love (both schemes will work). In either case, given any married couple $\langle m, w \rangle$, either $w = f(m)$ or $m = g(w)$.

A second proof of the Bernstein-Schröder theorem As we said, this proof uses a remarkable lemma, due to Knaster and Tarski (Tarski 1955). (Actually we use a special case of it—the proof generalizes to complete lattices.) First we need an additional piece of terminology. A function h is called *monotone* with respect to $\subseteq$—sometimes *monotone increasing*—if, for each X and Y in the domain of h, $X \subseteq Y$ implies $h(X) \subseteq h(Y)$. Note: monotone should not be confused with *progressing*, which says $X \subseteq h(X)$.

Lemma 2.2 (Fixed point lemma). *Let A be a set and let h be a monotone function from subsets of A to subsets of A. Then there is a subset B of A such that $h(B) = B$!*

Proof Let Σ be the set of all subsets X of A such that $X \subseteq h(X)$ and let $B = \cup \Sigma$. We will show that B is a fixed point of h (i.e., $h(B) = B$).

Step 1. We first show that $B \subseteq h(B)$. Well, suppose b is an element of B. Then for some X in Σ, $b \in X$. Since $X \in \Sigma$, then $X \subseteq h(X)$, hence $b \in h(X)$. But $X \subseteq B$, and since h is monotone, then $h(X) \subseteq h(B)$. Therefore $b \in h(B)$. This proves that $B \subseteq h(B)$.

Step 2. Since $B \subseteq h(B)$, then $h(B) \subseteq h(h(B))$ (since h is monotone), which means that $h(B)$ satisfies the condition of being in Σ! Since $h(B) \in \Sigma$ and $B = \cup \Sigma$, then $h(B) \subseteq B$. This, with $B \subseteq h(B)$, yields $h(B) = B$. This concludes the proof. ∎

Proof 2 of the Bernstein-Schröder theorem Let f be a 1-1 map from A into B and let g be a 1-1 map from B into A. We define the following map h from subsets of A to subsets of A: For any subset X of A, we let

$$h(X) = A - g''(B - f''(X)).$$

($f''X$ is a subset of B, hence $B - f''(X)$ is a subset of B, hence $g''(B - f''(X))$ is a subset of A, and so for $X \subseteq A$, $h(X)$ is indeed a subset of A.)

We now show that the function h is monotone. So suppose X, Y are subsets of A and that $X \subseteq Y$. Then $f''(X) \subseteq f''(Y)$, hence $B - f''(Y) \subseteq B - f''(X)$, hence $g''(B - f''(Y)) \subseteq g''(B - f''(X))$, hence $A - g''(B - f''(X)) \subseteq A - g''(B - f''(Y))$—i.e., $h(X) \subseteq h(Y)$. So h is monotone.

Then by the fixed point lemma, there is a subset X of A such that $X = h(X)$—hence $A - X = A - h(X)$. Now $A - h(X) = A - [A - g''(B - f''(X))] = g''(B - f''(X))$. So f is a 1-1 mapping from X onto $f''(X)$ and g is a 1-1 mapping from $B - f''(X)$ onto $A - X$.

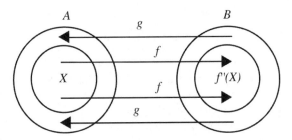

(Returning to our marriage example, if A is the set of men and B is the set of women, each man in X marries the woman he loves and each woman in $B - f''(X)$ marries the man she loves.)

Remark Both proofs yield the additional information that if f is a 1-1 mapping from A into B and g is a 1-1 mapping from B into A then there exists a 1-1 mapping φ from A onto B such that for any ordered pair $\langle x, y \rangle$ in φ, either $f(x) = y$ or $g(y) = x$.

EXERCISES

Exercise 2.1. The fixed point lemma 2.2 can easily be strengthened.

(a) Show the fixed point constructed in the proof is actually the *largest*. That is, if B is the fixed point of the proof, and if A is any other fixed point of h, then $A \subseteq B$.

(b) Give an alternative proof of the lemma that shows there is a *smallest* fixed point for h as well.

§3 Denumerable sets

We recall that for any natural number n, a set A is said to have n elements if $A \cong n$, and a set A is called *finite* if there exists a natural number n such that A has n elements; otherwise A is called infinite. We recall that A is called *denumerable* if $A \cong \omega$. We call A *countable* if A is either denumerable or finite—otherwise A is called *uncountable* or *non-denumerable*.

Theorem 3.1.

(1) Any denumerable set A can be put into a 1-1 correspondence with a proper subset of A.

(2) The union of any finite number of denumerable sets is denumerable.

(3) If A, B are denumerable, so is $A \times B$.

(4) If A is denumerable, then the set of all finite sequences of elements of A is denumerable.

(5) If A is denumerable and F is finite, then $A \cong (A \cup F)$.

EXERCISES

Exercise 3.1. Prove all parts of Theorem 3.1.

§4 Infinite sets and choice functions

We henceforth put a star before all theorems and exercises that require the axiom of choice.

* **Theorem 4.1.** Every infinite set has a denumerable subset.

More specifically we show the following stronger version.

Theorem 4.2. *If A is infinite and if there is a choice function for A, then A has a denumerable subset.*

Proof Suppose that A is infinite and that C is a choice function for A. Informally, the idea is to pick an element of A, then pick another element of A, and so forth, denumerably many times, but this involves making infinitely many choices, hence a choice function C is necessary.

More formally, it follows from Theorem 8.1 of Chapter 3 that there is a denumerable sequence $a_0, a_1, \ldots, a_n, \ldots$ of elements of A such that a_0 is the C-chosen element of A and for each n, a_{n+1} is the C-chosen element of $A - \{a_0, \ldots, a_n\}$. (Exercise 4.1 below.) ■

Second proof Since A has a choice function, then A can be well ordered. Then in any such well ordering, the set of elements that come before the first limit element is a denumerable set. ■

* Corollary 4.3. Any infinite set A can be put into a 1-1 correspondence with a proper subset of itself.

Proof By Theorem 4.1 A has a denumerable subset D. Then D can be put into a 1-1 correspondence with a proper subset D_0 of D; let f be such a 1-1 correspondence. For any x in A, let $h(x) = f(x)$, if $x \in D$, and let $h(x) = x$ if $x \in A - D$. Then h is a 1-1 correspondence from A onto $A - (D - D_0)$, which is a proper subset of A. ∎

Remarks No finite set can be put into a 1-1 correspondence with any of its proper subsets, hence if we assume the axiom of choice, then a set is infinite if and only if it can be put into a 1-1 correspondence with a proper subset of itself. Some treatments of set theory *define* a set to be infinite if it can be put into a 1-1 correspondence with a proper subset of itself.

* Corollary 4.4. If A is infinite and F is finite, then $A \cong (A \cup F)$.

Proof By Theorem 4.1, A has a denumerable subset D. By part 5 of Theorem 3.1, D can be put into a 1-1 correspondence φ_1 with $D \cup F$. Let φ_2 be the identity map from $A - D$ onto $A - D$. Then $\varphi_1 \cup \varphi_2$ is a 1-1 mapping from A onto $A \cup F$. ∎

EXERCISES

Exercise 4.1. Complete the first proof of Theorem 4.2. Hint: For any proper subset x of A, let $g(x) = A \cup \{C(A - x)\}$, and let $g(A) = A$. By Theorem 8.1 of Chapter 3, reading "$\mathcal{P}(A)$ for "A," there is a function h from ω into $\mathcal{P}(A)$ such that $h(0) = g(\emptyset)$ and for any n, $h(n^+) = g(h(n))$. Now take a_0 to be the sole element of $h(0)$ and for each n, take a_{n^+} to be the sole element of $h(n^+) - h(n)$. Now show that a_0 is the C-chosen element of A and that a_{n^+} is the C-chosen element of $A - \{a_0, \ldots, a_n\}$.

Part II—Hartog's Theorem

§5 Hartog's theorem

We have promised to show that the axiom of choice can be derived from the principle that any two sets are comparable in size. This is an easy corollary of the following theorem, one that will be seen to have many other interesting applications.

Theorem 5.1 (Hartog's theorem). *For any set A there is an ordinal α that cannot be 1-1 mapped into A.*

Remark Of course if we use the axiom of choice, Theorem 5.1 becomes trivial. [Given A, the set $\mathcal{P}(A)$ cannot be 1-1 mapped into A (by Cantor's theorem) and $\mathcal{P}(A)$ can be well ordered, hence put into a 1-1 correspondence with some ordinal α, hence α cannot be put into a 1-1 correspondence with any subset of A.] The whole point now is that Hartog's theorem can be proved *without* the axiom of choice.

§5. HARTOG'S THEOREM

Proof of Hartog's theorem Given a set A, let B be the class of all x such that x is a well ordering of some subset of A. Now, B is a set (because every element x of B is a set of ordered pairs of elements of A, hence is a subset of $A \times A$, hence an element of $\mathcal{P}(A \times A)$, so $B \subseteq \mathcal{P}(A \times A)$).

Each element x of B is order isomorphic to one and only one ordinal; let $F(x)$ be this ordinal. Then $F''(B)$ is a *set* of ordinals (by the axiom of substitution) so let α be any ordinal greater than all ordinals in $F''(B)$. Then α is not order isomorphic to any element of B, hence α cannot be put into a 1-1 correspondence with any subset y of A (because if it could, it would induce a well ordering x of y, hence x would be in B, hence some element of B would be order isomorphic to α). ∎

Trichotomy of cardinals The principle that for any two sets A and B, either $A \leq B$ or $B \leq A$ can be equivalently stated: For any two sets A and B, either $A < B$ or $A \cong B$ or $B < A$, and is accordingly known as the *trichotomy of cardinals*. Hartog's theorem at once yields the following.

Theorem 5.2. *The trichotomy of cardinals implies the axiom of choice.*

Proof Assume the trichotomy of cardinals. Take any set A. Then A is comparable in size with every ordinal α—either $A \leq \alpha$ or $\alpha < A$. By Hartog's theorem it cannot be that for *every* α, $\alpha < A$, hence for *some* ordinal α, $A \leq \alpha$. Thus A can be put into a 1-1 correspondence with a subset of α, and since α is well ordered, then A can be well ordered. Thus every set A can be well ordered, and so the axiom of choice follows. ∎

We now consider an interesting question: Without the axiom of choice, can we prove the existence of a non-denumerable ordinal, or better yet can we prove that for every ordinal α, there exists an ordinal β of higher cardinality? With the axiom of choice, we of course can, since then $\mathcal{P}(\alpha)$, which is of higher cardinality than α, can be well ordered, hence can be put into a 1-1 correspondence with an ordinal β. But now, due to Hartog's theorem, we don't need the axiom of choice for this, as we will now see.

Theorem 5.3. *For any ordinal α there is an ordinal β of higher cardinality.*

Proof Given α, by Hartog's theorem there is an ordinal β that cannot be mapped 1-1 into α. Therefore β is not a subset of α, so α must be a subset of β, and so $\alpha \leq \beta$. But $\beta \leq \alpha$ is false, and so $\alpha < \beta$. ∎

Remark It is of interest that Cantor's theorem and Hartog's theorem provide two very different methods of obtaining non-denumerable sets—in fact sets of arbitrarily high cardinality.

Hartog's function For any set x, by $\mathcal{H}(x)$ we shall mean the least ordinal that is not similar to any subset of x. This function $\mathcal{H}$ is known as *Hartog's function*, and will have a surprising use in Part V of this chapter!

We now define

$$\mathcal{P}^0(x) = x, \mathcal{P}^1(x) = \mathcal{P}(x), \mathcal{P}^2(x) = \mathcal{P}(\mathcal{P}(x)), \ldots, \mathcal{P}^{n+1}(x) = \mathcal{P}(\mathcal{P}^n(x)).$$

Theorem 5.4. $\mathcal{H}(x) \leq \mathcal{P}^4(x)$.

Proof Let $\beta = \mathcal{H}(x)$. Since β is the least ordinal that is not similar to any subset of x, then for every $\alpha < \beta$, α is similar to some subset of x. For any $\alpha < \beta$, let $h(\alpha)$ be the set of all well orderings w of subsets of x such that w is order isomorphic to α. The function h is 1-1 and its domain is β.

Now suppose w is an element of $h(\alpha)$ (where $\alpha \in \beta$). Since w is a well ordering of a subset of x, then w is a set of ordered pairs of elements of x, hence $w \subseteq x \times x$, hence $w \in \mathcal{P}(x \times x)$. Thus each element of $h(\alpha)$ is an element of $\mathcal{P}(x \times x)$, and so $h(\alpha) \subseteq \mathcal{P}(x \times x)$, hence $h(\alpha) \in \mathcal{PP}(x \times x)$. Also $x \times x \subseteq \mathcal{PP}(x)$, hence $\mathcal{PP}(x \times x) \subseteq \mathcal{PPPP}(x)$, and therefore $h(\alpha) \in \mathcal{PPPP}(x)$, and so $h''(\beta) \subseteq \mathcal{PPPP}(x)$, hence $h''(\beta) \leq \mathcal{PPPP}(x)$. But h is 1-1, and so $\beta \leq \mathcal{PPPP}(x)$. ∎

Part III—The Fundamental Theorem of Cardinal Arithmetic

§6 A fundamental theorem

A result known as the *fundamental theorem of cardinal arithmetic* is that for any infinite set A, we have $A \times A \cong A$. Of course we already know this for a *denumerable* set A — and the axiom of choice is not needed for this. But for sets of larger (higher) cardinality than ω, the axiom of choice is required for the proof.

In this chapter we shall give two quite different proofs of this basic result. Our first proof utilizes the following theorem (of considerable interest in its own right) whose proof requires the axiom of choice. Recall, we star theorems that use the axiom of choice.

* **Theorem 6.1.** For any infinite sets A, A' of the same size, $A \cong (A \cup A')$.

Proof We already know that this is true if A is a *denumerable* set (since the union of two denumerable sets is denumerable). Using this fact, together with the axiom of choice, we obtain the result for non-denumerable sets as follows.

We first consider the case that A is disjoint from A'. Let π be a 1-1 map from A onto A'. For any element x of A, we let $x' = \pi(x)$, and for any subset S of A, we let $S' = \pi''(S)$ (obviously $S' \subseteq A'$).

We let Σ be the set of all 1-1 maps, each of which is from some (necessarily infinite) subset S of A onto $S \cup S'$. Now, Σ is non-empty, because A has a denumerable subset D (by Theorem 4.1) and D can be mapped 1-1 onto $D \cup D'$. So Σ is non-empty. It is a routine matter to verify that for any nest $\mathcal{N}$ of such 1-1 correspondences, $\cup \mathcal{N}$ is again such a 1-1 correspondence, and so Σ is closed under chain unions. Then by Kuratowski's maximal principle (derived from the axiom of choice), Σ must contain a *maximal* element φ. Thus φ is a 1-1 correspondence between some subset M of A and $M \cup M'$—and φ cannot be further extended to any element of Σ. It may well be tempting at this point to conclude that M must be the whole of A, but this is not necessarily so! (It could be, for example, that A contains just one element x not in M, hence the only way to extend φ would be to put the one element set $\{x\}$ into a 1-1 correspondence with the two element set $\{x, x'\}$, which is obviously impossible!) However, we will show that the set $A - M$—call it "F"—must be finite! For if F

were infinite, it would include a denumerable subset D, and there would then be a 1-1 mapping ψ from D onto $D \cup D'$, hence $\varphi \cup \psi$ would be a proper extension of φ in Σ, contrary to the fact that φ is a *maximal* element of Σ. Therefore F must be finite.

Since F is finite, then $M \cup F \cong M$ (by Corollary 4.4) and so $A \cong M$. (Thus although M may not be the whole of A, it is of the same *size* as A.) Since $A \cong M$ then $A' \cong M'$, and since $M \cong (M \cup M')$ it obviously follows that $A \cong (A \cup A')$.

This proves that in the case that A is disjoint from A', $A \cong (A \cup A')$. Now suppose that A is not disjoint from A'. By Proposition 1.1, we can take disjoint sets A_1 and A_2 of the same size as A. Then $A \cong A_1$ and $A' \cong A_2$, so by Proposition 1.2, $A \cup A' \leq A_1 \cup A_2$, and $A_1 \cup A_2 \cong A_1$ (as we have shown, for *disjoint* sets A_1 and A_2), hence $A_1 \cup A_2 \cong A$, hence $A \cup A' \leq A$. Also, of course, $A \leq A \cup A'$, and so $A \cong A \cup A'$. ∎

As a corollary we easily get the following.

* Theorem 6.2 (Law of additive absorption). For any infinite set A and any set B, if $B \leq A$ then $A \cup B \cong A$.

Proof Suppose A is infinite and $B \leq A$. By Proposition 1.1 we can take two disjoint sets A_1 and A_2 of the same size as A. Then $B \leq A_2$ and of course $A \leq A_1$, so by Proposition 1.2 $A \cup B \leq A_1 \cup A_2$, but $A_1 \cup A_2 \cong A_1$ (by Theorem 6.1), hence $A_1 \cup A_2 \cong A$, hence $A \cup B \leq A$, and of course $A \leq A \cup B$, so $A \cup B \cong A$ (Theorem 2.1). ∎

Next we need the following.

* Lemma 6.3. Suppose M is infinite, $M \cong M'$, M' is disjoint from M, and $M \cong M \times M$. Then $M \cup M' \cong (M \cup M') \times (M \cup M')$. Moreover any 1-1 mapping of M onto $M \times M$ can be extended to a 1-1 mapping of $M \cup M'$ onto $(M \cup M') \times (M \cup M')$.

Proof Let φ be a 1-1 map from M onto $M \times M$ and suppose M is infinite and M' is of the same size as M. Then it is obvious that the six sets M, M', $M \times M$, $M \times M'$, $M' \times M$, and $M' \times M'$ are all of the same size. Also, $(M \cup M') \times (M \cup M') = (M \times M) \cup (M \times M') \cup (M' \times M) \cup (M' \times M')$ (as can be easily verified). Let N be the union of the three sets $M \times M'$, $M' \times M$, $M' \times M'$. Then by two applications of Theorem 6.1, N is the same size as $M \times M'$, hence is the same size as M', so there is a 1-1 mapping ψ from M' onto N. Hence $\varphi \cup \psi$ is a 1-1 mapping from $M \cup M'$ onto $(M \times M) \cup N$, which is $(M \cup M') \times (M \cup M')$. And so φ can be extended to a 1-1 mapping from $M \cup M'$ onto $(M \cup M') \times (M \cup M')$. ∎

Now we can prove the following important result.

* Theorem 6.4 (Fundamental theorem of cardinal arithmetic). For any infinite set A, $A \cong A \times A$.

Proof Let Σ be the set of all 1-1 maps, each of which is from a subset S of A onto $S \times S$ (which is a subset of $A \times A$). The set Σ is non-empty, since A includes a denumerable set D and $D \cong (D \times D)$. The set Σ is closed under chain unions (as can be easily verified) and so Σ contains a maximal element φ. Let M be the domain of φ. As in the proof of Theorem 6.1, M need not be the whole of A, but we will show that $A - M$ must be smaller than M (though not necessarily finite, as in the proof of Theorem 6.1).

Well, suppose $A - M$ were not smaller than M. Then by the Bernstein-Schröder theorem 2.1, $A - M$ would include a subset M' of the same size as M, hence by Lemma 6.3, φ could be further extended to a 1-1 mapping from $M \cup M'$ onto $(M \cup M') \times (M \cup M')$, contrary to the fact that φ is a *maximal* member of Σ. Thus $A - M$ must be smaller than M.

Since $A - M < M$, then of course $A - M \leq M$, and so by the law of additive absorption (Theorem 6.2) $M \cong (A - M) \cup M$, but $(A - M) \cup M = A$ (since $M \subseteq A$), and so $M \cong A$. Then, since $M \cong M \times M$, it follows that $A \cong A \times A$. ∎

* **Corollary 6.5 (Law of multiplicative absorption).** *If $B \leq A$, A is infinite, and B is non-empty, then $B \times A \cong A$.*

Proof Assume the hypothesis. Then B is similar to some subset B_1 of A, hence $B \times A \cong B_1 \times A$, but $B_1 \times A \leq A \times A$ (since $B_1 \times A \subseteq A \times A$), so $B \times A \leq A \times A$, and since $A \times A \cong A$ (Theorem 6.4) then $B \times A \leq A$. Also $A \leq B \times A$ (since $B \neq \emptyset$), and so $B \times A \cong A$. ∎

Part IV—Cardinal Numbers

§7 Preliminaries

We recall that by a *cardinal*, or *cardinal number* is meant an ordinal that is not the same size as any lesser ordinal. The first infinite cardinal is ω—also called $\aleph_0$. The next infinite cardinal (the first non-denumerable cardinal) is called $\aleph_1$; the next is called $\aleph_2$, and so forth—that is, for each ordinal α, $\aleph_\alpha$ is the αth *infinite* cardinal in the well ordering of the cardinals according to magnitude. Of course $\aleph_0$ and ω are the same object—we use the $\aleph_0$ notation when we wish to emphasize its role as a *cardinal*, and ω when its role as an *ordinal* is important.

As we have noted, using the axiom of choice, each set A is similar to one and only one cardinal.

Definition 7.1. The *cardinal* of A is the cardinal that is similar to A. It is denoted $\overline{\overline{A}}$.

If c is a cardinal then obviously $\overline{\overline{c}} = c$. Also, of course, for any ordinal α, since $\overline{\overline{\alpha}}$ is a cardinal, then $\overline{\overline{\alpha}} \leq \alpha$.

The following little lemma—reminiscent of Theorem 2.1, Chapter 6, will be helpful.

Lemma 7.2. *Suppose φ is an order isomorphism from an ordinal α onto a set S of ordinals. Then for every ordinal γ in α, $\gamma \leq \varphi(\gamma)$.*

Proof As with the proof of Theorem 2.1, Chapter 6, let us say that an element γ of α moves backwards (under φ) if $\varphi(\gamma) < \gamma$. Well, if γ moves backwards, then $\varphi(\gamma)$ is also in α, hence $\varphi(\gamma)$ is also in the domain of φ, and then from $\varphi(\gamma) < \gamma$ it follows that $\varphi(\varphi(\gamma)) < \varphi(\gamma)$ (since φ is an isomorphism), and so $\varphi(\gamma)$ also moves backwards. Thus if γ moves backwards, so does some lesser element of α—namely $\varphi(\gamma)$—hence there cannot be a least element of α that moves backwards, and hence no element of α moves backwards. ∎

Remark Both the lemma above and Theorem 2.1 of Chapter 6 are but special cases of the following fact: If L is a lower section of a well ordered system $\langle A, \leq \rangle$ and φ is an isomorphism (with respect to $\leq$) from L into A, then for every $x \in L$, $x \leq \varphi(x)$. We leave the proof of this as Exercise 7.1 below.

From the lemma above we have the following.

Proposition 7.3. *If $S \subseteq \beta$, where β is an ordinal, and S is isomorphic to α, then $\alpha \leq \beta$.*

Proof Assume the hypothesis. Then α is isomorphic to S under some mapping φ. By the lemma above, for each $\gamma \in \alpha$, $\gamma \leq \varphi(\gamma)$, and since $\varphi(\gamma) \in S$, then $\varphi(\gamma) < \beta$, hence $\gamma < \beta$, hence $\gamma \in \beta$, and so $\alpha \subseteq \beta$. Then $\alpha \leq \beta$. ∎

We now apply this proposition to part of the following proposition about cardinals.

Proposition 7.4. *For any cardinals c and d:*

(1) $c \cong d$ iff $c = d$;

(2) $c \prec d$ iff $c < d$;

(3) $c \preceq d$ iff $c \leq d$.

Proof
(1) If $c = d$, then of course $c \cong d$. Now suppose $c \cong d$. Then $c < d$ is not possible, since d is a cardinal. Also, since $d \cong c$ and c is a cardinal, then $d < c$ is not possible. Hence $c = d$.

(2) Suppose $c < d$. Then c is a subset of d, hence $c \preceq d$, but $c \cong d$ is not possible, since d is a cardinal. Therefore $c < d$ implies $c \prec d$.

To show the converse is a little more tricky; here is one way. Suppose $c \prec d$. Then $c \cong S$ for some subset S of d. The set S is order isomorphic to some ordinal α (by the counting theorem—Theorem 4.1 of Chapter 6). By Proposition 7.3, $\alpha < d$. Also $c \cong \alpha$ (since $c \cong S$ and $S \cong \alpha$), so $c \leq d$. But $c \neq d$ (since $c \prec d$), hence $c < d$. Thus $c \prec d$ implies $c < d$.

(3) This is immediate from (1) and (2). ∎

Remarks Another way to show that $c \prec d$ implies $c < d$ is this. Suppose $c \prec d$. Then $d \prec c$ is not possible, by the Bernstein-Schröder theorem, hence $d < c$ doesn't hold (because $d < c$ implies $d \prec c$, as we have shown), and also $c = d$ doesn't hold (since $c \prec d$), hence $c < d$.

The first proof, however, though more elaborate, yields an alternative proof of the Bernstein-Schröder theorem, if we use the axiom of choice, as we will shortly see.

Assuming the axiom of choice now, each set A has a cardinal number $\overline{\overline{A}}$, and so the following is immediate from Proposition 7.4.

* **Proposition 7.5.** *For any sets A and B:*

(1) $A \cong B$ iff $\bar{\bar{A}} = \bar{\bar{B}}$;

(2) $A \prec B$ iff $\bar{\bar{A}} < \bar{\bar{B}}$;

(3) $A \preceq B$ iff $\bar{\bar{A}} \leq \bar{\bar{B}}$.

Remarks Assuming the axiom of choice, Proposition 7.5 gives the following simple proof of the Bernstein-Schröder theorem: If $A \preceq B$ and $B \preceq A$ then by the proposition above, $\bar{\bar{A}} \leq \bar{\bar{B}}$ and $\bar{\bar{B}} \leq \bar{\bar{A}}$, hence $\bar{\bar{A}} = \bar{\bar{B}}$, hence $A \cong B$.

EXERCISES

Exercise 7.1. Prove the fact stated in the remark following the proof of Lemma 7.2 and show how it yields both Theorem 2.1, Chapter 6 and Lemma 7.2 above.

§8 Cardinal arithmetic

In §6 of Chapter 6 we took a brief look at ordinal arithmetic, and saw that it generalized the arithmetic of the natural numbers. Now we see there is a second way of generalizing elementary arithmetic that is equally natural—to cardinal arithmetic.

Recall, one way of characterizing ordinal addition was this. To add ordinals α and β, take two well ordered sets, a, ordered like α (that is, isomorphic to it), a disjoint set b, ordered like β, and combine them, $a \cup b$, giving the result an ordering that amounts to "following" a with b. The result is well ordered, so is isomorphic to an ordinal, and that ordinal is the sum of α and β. Now, cardinal addition is like this except that only sizes of sets matter, and not ordering relations.

Definition 8.1. For any cardinals c_1 and c_2, by $c_1 + c_2$ is meant the cardinality of $C_1 \cup C_2$, where C_1 is any set of cardinality c_1 and C_2 is any set *disjoint from C_1* of cardinality c_2 (which sets C_1, C_2 are chosen obviously makes no difference, providing, of course, that they are disjoint).

The cardinal $c_1 + c_2$ is called the *cardinal sum* of c_1 and c_2. If c_1 and c_2 are finite cardinals, i.e., natural numbers, the cardinal sum $c_1 + c_2$ is their ordinary sum. It takes some work to prove this—we omit the formal proof. But this notion of addition, when applied to natural numbers, is an obvious abstraction of the common understanding of what ordinary addition is about.

We saw that ordinal addition was not commutative (except in the finite case). On the other hand, for cardinal addition, $c_1 + c_2 = c_2 + c_1$ (since $C_1 \cup C_2 = C_2 \cup C_1$). Cardinal addition is likewise associative. But real differences between natural number addition and cardinal addition exist. By Theorem 6.2, if c and d are cardinals, d is infinite, and $c \leq d$, then $c + d = d$ (law of additive absorption). In particular, $\aleph_0 + n = n + \aleph_0 = \aleph_0$, for any natural number n.

Definition 8.2. For any cardinals c and d, by $c \cdot d$ is meant the cardinality of $c \times d$.

$c \cdot d$ is called the *cardinal product* of c and d. It is not hard to show it is commutative and associative. We leave this as an exercise. On the other hand, once again

§8. CARDINAL ARITHMETIC   119

differences with ordinary multiplication show up once infinite cardinals are involved. Theorem 6.4 is for any infinite set so as a special case, for any infinite cardinal c, $c \cdot c = c$ (fundamental theorem of cardinal arithmetic)!

The proof we gave of Theorem 6.4 required the axiom of choice, but the axiom of choice is *not* required to show that if c is an infinite *cardinal*, then $(c \times c) \cong c$. We shall now do this, and when we have, we will have an alternative proof of Theorem 6.4. But first we need some antecedent material.

A proper well ordering of On^2 By On^2 we mean $On \times On$—i.e., the class of all ordered pairs $\langle \alpha, \beta \rangle$ of ordinals. We introduce a proper well ordering of On^2 in the following manner: By $\max(\alpha, \beta)$ we mean the maximum of α, β (which is $\alpha \cup \beta$). Then, given two pairs $\langle \alpha_1, \beta_1 \rangle$ and $\langle \alpha_2, \beta_2 \rangle$ we define $\langle \alpha_1, \beta_1 \rangle < \langle \alpha_2, \beta_2 \rangle$ ($\langle \alpha_1, \beta_1 \rangle$ is *less than* $\langle \alpha_2, \beta_2 \rangle$) if either:

(1) $\max(\alpha_1, \beta_1) < \max(\alpha_2, \beta_2)$, or

(2) $\max(\alpha_1, \beta_1) = \max(\alpha_2, \beta_2)$ and $\alpha_1 < \alpha_2$, or

(3) $\max(\alpha_1, \beta_1) = \max(\alpha_2, \beta_2)$ and $\alpha_1 = \alpha_2$ and $\beta_1 < \beta_2$.

Thus we order the pairs first according to their maximums; if these are the same, we order the pairs according to their first members; if these are also the same, then we order the pairs according to their second members.

The ordering begins like this: $\langle 0,0 \rangle, \langle 0,1 \rangle, \langle 1,0 \rangle, \langle 1,1 \rangle, \langle 0,2 \rangle, \langle 1,2 \rangle, \langle 2,0 \rangle, \langle 2,1 \rangle,$ $\langle 2,2 \rangle, \langle 0,3 \rangle, \cdots, \langle \alpha, \alpha+2 \rangle, \langle \alpha+1, \alpha+2 \rangle, \langle \alpha+2, 0 \rangle, \cdots, \langle \alpha+2, \alpha+2 \rangle, \langle 0, \alpha+3 \rangle, \cdots$.

It is obvious that this ordering $\leq$ of On^2 is a linear ordering. To see that it is a well ordering, take any non-empty subclass A of On^2. Let γ be the least ordinal that is the maximum of some pair in A; let α be the least ordinal that is the first element of some pair in A whose maximum is γ; then let β be the least ordinal that is the second member of some pair in A whose maximum is γ and whose first member is α. Then $\langle \alpha, \beta \rangle$ is clearly the least element of A. And so A does have a least member. Thus On^2 is well ordered under $\leq$.

In fact, On^2 is *properly* well ordered under $\leq$ (i.e., every proper lower section is a set), for suppose L is a proper lower section. Then there is some pair $\langle \alpha, \beta \rangle$ such that every element of L precedes $\langle \alpha, \beta \rangle$ in the ordering. Now let γ be any ordinal greater than both α and β. Then $\max(\alpha, \beta) < \gamma$. Now, for any pair $\langle \alpha_1, \beta_1 \rangle$ in L, $\langle \alpha_1, \beta_1 \rangle$ precedes $\langle \alpha, \beta \rangle$, hence $\max(\alpha_1, \beta_1) \leq \max(\alpha, \beta) < \gamma$, and so α_1, β_1 are both $< \gamma$, hence they are both elements of γ, so $\langle \alpha_1, \beta_1 \rangle \in \gamma \times \gamma$. Thus $L \subseteq \gamma \times \gamma$, but $\gamma \times \gamma$ is a set, and so L is a set. This proves that we have a *proper* well ordering of On^2.

It then follows that On^2 is *order isomorphic* to the class On; we let P be this isomorphism (there is only one!). Thus P maps On^2 1-1 onto On. For any ordinals α, β we let $P(\alpha, \beta) = P(\langle \alpha, \beta \rangle)$. Then $\langle \alpha, \beta \rangle \leq \langle \alpha', \beta' \rangle$ if and only if $P(\alpha, \beta) \leq P(\alpha', \beta')$.

Now we shall establish some facts about this P operation.

Lemma 8.3. *Let $\gamma = \max(\alpha, \beta)$. Then $P(\alpha, \beta) \subseteq P''((\gamma+1) \times (\gamma+1))$.*

Proof Let $\gamma = \max(\alpha,\beta)$, and let δ be any member of $P(\alpha,\beta)$. Then $\delta < P(\alpha,\beta)$. Now, $\delta = P(\alpha_1,\alpha_2)$ for unique ordinals α_1, α_2, and so $P(\alpha_1,\alpha_2) < P(\alpha,\beta)$. Hence $\max(\alpha_1,\alpha_2) \leq \max(\alpha,\beta)$. Hence $\alpha_1 \leq \gamma$ and $\alpha_2 \leq \gamma$, so $\alpha_1 \in \gamma + 1$ and $\alpha_2 \in \gamma + 1$, so $\langle \alpha_1, \alpha_2 \rangle \in (\gamma+1) \times (\gamma+1)$, so $P(\alpha_1,\alpha_2) \in P''((\gamma+1) \times (\gamma+1))$, hence $\delta \in P''((\gamma+1) \times (\gamma+1))$. Thus for every member δ of $P(\alpha,\beta)$, we have $\delta \in P''((\gamma+1) \times (\gamma+1))$, and so $P(\alpha,\beta) \subseteq P''((\gamma+1) \times (\gamma+1))$. ∎

Proposition 8.4. *If $\gamma = \max(\alpha,\beta)$ and at least one of α, β is infinite, then $P(\alpha,\beta) \leq \bar{\bar{\gamma}} \cdot \bar{\bar{\gamma}}$.*

Proof Suppose $\gamma = \max(\alpha,\beta)$ and γ is infinite. By Lemma 8.3, $P(\alpha,\beta) \subseteq P''((\gamma+1) \times (\gamma+1))$. Now, $P''((\gamma+1) \times (\gamma+1))$ is the same size as $(\gamma+1) \times (\gamma+1)$ (since P is 1-1), which in turn is the same size as $\gamma \times \gamma$ (since γ is infinite and so is the same size as $\gamma + 1$). Therefore $P(\alpha,\beta) \leq \gamma \times \gamma$, and since $\gamma \times \gamma$ is the same size as $\bar{\bar{\gamma}} \cdot \bar{\bar{\gamma}}$, then $P(\alpha,\beta) < \bar{\bar{\gamma}} \cdot \bar{\bar{\gamma}}$. But $\bar{\bar{\gamma}} \cdot \bar{\bar{\gamma}}$ is a cardinal, and so $P(\alpha,\beta) \leq \bar{\bar{\gamma}} \cdot \bar{\bar{\gamma}}$. ∎

Next we need the following.

Lemma 8.5. *Suppose c is a cardinal such that for every α, β less than c, $P(\alpha,\beta) < c$. Then $c \cdot c = c$.*

Proof Assume the hypothesis. Then for any α and β, if α, β are both elements of c, then $P(\alpha,\beta) \in c$, so $P''(c \times c) \subseteq c$. Also, since P is 1-1, then $P''(c \times c)$ is the same size as $c \times c$, and so $c \times c \leq c$. Also, of course, $c \leq c \times c$, so $c \times c$ is the same size as c, hence $c \cdot c = c$. ∎

Now, *without the axiom of choice*, we can prove the following.

Theorem 8.6. *For any infinite cardinal c, $c \cdot c = c$.*

Proof We prove this by transfinite induction on the infinite cardinals. Suppose c is an infinite cardinal such that for every infinite cardinal $c' < c$, $c' \cdot c' = c'$. We show that $c \cdot c = c$. We do this by showing that c obeys the hypothesis of the lemma above.

Well, suppose α and β are both less than c. If α and β are both finite, then of course $P(\alpha,\beta) < c$. So suppose that at least one of α, β is infinite and let γ be their max. Then γ is infinite, and so by Proposition 8.4, $P(\alpha,\beta) \leq \bar{\bar{\gamma}} \cdot \bar{\bar{\gamma}}$. Since α, β are both less than c, so is γ, hence $\bar{\bar{\gamma}} < c$ (since $\bar{\bar{\gamma}} \leq \gamma$). Then by the induction hypothesis (taking $\bar{\bar{\gamma}}$ for c'), $\bar{\bar{\gamma}} \cdot \bar{\bar{\gamma}} = \bar{\bar{\gamma}}$, and so $P(\alpha,\beta) \leq \bar{\bar{\gamma}}$. Hence $P(\alpha,\beta) < c$. And so by the lemma above, $c \cdot c = c$. ∎

Corollary 8.7. *For any infinite cardinal c and for any non-zero cardinal $d \leq c$, $d \cdot c = c$.*

Proof Exercise 8.5 below. ∎

Discussion Theorem 8.6 was proved without the axiom of choice. With the axiom of choice, Theorem 8.6 gives the following alternative proof of Theorem 6.4: Suppose A is an infinite set. Then (using the axiom of choice) A does have a cardinal c (and of course c is also infinite). Then $A \times A$ has cardinality $c \cdot c$. And since $c \cdot c = c$ (by Theorem 8.6) then $A \times A$ also has cardinality c. Hence $(A \times A) \cong A$.

Exponentiation also generalizes to infinite cardinals.

Definition 8.8. For cardinals c and d, by c^d is meant the cardinality of the set of all functions from d to c.

It can be shown that this agrees with ordinary exponentiation on the natural numbers, and that several of the familiar laws of exponentiation still hold. We do not carry the investigation of cardinal exponentiation any further since it would take us too far afield.

EXERCISES

Exercise 8.1. Show cardinal addition is well defined. That is, show that if a and a' are the same size, b and b' are the same size, a and b are disjoint, and a' and b' are disjoint, then $a \cup b$ and $a' \cup b'$ are the same size.

Exercise 8.2. Show cardinal multiplication is well defined. Assume a and a' are the same size, and b and b' are the same size, and show $a \times b$ and $a' \times b'$ are the same size.

Exercise 8.3. Prove the following facts about cardinal multiplication. For any cardinals a, b, and c:

(a) $a \cdot b = b \cdot a$

(b) $a \cdot (b \cdot c) = (a \cdot b) \cdot c$

(c) $a \cdot 0 = 0$

(d) $a \cdot 1 = a$

(e) if $a \le b$ then $a \cdot c \le b \cdot c$. (Can $\le$ be replaced by $<$ here?)

Exercise 8.4. Suppose a, b, and c are cardinals. Using cardinal multiplication and exponentiation, show $(a^b)^c = a^{(b \cdot c)}$.

Exercise 8.5. Prove Corollary 8.7.

Part V—G.C.H. implies A.C.

§9 Sierpiński's theorem

G.C.H. (the generalized continuum hypothesis) is that given any set x, there is no set y such that $x \prec y \prec \mathcal{P}(x)$—or equivalently, that for any sets x and y, if $x \preceq y \preceq \mathcal{P}(x)$, then either $x \cong y$ or $y \cong \mathcal{P}(x)$.

We now wish to prove the lovely result of Sierpiński (1947) that the generalized continuum hypothesis implies the axiom of choice (A.C.). Our presentation is an adaptation of that in (Mendelson 1987), which in turn is a modified form of the proof in (Cohen 1966). We shall break the proof into several small pieces, which are relatively easy to handle.

For any sets x and y we define $x + y$ to be the set $x^* \cup y^*$, where $x^* = x \times \{0\}$ and $y^* = y \times \{1\}$. The sets x^* and y^* are disjoint from each other, and $x^* \cong x$ and $y^* \cong y$.

Let us note that to say that $x + y \cong z$ is equivalent to saying that for any set x_1 similar to x and any set y_1 similar to y, if x_1 is disjoint from y_1, then $x_1 \cup y_1 \cong z$. We also note that if x is disjoint from y then $x+y \cong x \cup y$. Finally, $x+x$ is the set $(x \times \{0\}) \cup (x \times \{1\})$, hence is the set $x \times 2$. Thus $x + x = x \times 2$.

Now we establish some basic properties of the + operation (which do not require the axiom of choice).

Proposition 9.1.

(1) If $\omega \leq x$ then $x \cong x + 1$.

(2) If $x + x \cong x$ and x is non-empty, then $x \cong x + 1$.

Proof

(1) Suppose $\omega \leq x$. Then x has a denumerable subset D. Let a be any element outside x. Then by the last part of Theorem 3.1, $D \cong D \cup \{a\}$. Then $(x-D) \cup D \cong (x-D) \cup (D \cup \{a\})$, but $(x-D) \cup D = x$ and $(x-D) \cup (D \cup \{a\}) = x \cup \{a\}$, and so $x \cong x \cup \{a\}$, hence $x \cong x + 1$.

(2) Suppose $x + x \cong x$. If also $x \neq \emptyset$, then $1 \leq x$, hence $x + 1 \leq x + x$, hence $x + 1 \leq x$. Also, of course, $x \leq x + 1$, so by the Bernstein-Schröder theorem, $x \cong x + 1$. □

Proposition 9.2 (Important!). $\mathcal{P}(x + y) \cong \mathcal{P}(x) \times \mathcal{P}(y)$.

Proof It suffices to prove that for any disjoint sets x and y, $\mathcal{P}(x \cup y) \cong \mathcal{P}(x) \times \mathcal{P}(y)$. So suppose that x is disjoint from y. Then every subset of x is disjoint from every subset of y. Now, a set z is an element of $\mathcal{P}(x \cup y)$ if and only if it is the union of some subset x_1 of x with some subset y_1 of y. Moreover, given z, there can be only *one* pair $\langle x_1, y_1 \rangle$ of subsets of x, y respectively such that $x_1 \cup y_1 = z$. (Reason: suppose $\langle x_2, y_2 \rangle$ is another such pair. Then $x_1 \cup y_1 = x_2 \cup y_2$, hence $x_1 \subseteq x_2 \cup y_2$—since $x_1 \subseteq x_1 \cup y_1$—hence $x_1 \subseteq x_2$—since x_1 is disjoint from y_2. Similarly $x_2 \subseteq x_1$, so $x_1 = x_2$. Similarly $y_1 = y_2$.) Since now, every element of $\mathcal{P}(x \cup y)$ is $x_1 \cup y_1$ for a *unique* pair $\langle x_1, y_1 \rangle$ in $\mathcal{P}(x) \times \mathcal{P}(y)$, then $\mathcal{P}(x \cup y)$ is in 1-1 correspondence with $\mathcal{P}(x) \times \mathcal{P}(y)$. ∎

Corollary 9.3. $\mathcal{P}(x + x) \cong \mathcal{P}(x) \times \mathcal{P}(x)$.

Corollary 9.4. $\mathcal{P}(x + 1) \cong \mathcal{P}(x) + \mathcal{P}(x)$.

Proof By Proposition 9.2, $\mathcal{P}(x + 1) \cong \mathcal{P}(x) \times \mathcal{P}(1)$. Now, $\mathcal{P}(1) = 2$, so $\mathcal{P}(x + 1) \cong \mathcal{P}(x) \times 2$, but $\mathcal{P}(x) \times 2$ is the set $\mathcal{P}(x) + \mathcal{P}(x)$. ∎

Corollary 9.5. Suppose $x \cong x + 1$. Then:

(1) $\mathcal{P}(x) \cong \mathcal{P}(x) + \mathcal{P}(x)$;

(2) $\mathcal{P}(x) \cong \mathcal{P}(x) + x$.

Proof Suppose $x \cong x + 1$.

(1) Then $\mathcal{P}(x) \cong \mathcal{P}(x + 1)$, but $\mathcal{P}(x + 1) \cong \mathcal{P}(x) + \mathcal{P}(x)$ (Corollary 9.4), hence $\mathcal{P}(x) \cong \mathcal{P}(x) + \mathcal{P}(x)$.

(2) Since $x \leq \mathcal{P}(x)$, then $\mathcal{P}(x) + x \leq \mathcal{P}(x) + \mathcal{P}(x)$, hence $\mathcal{P}(x) + x \leq \mathcal{P}(x)$ (because $\mathcal{P}(x) + \mathcal{P}(x) \cong \mathcal{P}(x)$, by the first part). Also, of course, $\mathcal{P}(x) \leq \mathcal{P}(x) + x$, and so by the Bernstein-Schröder theorem, $\mathcal{P}(x) \cong \mathcal{P}(x) + x$. ∎

Proposition 9.6. *If $\omega \leq x$ then $\mathcal{P}(x) \cong \mathcal{P}(x) + \mathcal{P}(x)$.*

Proof If $\omega \leq x$ then $x \cong x + 1$ (by the first part of Proposition 9.1), hence by the first part of Corollary 9.5, $\mathcal{P}(x) \cong \mathcal{P}(x) + \mathcal{P}(x)$. ∎

Remarks Let us pause and note the significance of Proposition 9.6. Without the axiom of choice we cannot prove of an arbitrary infinite set y that $y \cong y + y$. But by Proposition 9.6, if y is the power set of a set x that includes a denumerable subset, then it follows without the axiom of choice that $y \cong y + y$.

Proposition 9.6 has the following corollary.

Proposition 9.7. *If $z = \mathcal{P}(y \cup \omega)$ for some set y, then $\mathcal{P}^n(z) + \mathcal{P}^n(z) \cong \mathcal{P}^n(z)$, for every natural number n.*

Proof Suppose $z = \mathcal{P}(y \cup \omega)$. Let $v = y \cup \omega$. Then $\omega \leq v$ and $z = \mathcal{P}(v)$. Since $\omega \leq v$ then $\omega \leq \mathcal{P}^n(v)$, since $v \leq \mathcal{P}^n(v)$, hence by Proposition 9.6, taking $\mathcal{P}^n(v)$ for x, we have $\mathcal{P}(\mathcal{P}^n(v)) + \mathcal{P}(\mathcal{P}^n(v)) \cong \mathcal{P}(\mathcal{P}^n(v))$, and thus $\mathcal{P}^n(z) + \mathcal{P}^n(z) \cong \mathcal{P}^n(z)$. ∎

Proposition 9.8. *If $x + x \cong x$ and $y \leq \mathcal{P}(x)$ then $x + y \leq \mathcal{P}(x)$.*

Proof If x is empty and $y \leq \mathcal{P}(c)$, y must either be empty or have one member. Either way, $x + y \leq \mathcal{P}(x)$ follows easily. So now assume x is not empty, $x + x \cong x$, and $y \leq \mathcal{P}(x)$. Since $x + x \cong x$ then $x + 1 \cong x$ (Proposition 9.1 part (2)), hence $\mathcal{P}(x) + x \cong \mathcal{P}(x)$ (Corollary 9.5, part (2)). Then, since $y \leq \mathcal{P}(x)$ it follows that $y + x \leq \mathcal{P}(x) + x$, and since $\mathcal{P}(x) + x \cong \mathcal{P}(x)$, then $y + x \leq \mathcal{P}(x)$. ∎

The next lemma and its consequence (Proposition 9.10) are quite interesting.

Lemma 9.9. *Suppose $x \cup y \cong z \times z$. Then either there is a function h such that $h''(x) = z$, or $z \leq y$.*

Proof Suppose $x \cup y \cong z \times z$. Suppose also that there is no function h such that $h''(x) = z$. We will show that $z \leq y$.

Let f be a 1-1 mapping from $x \cup y$ onto $z \times z$. For any element $a \in x$, $f(a)$ is an ordered pair $\langle z_1, z_2 \rangle$ of elements of z; let $h(a)$ be its first member z_1. Then $h''(x) \subseteq z$. But $h''(x) \neq z$ (by our supposition) and so $h''(x)$ is a *proper* subset of z, hence z contains an element b not in $h''(x)$, so $b \neq h(a)$ for every $a \in x$. Therefore for any element z_1 of z, there is no a in x such that $f(a) = \langle b, z_1 \rangle$ (for if $f(a) = \langle b, z_1 \rangle$, then $h(a) = b$). Yet there is one (and only one) element a in $x \cup y$ such that $f(a) = \langle b, z_1 \rangle$, so such an a must be in y. Let g be f^{-1} (the inverse of f). Then for every z_1 in z, the element $g(\langle b, z_1 \rangle)$ is in y, and so $g''(\{b\} \times z)$ is a subset y_1 of y. Then y_1 is the same size as $\{b\} \times z$, which in turn is the same size as z, and so z is the same size as the subset y_1 of y, and so $z \leq y$. ∎

The following is the main fact behind the proof of Sierpiński's theorem.

Proposition 9.10. *If $x + y \cong \mathcal{P}(x + x)$ then $\mathcal{P}(x) \leq y$.*

Proof It follows from Lemma 9.9 that if $x + y \cong z \times z$, then either there is a function h such that $h''(x) = z$, or $z \leq y$. We take $\mathcal{P}(x)$ for z, and since there is no function h such that $h''(x) = \mathcal{P}(x)$ (see note below) then if $x + y \cong \mathcal{P}(x) \times \mathcal{P}(x)$ then $\mathcal{P}(x) \leq y$. But $\mathcal{P}(x) \times \mathcal{P}(x) \cong \mathcal{P}(x + x)$ (by Corollary 9.3) and so if $x + y \cong \mathcal{P}(x + x)$ then $\mathcal{P}(x) \leq y$. ∎

Note By Cantor's theorem, if h is 1-1, then $h''(x) \neq \mathcal{P}(x)$, but Cantor's argument shows this just as well if h is not 1-1, for again if h is any function from x *into* $\mathcal{P}(x)$, then the set of all y in x such that $y \notin h(y)$ cannot be in the range of h, so $h''(x)$ is a *proper* subset of $\mathcal{P}(x)$.

Corollary 9.11. *If $x + x \cong x$ and $x + y \cong \mathcal{P}(x)$ then $\mathcal{P}(x) \leq y$.*

Proof Assume the hypothesis. Since $x + x \cong x$ then $\mathcal{P}(x + x) \cong \mathcal{P}(x)$. Then $x + y \cong \mathcal{P}(x + x)$ (since $x + y \cong \mathcal{P}(x)$). Then $\mathcal{P}(x) \leq y$ by Proposition 9.10. ∎

Proposition 9.12. *Suppose G.C.H. holds and that x, y are sets such that $x + x \cong x$ and $y \leq \mathcal{P}(x)$. Then x is comparable with y ($x \leq y$ or $y \leq x$).*

Proof Assume the hypothesis. Since $x + x \cong x$ and $y \leq \mathcal{P}(x)$ then by Proposition 9.8, $x + y \leq \mathcal{P}(x)$. Also, of course $x \leq x + y$. Thus $x \leq x + y \leq \mathcal{P}(x)$. Then by G.C.H. either $x \cong x + y$ or $x + y \cong \mathcal{P}(x)$. If the former, then $y \leq x$ (since $y \leq x + y$). If the latter, then $\mathcal{P}(x) \leq y$ (by Corollary 9.11, since $x + x \cong x$), hence $x \leq y$. ∎

Lemma 9.13 (Main lemma). *Suppose that G.C.H. holds and $z = \mathcal{P}(y \cup \omega)$, for some set y, and b is a set such that $z \leq b$ doesn't hold. Then for every natural number n, $b \leq \mathcal{P}^n(z)$ implies $b \leq z$.*

Proof Assume the hypothesis. We show by induction on n that $b \leq \mathcal{P}^n(z)$ implies $b \leq z$. For $n = 0$ the case is trivial (since $\mathcal{P}^0(z) = z$). Now suppose that n is such that $b \leq \mathcal{P}^n(z)$ implies $b \leq z$. Suppose also that $b \leq \mathcal{P}^{n+1}(z)$. We are to show that $b \leq z$.

Since $b \leq \mathcal{P}^{n+1}(z)$ then $b \leq \mathcal{P}(\mathcal{P}^n(z))$. Since $z = \mathcal{P}(y \cup \omega)$, then $\mathcal{P}^n(z) + \mathcal{P}^n(z) \cong \mathcal{P}^n(z)$ (by Proposition 9.7). Then by Proposition 9.12 (taking $\mathcal{P}^n(z)$ for x and b for y) b is comparable with $\mathcal{P}^n(z)$. But $\mathcal{P}^n(z) \leq b$ doesn't hold, since $z \leq b$ doesn't hold (by hypothesis), and so $b \leq \mathcal{P}^n(z)$. Then by the induction hypothesis, $b \leq z$. This completes the induction. ∎

Theorem 9.14 (Sierpiński). *G.C.H. implies every set can be well ordered (and hence that A.C. holds).*

Proof Assume G.C.H. Now suppose that some set y cannot be well ordered. We then get a contradiction as follows.

Let $z = \mathcal{P}(y \cup \omega)$. Since y cannot be well ordered then there is no ordinal β such that $y \leq \beta$. Therefore there is no ordinal β such that $z \leq \beta$ (since $y \leq z$). Now, by Hartog's theorem (Theorem 5.1) and Theorem 5.4 there is an ordinal β that cannot be 1-1 mapped into z but $\beta \leq \mathcal{P}^4(z)$. And so $z = (y \cup \omega)$, $z \not\leq \beta$, and $\beta \leq \mathcal{P}^4(z)$. Then by Lemma 9.13, taking β for b, it follows that $\beta \leq z$, contrary to the fact that β cannot be 1-1 mapped into z. ∎

PART II

CONSISTENCY OF THE CONTINUUM HYPOTHESIS

CHAPTER 10

MOSTOWSKI-SHEPHERDSON MAPPINGS

We now embark on a journey over the next several chapters leading to Gödel's proof of the relative consistency of the continuum hypothesis. This chapter contains some key material that will be needed for this. Some of what we now do generalizes results of Chapters 6, 7 and 8.

§1 Relational systems

By a *relational system* Γ we shall mean a class A together with a relation R on A. Par abuse of notation, we write $\Gamma = (A, R)$.

Remarks If A or R is a proper class (i.e., not a set) then we cannot *formally* have the ordered pair $\langle A, R \rangle$—that is, such an entity is not an element, nor a subclass of V, but we can *informally* speak of A together with R, which we denote by (A, R). (Note the round parentheses, in contrast to the angular ones we use for ordered pair, $\langle x, y \rangle$ of sets.)

Given a relational system (A, R), for any element $a \in A$, by a^* we shall mean the class of all x in A such that xRa holds. We refer to the elements of a^* as the *components* of a.

We note that if A is the class V and R is the $\in$-relation of membership, then in the relational system $(V, \in)$, for any $a \in V$, a^* is simply a itself (since the components of a are then the members of a). Indeed, if A is any *transitive* class, then in the relational system $(A, \in)$, for any $a \in A$, $a^* = a$ — more generally, whether A is transitive or not, $a^* = a \cap A$ (and hence $a^* \subseteq a$).

We shall call (A, R) a *proper* relational system if, for every $a \in A$, the class a^* is a *set*. This generalizes the notion of a proper well ordering. Indeed, a well ordering $\leq$ of a class A is a proper well ordering if and only if the relational system $(A, \leq)$ is a proper relational system, because a^* is the class $L_\leq(a)$ (the class of all $x \in A$ such that $x \leq a$). Given a well ordering $\leq$ of A, we shall usually be more interested in the relational system $(A, <)$ than the system $(A, \leq)$ (for reasons that will appear later), but we note now that a well ordering $\leq$ of A is proper if and only if the relational system $(A, <)$ is proper (because for any $x \in A$, x^* is now $L_<(x)$). Let us record this obvious fact.

Proposition 1.1. *A well ordering $\leq$ of A is proper if and only if the relational system $(A, <)$ is proper.*

Of course, for any *set* A, the system (A, R) is automatically proper. Also, we have the following.

Proposition 1.2. *For any class A, whether a set or not, the relational system $(A, \in)$ is proper.*

Proof For $x^* = x \cap A$, which is a subclass of x, and hence a set. ∎

Extensionality $\Gamma = (A, R)$ is called *extensional* iff the following condition holds: for any elements x, y in A, if $x \neq y$, then there is at least one element z in A such that z is a component of one of x or y, but not of both. Stated otherwise, extensionality means that for all x, y in A, if x, y have the same components, then x, y are identical—thus $x^* = y^*$ implies $x = y$.

One sometimes says that A is extensional with respect to R, meaning that (A, R) is extensional. Also, a class A is said to be extensional (without reference to a relation R) if it is extensional with respect to the relation $\in$.

Examples

(1) The set $\{2, 5\}$ is extensional (with respect to $\in$), since the only two distinct members of the set are 2 and 5, and one of them contains an element of the set which the other does not contain—namely 5 contains 2, but 2 does not contain 2.

(2) However, if A is the set $\{\{2\}, \{5\}\}$, then A is *not* extensional, because the elements $\{2\}, \{5\}$ contain exactly the same elements of A (namely, no elements at all!), yet they are distinct.

(3) Let A be the set $\{7, \{2, 7\}, \{3, 7\}\}$. This set is not extensional, because $\{2, 7\}$ and $\{3, 7\}$ contain exactly the same elements of A—namely 7—but they are different sets.

(4) Any transitive class is automatically extensional (why?).

(5) Any class of ordinals is extensional (why?).

Proposition 1.3. *If $\leq$ is a well ordering (or even a linear ordering) of A, then $(A, <)$ is extensional.*

Proof Suppose x and y are *distinct* elements of A; then one of them—say x—is less than the other. Since $x < y$, then $x \in y^*$, but $x \notin x^*$ (since $x < x$ doesn't hold), so $x^* \neq y^*$ (since y^* contains x and x^* doesn't). ∎

Corollary 1.4. *Any class A of ordinals is extensional.*

Proof For $(A, \in)$ is then $(A, <)$. ∎

Well foundedness Returning to the more general case, let Γ be a relational system (A, R) where R is an arbitrary relation. For any subclass B of A, an element x of B is called an *initial* element of B if x has no components in B (thus all components of x lie in $A - B$), or equivalently, $x^* \cap B = \emptyset$. We now call Γ *well founded* iff every non-empty subclass of A contains an initial element. This obviously generalizes the notion of well foundedness of Chapter 8, since a class A is well founded as defined in Chapter 8 iff the relational system $(A, \in)$ is well founded in the sense above. (As we will see, several of the notions and results of Chapter 8 that were stated for the particular relation $\in$ generalize to an arbitrary relation R.)

Let us note that if $\leq$ is a well ordering of A, then the relational system $(A, \leq)$ is definitely *not* well founded, for A itself has no initial element x (since $x \leq x$, hence $x \in x^*$, and thus x^* has a component in A—namely x). However, if $\leq$ is a well ordering of A, then the relational system $(A, <)$ is well founded, because for any non-empty subclass B of A, the least element of A (under the well ordering $\leq$) is an initial element of B (in fact the only initial element of B). And so we now know (by Propositions 1.1 and 1.3) the following.

Theorem 1.5. *If A is properly well ordered under $\leq$ then the relational system $(A, <)$ is proper, extensional, and well founded.*

Isomorphisms By an *isomorphism* from a relational system (A_1, R_1) to a relational system (A_2, R_2) is meant a 1-1 function F from A_1 onto A_2 such that for any elements x and y of A_1,
$$xR_1y \text{ iff } F(x)R_2F(y).$$
Two relational systems are called *isomorphic* if there is an isomorphism from one to the other. The main theorem of this chapter — the Mostowski-Shepherdson mapping theorem—is that if (A, R) is extensional, well founded, and proper, then there is a *transitive* class M such that (A, R) is isomorphic to $(M, \in)$ (and thus there is a 1-1 function F mapping A onto M such that for any x and y in A, xRy iff $F(x) \in F(y)$). This result is one of the basic tools we will need for the proof of the consistency of the continuum hypothesis. We shall lead up to the proof gradually and establish some other interesting things along the way.

EXERCISES

Exercise 1.1. Assume that (A, R) is isomorphic to (A', R'). Show the following.

(a) (A, R) is proper if and only if (A', R') is proper.

(b) (A, R) is extensional if and only if (A', R') is extensional.

(c) (A, R) is well founded if and only if (A', R') is well founded.

§2 Generalized induction and Γ-rank

The following theorem generalizes Theorem 3.2, Chapter 8.

Theorem 2.1 (Generalized induction). *Suppose (A, R) is well founded. Then a sufficient condition for a property P to hold for all elements of A is that for every $x \in A$, if P holds for all components of x, then P holds for x.*

Proof Suppose (A, R) is well founded and that for every $x \in A$, if P holds for all elements of x^*, then P holds for x. Suppose that P failed to hold for some element. Let B be the class of all elements of A for which P fails. Then B is non-empty, hence by well foundedness, B contains an initial element x. Then no component of x is in B, hence all components of x have property P, and so by hypothesis x has property P, which is a contradiction. ∎

The function $p(x)$ Given a relational system (A, R), for any subset x of A, by $p(x)$ we shall mean the class of all elements a of A such that $a^* \subseteq x$. Thus $a \in p(x)$ iff $a^* \subseteq x$.

Let us note that for a relational system $(A, \in)$ where A is transitive, $p(x)$ is $\mathcal{P}(x) \cap A$ ($\mathcal{P}(x)$ is the power set of x). Of course for the relational system $(V, \in)$, $p(x) = \mathcal{P}(x)$.

Lemma 2.2. *If (A, R) is extensional and proper, then for every subset x of A, $p(x)$ is a set.*

Proof Suppose that (A, R) is extensional and proper. For any element a of A, let $h(a) = a^*$. Since (A, R) is proper, then $h(a)$ is a set. Since (A, R) is extensional, then the function h is 1-1. For any subset x of A, $p(x)$ is the class of all $a \in A$ such that $a^* \subseteq x$, and so for any $a \in A$, $a \in p(x)$ iff $h(a) \in \mathcal{P}(x)$. Thus h maps $p(x)$ 1-1 onto some subset of $\mathcal{P}(x)$, and hence $p(x)$ is a set, by the axiom of substitution. ∎

Γ-rank We now let Γ be a relational system (A, R) such that for every subset x of A, $p(x)$ is a set. By transfinite induction we define the following On-sequence, Γ_0, Γ_1, ..., Γ_α, ... of subsets of A:

$$\begin{aligned}
\Gamma_0 &= \emptyset \\
\Gamma_{\alpha+1} &= p(\Gamma_\alpha) \\
\Gamma_\lambda &= \cup_{\alpha < \lambda} \Gamma_\alpha \ (\lambda \text{ a limit ordinal}).
\end{aligned}$$

(Since for every subset x of A, $p(x)$ is a set, then if Γ_α is a set, so is $\Gamma_{\alpha+1}$, and hence by transfinite induction, for every α, Γ_α is a set.)

An element x of A is said to have Γ-*rank* if it is in some Γ_α, in which case we define its Γ-rank to be the first ordinal α such that $x \in \Gamma_{\alpha+1}$.

We note that if Γ is the relational system $(V, \in)$, then for each α, Γ_α is R_α, as defined in Chapter 7 (since $p(x)$ is then $\mathcal{P}(x)$).

Proposition 2.3. *For each ordinal α, $\Gamma_\alpha \subseteq \Gamma_{\alpha+1}$.*

Proof Call α *good* if $\Gamma_\alpha \subseteq \Gamma_{\alpha+1}$. We show by transfinite induction that every α is good.

(1) Since Γ_0 is empty, $\Gamma_0 \subseteq \Gamma_1$, so 0 is good.

(2) Suppose α is good; we wish to show that $\alpha + 1$ is good. Well, suppose x is any element of $\Gamma_{\alpha+1}$. Then $x^* \subseteq \Gamma_\alpha$. Also $\Gamma_\alpha \subseteq \Gamma_{\alpha+1}$ (by the induction hypothesis, α is good). Hence $x^* \subseteq \Gamma_{\alpha+1}$, and so $x \in \Gamma_{\alpha+2}$. Thus every x in $\Gamma_{\alpha+1}$ is also in $\Gamma_{\alpha+2}$, which means that $\alpha + 1$ is good.

(3) Suppose λ is a limit ordinal such that every ordinal less than λ is good; we are to show that λ must be good. Suppose x is any element of Γ_λ. Then for some $\alpha < \lambda$,

§2. GENERALIZED INDUCTION AND Γ-RANK

$x \in \Gamma_\alpha$. But $\Gamma_\alpha \subseteq \Gamma_{\alpha+1}$ ($\alpha < \lambda$, so by the induction hypothesis α is good). Hence $x \in \Gamma_{\alpha+1}$. Hence $x^* \subseteq \Gamma_\alpha$, and since $\Gamma_\alpha \subseteq \Gamma_\lambda$, then $x^* \subseteq \Gamma_\lambda$, and so $x \in \Gamma_{\lambda+1}$. Thus $\Gamma_\lambda \subseteq \Gamma_{\lambda+1}$, which means that λ is good.

This completes the induction. ∎

By virtue of Proposition 2.3, the *On* sequence $\Gamma_0, \Gamma_1, \ldots, \Gamma_\alpha, \ldots$ is an *ordinal hierarchy*, as defined in §2 of Chapter 7, and so by $\mathbf{O}_1, \mathbf{O}_3, \mathbf{O}_4$, and $\mathbf{O}_5$ of Theorem 2.1 of Chapter 7 we see that the following four properties hold for the Γ_α sequence.

P_1 $\alpha \leq \beta$ implies $\Gamma_\alpha \subseteq \Gamma_\beta$.

P_2 $x \in \Gamma_\alpha$ iff $x \in \Gamma_{\beta+1}$, for some $\beta < \alpha$.

P_3 $x \in \Gamma_\alpha$ iff x has Γ-rank $< \alpha$.

P_4 x has Γ-rank α iff $x \in \Gamma_{\alpha+1} - \Gamma_\alpha$.

As obvious corollaries of P_3 we have the following.

P_5 For any subclass B of A and any ordinal α, $B \subseteq \Gamma_\alpha$ iff all elements of B have Γ-rank $< \alpha$.

P_6 For any $x \in A$, $x^* \subseteq \Gamma_\alpha$ iff x has Γ-rank $\leq \alpha$.

Proof (Of P_6) $x^* \subseteq \Gamma_\alpha$ iff $x \in \Gamma_{\alpha+1}$ iff x has Γ-rank $< \alpha + 1$ (by P_3) iff x has Γ-rank $\leq \alpha$. ∎

From P_5 we get the following.

P_7 If x has Γ-rank, then every element of x^* has a lower Γ-rank.

Proof Suppose x has Γ-rank α. Then $x^* \subseteq \Gamma_\alpha$ (since $x \in \Gamma_{\alpha+1}$), hence by P_5 every element of x^* has Γ-rank $< \alpha$. ∎

P_8 If x has Γ-rank, then its Γ-rank is the least ordinal greater than the Γ-ranks of all the elements of x^*.

Proof Suppose x has Γ-rank α. Then by P_7, α is greater than the Γ-ranks of all the elements of x^*.

Now suppose that β is greater than the Γ-ranks of all the elements of x^*. Since x has Γ-rank, so does every element of x^* (by P_7) and so every element of x^* has Γ-rank $< \beta$, hence by P_5, $x^* \subseteq \Gamma_\beta$, and so by P_6, x has Γ-rank $\leq \beta$, and thus $\alpha \leq \beta$. Therefore α is the least ordinal that is greater than the Γ-ranks of all the elements of x^*. ∎

As a special case of P_8, consider the relational system $(V, \in)$. Then the $(V, \in)$-rank of x is simply the rank of x, as defined in Chapter 7, and for this relational system, x^* is simply x, and so by P_8 we have the following.

P_9 If x has rank, then its rank is the least ordinal greater than the ranks of all the elements of x.

Now let us return to the general case of a relational system (A, R), but one in which for every subset x of A, $p(x)$ is a set (and so (A, R)-rank is then well defined).

P_{10} For any element x of A, if x^* is a set and if every element of x^* has Γ-rank, then x has Γ-rank.

Proof Suppose that every element of x^* has Γ-rank and that x^* is a set. Then by the axiom of substitution, the class of Γ-ranks of the elements of x^* is a set, hence there is an ordinal α greater than the Γ-ranks of all the elements of x^*, and so every element of x^* has a Γ-rank $< \alpha$. Then by P_5, $x^* \subseteq \Gamma_\alpha$, hence by P_6, x has Γ-rank $\leq \alpha$. ∎

Now we have the following.

Proposition 2.4. *Suppose that Γ is a proper well founded relational system (A, R) such that for every subset x of A, $p(x)$ is a set. Then every element of A has Γ-rank.*

Proof Assume the hypothesis. Since for every subset x of A, $p(x)$ is a set, the notion of Γ-rank is well defined. Since Γ is well founded, then by the generalized induction principle (Theorem 2.1), to prove that every element of A has Γ-rank, it suffices to show that for every $x \in A$, if all elements of x^* have Γ-rank, then x has Γ-rank. So suppose that all elements of x^* have Γ-rank. Now, x^* is a set (since Γ is proper) and so x then has Γ-rank by P_{10}. ∎

From Proposition 2.4 and Lemma 2.2 we have the following key result.

Theorem 2.5. *If Γ is a proper, extensional, and well founded system (A, R), then every element of A has Γ-rank.*

Proposition 2.4 also yields the following.

Proposition 2.6. *If Γ is a well founded system (A, R) and A is a set, then every element of A has Γ-rank.*

Proof If A is a set, then it is automatic that for every subset x of A, $p(x)$ is a set, and so the result follows from Proposition 2.4. ∎

A-ranks Suppose Γ is a relational system $(A, \in)$ and A is transitive and well founded (i.e., $(A, \in)$ is well founded). It is automatic that for every subset x of A, $p(x)$ is a set, since $p(x)$ is then $p(x) \cap A$, and so by Proposition 2.4, every element x of A has Γ-rank, which we more simply call the *A-rank* of x. Thus the A-rank of x is the $(A, \in)$-rank of x. In particular, the V-rank of a set x is simply the rank of x, as defined in Chapter 7 (the least ordinal α such that $x \in R_{\alpha+1}$).

Theorem 2.7. *Suppose every element of A has V-rank. Then every element of A also has A-rank and for each $x \in A$, the A-rank of x is $\leq$ the V-rank of x.*

Proof Since every element of A has V-rank then A is well founded (Theorem 3.4 of Chapter 8), hence every element of A has A-rank, as we have seen above. Since A is well founded, we can use generalized induction on A. So suppose that x is an element

§3. GENERALIZED TRANSFINITE RECURSION

of A such that for every y in $x \cap A$, the A-rank of y is $\leq$ the V-rank of y. We are to show that the A-rank of x is $\leq$ the V-rank of x.

Let α be the A-rank of x and β be the V-rank of x. We are to show that $\alpha \leq \beta$. By P_8, α is the least ordinal greater than the A-ranks of all the elements of $x \cap A$ (since in the relational system $(A, \in)$, x^* is $x \cap A$). Also, since every element of $x \cap A$ is an element of x, then by P_6 of Chapter 7, β is greater than the V-ranks of all the elements of $x \cap A$, and hence by the induction hypothesis, β is greater than the A-ranks of all the elements of $x \cap A$, and since α is the least such ordinal, then $\alpha \leq \beta$. This completes the induction. ∎

As a corollary we have the following.

Theorem 2.8. *For any class A of elements with V-rank, each ordinal α in A has A-rank, and its A-rank is $\leq \alpha$.*

Proof By P_7 of Chapter 7, each ordinal α has V-rank α, and by Theorem 2.7 each ordinal α in A has A-rank, and the A-rank of α is $\leq$ the V-rank of α, which is α. ∎

§3 Generalized transfinite recursion

In Chapter 6, Theorem 5.8, we proved that for any function h (defined on all sets) there is a function F on On such that for every ordinal α, $F(\alpha) = h(F''(\alpha))$. We now turn to a generalization of this.

Theorem 3.1 (Generalized transfinite recursion theorem). *Let Γ be a relational system (A, R) such that every element of A has Γ-rank. Then for every function g defined on V there is a unique function F defined on A such that for every x in A, $F(x) = g(F''(x^*))$.*

Proof By transfinite recursion (Theorem 5.8, Chapter 6) we can define the following On sequence, $f_0, f_1, \ldots, f_\alpha, \ldots$, where for each α, f_α is a function on Γ_α:

(1) $f_0 = \emptyset$

(2) $f_{\alpha+1} = f_\alpha$ (the set of all ordered pairs $\langle x, g(f_\alpha''(x^*))\rangle$ such that $x \in \Gamma_{\alpha+1} - \Gamma_\alpha$)

(3) $f_\lambda = \cup_{\alpha < \lambda} f_\alpha$ (for λ a limit ordinal).

Concerning (2), we note that for $x \in \Gamma_{\alpha+1} - \Gamma_\alpha$, since $x \in \Gamma_{\alpha+1}$, then $x^* \subseteq \Gamma_\alpha$, hence f_α is defined on all elements of x^*. Thus $f_{\alpha+1}$ is defined on $\Gamma_{\alpha+1}$, and for any $x \in \Gamma_\alpha$, $f_{\alpha+1}(x) = f_\alpha(x)$, but for $x \in \Gamma_{\alpha+1} - \Gamma_\alpha$, $f_{\alpha+1}(x) = g(f_\alpha''(x^*))$.

Since $f_\alpha \subseteq f_{\alpha+1}$, then also (by (1) and (3)), the On-sequence $f_0, f_1, \ldots, f_\alpha, \ldots$ is an ordinal hierarchy, and so for $\alpha < \beta$, $f_\alpha \subseteq f_\beta$ (which means that for every $x \in \Gamma_\alpha$, $f_\alpha(x) = f_\beta(x)$).

We now take $F(x) = f_\beta(x)$, where β is the first ordinal such that $x \in \Gamma_\beta$, and so for any ordinal α, $x \in \Gamma_\alpha$ implies $F(x) = f_\alpha(x)$ (because then $\beta \leq \alpha$, hence $f_\alpha(x) = f_\beta(x)$).

Now we show that for every x in A, $F(x) = g(F''(x^*))$. Well, let α be the Γ-rank of x. Then $F''(x^*) = f_\alpha''(x^*)$ (because $x \in \Gamma_{\alpha+1}$, hence $x^* \subseteq \Gamma_\alpha$, hence for each y in x^*, $y \in \Gamma_\alpha$, hence $F(y) = f_\alpha(y)$, and so $F''(x^*) = f_\alpha''(x^*)$). Since $F''(x^*) = f_\alpha''(x^*)$, then $g(F''(x^*)) = g(f''_\alpha(x^*))$, but $g(f''_\alpha(x^*)) = f_{\alpha+1}(x)$ (since $x \in \Gamma_{\alpha+1} - \Gamma_\alpha$), hence $g(F''(x^*)) = f_{\alpha+1}(x)$, but also $f_{\alpha+1}(x) = F(x)$, and so $F(x) = g(F''(x^*))$.

As for uniqueness, suppose F_1 is a function on A such that for every $x \in A$, $F_1(x) = g(F''_1(x^*))$. We are to show that for all x in A, $F_1(x) = F(x)$. We do this by generalized induction on x (Theorem 2.1). And so suppose that for each $y \in x^*$, $F_1(y) = F(y)$. We are to show that $F_1(x)$ then equals $F(x)$. Well, since for each y in x^*, $F_1(y) = F(y)$, then $F''_1(x^*) = F''(x^*)$, hence $g(F''_1(x^*)) = g(F''(x^*))$, and hence $F_1(x) = F(x)$. ∎

Suppose now that Γ is an extensional, well founded, proper system (A, R). Then by Theorem 2.5, every element of A has Γ-rank, hence by Theorem 3.1 we have the following.

Theorem 3.2. *If (A, R) is proper, extensional, and well founded, then for any function g defined on all sets there is a unique function F on A such that for all $x \in A$, $F(x) = g(F''(x^*))$.*

As a special case, taking the identity function for g, we have this.

Theorem 3.3. *If (A, R) is proper, extensional, and well founded, then there is a unique function F on A such that for all x in A, $F(x) = F''(x^*)$.*

Definition 3.4. A function F that satisfies the conclusion of Theorem 3.3, $F(x) = F''(x^*)$ for all $x \in A$, is called a *Mostowski-Shepherdson map* for (A, R).

The significance of these maps will be seen in the next section. Theorem 3.3 thus says that a proper, extensional, well founded system (A, R) has one and only one Mostowski-Shepherdson map.

By Theorem 1.5, if A is properly well ordered under $\leq$ then the relational system $(A, <)$ is proper, extensional, and well founded. Hence Theorem 3.3 immediately yields the following.

Corollary 3.5. *If A is properly well ordered under $\leq$ then the relational system $(A, <)$ has a Mostowski-Shepherdson map.*

EXERCISES

Exercise 3.1. Why is Theorem 5.8 of Chapter 6 a special case of Theorem 3.1 of this chapter?

§4 Mostowski-Shepherdson maps

Theorem 4.1 (Mostowski-Shepherdson mapping theorem). *Suppose (A, R) is a relational system that is proper, extensional, and well founded. Then:*

(1) there is a Mostowski-Shepherdson map F for (A, R) (a function on A such that $F(x) = F''(x^*)$ for every x in A);

(2) F maps A onto some transitive class M;

(3) (A, R) is isomorphic to $(M, \in)$ under F (for any x and y in A, xRy iff $F(x) \in F(y)$, and F is 1-1).

Proof Assume the hypothesis. By Theorem 3.3 there is a Mostowski-Shepherdson map F for (A, R).

(1) We show that F is 1-1. Call an element x of A *good* if for every $y \in A$, $F(x) = F(y)$ implies $x = y$. We will show by generalized induction (Theorem 2.1) that every x in A is good. So suppose x is such that all elements of x^* are good; we are to show that x is good. Assume $F(x) = F(y)$, where y is an element of A; we are to show that $x = y$. To do this it suffices, by extensionality, to show that $x^* = y^*$, which we do by showing that $x^* \subseteq y^*$ and $y^* \subseteq x^*$.

To prove that $x^* \subseteq y^*$, let z be any element of x^*. Then $F(z) \in F''(x^*)$, but $F''(x^*) = F''(y^*)$ (since $F''(x^*) = F(x) = F(y) = F''(y^*)$), hence $F(z) \in F''(y^*)$, so $F(z) = F(w)$ for some $w \in y^*$. But z, being an element of x^*, is good by the inductive hypothesis, hence $z = w$. Therefore $z \in y^*$ (since $z = w$ and $w \in y^*$). This proves that $z \in x^*$ implies $z \in y^*$, and so $x^* \subseteq y^*$.

To prove that $y^* \subseteq x^*$, let z be any element of y^*. Then $F(z) \in F''(y^*)$, hence $F(z) \in F''(x^*)$ (since $F''(x^*) = F''(y^*)$, as we have already seen), and so $F(z) = F(w)$ for some $w \in x^*$. Then w is good, hence $z = w$, so $z \in x^*$. This proves that $y^* \subseteq x^*$, and so $x^* = y^*$. This completes the proof that F is a 1-1 map.

(2) We now show that F is an isomorphism from (A, R) to $(F''(A), \in)$. We already know that F is 1-1. Let x and y be any elements of A. We are to show that xRy is equivalent to $F(x) \in F(y)$.

Suppose xRy. Then $x \in y^*$. Hence $F(x) \in F''(y^*)$, so $F(x) \in F(y)$ (since $F''(y^*) = F(y)$). Thus xRy implies $F(x) \in F(y)$.

Conversely, suppose $F(x) \in F(y)$. Then $F(x) \in F''(y^*)$, so $F(x) = F(z)$ for some $z \in y^*$. But F is 1-1, hence $x = z$, hence $x \in y^*$, so xRy. Thus $F(x) \in F(y)$ implies xRy, which completes the proof that F is an isomorphism.

(3) Lastly we show that $F''(A)$ is transitive. Well, suppose $x \in F''(A)$; we must show that $x \subseteq F''(A)$. Since $x \in F''(A)$ then $x = F(z)$ for some $z \in A$. Hence $x = F''(z^*)$. Since $z^* \subseteq A$ then $F''(z^*) \subseteq F''(A)$, and so $x \subseteq F''(A)$. ∎

Our main application of Theorem 4.1 is the case that R is the $\in$-relation. Two classes A and B are called $\in$-*isomorphic* if the two corresponding relational systems $(A, \in)$ and $(B, \in)$ are isomorphic—which means that there is a 1-1 function F from A onto B such that for all x and y in A, $x \in y$ iff $F(x) \in F(y)$. And such a function F is called an $\in$-*isomorphism* from A to B. We also call F a Mostowski-Shepherdson map for A if it is a Mostowski-Shepherdson map for the relational system $(A, \in)$. If A is extensional and well founded, then there is a Mostowski-Shepherdson map for A by Theorem 3.3 (since $(A, \in)$ is automatically proper), and by Theorem 4.1, F is an $\in$-isomorphism from A onto a *transitive* class $F''(A)$. Also, a Mostowski-Shepherdson map F on A is a map such that for every $x \in A$, $F(x) = F''(x \cap A)$ (because in the relational system $(A, \in)$, x^* is the class of all members of A that are in x, and is thus the set $x \cap A$). And so we have the following.

Theorem 4.2. *Every extensional, well founded class A is $\in$-isomorphic to a transitive class. More specifically, for any well founded, extensional class A, there is a unique map F with domain A such that for every $x \in A$, $F(x) = F''(x \cap A)$, and this F is an $\in$-isomorphism from A onto a transitive class.*

It will be helpful to consider a simple Mostowski-Shepherdson mapping example. Let A be the finite set $\{3, 17, 54, \omega \cdot 2\}$ and let F be the Mostowski-Shepherdson map on A. Let us compute $F(3)$, $F(17)$, $F(54)$, and $F(\omega \cdot 2)$. Let us first note that $3 \cap A = \emptyset$; $17 \cap A = \{3\}$; $54 \cap A = \{3, 17\}$; and $\omega \cdot 2 \cap A = \{3, 17, 54\}$, and so $F(3) = F''(3 \cap A) = F''(\emptyset) = 0$. Then $F(17) = F''(17 \cap A) = F''(\{3\}) = \{F(3)\} = \{0\} = 1$. Then also, $F(54) = F''(54 \cap A) = F''(\{3, 17\}) = \{F(3), F(17)\} = \{0, 1\} = 2$. Further, $F(\omega \cdot 2) = F''(\omega \cdot 2 \cap A) = F''(\{3, 17, 54\}) = \{F(3), F(17), F(54)\} = \{0, 1, 2\} = 3$. Then also, $F''(A) = \{0, 1, 2, 3\} = 4$.

EXERCISES

Exercise 4.1. Show that for a finite set A of ordinals, the Mostowski-Shepherdson mapping on A maps A onto the natural number n, where n is the number of elements in A.

Exercise 4.2. Show that if A is an infinite set of natural numbers and F is the Mostowski-Shepherdson mapping on A, then $F''(A) = \omega$.

Exercise 4.3. Show that if A is a class of ordinals and A is not a set, then the Mostowski-Shepherdson mapping on A maps A onto the class On of all ordinals.

§5 More on Mostowski-Shepherdson mappings

Theorem 5.1. *Suppose that the relational system $A = (A, \in)$ is well founded and extensional, T is a transitive subclass of A, and F is the Mostowski-Shepherdson map on A. Then for each element x of T, $F(x) = x$.*

Proof Assume the hypothesis. Let x be any element of T. Since T is transitive then $x \subseteq T$, hence $x \subseteq A$, hence $x \cap A = x$, and so $F''(x) = F''(x \cap A)$, but also $F''(x \cap A) = F(x)$ (since F is a Mostowski-Shepherdson map on A), and so $F''(x) = F(x)$. And so for every x in T, $F(x) = F''(x)$.

Now suppose x is an element of T such that for every y in x, $F(y) = y$. Then $F''(x) = $ (set of all elements $F(y)$ such that $y \in x$) = (set of all elements y such that $y \in x$) = x. Thus $F(x) = F''(x) = x$. This proves that for any $x \in T$, if every element y of x is such that $F(y) = y$, then $F(x) = x$. Then by generalized induction for T (which is well founded, since it is a subclass of the well founded class A) it follows that $F(x) = x$ for every x in T. ∎

Corollary 5.2. *If A is extensional, well founded, and contains all the natural numbers, then the Mostowski-Shepherdson map F on A maps each natural number n to itself, $F(n) = n$.*

Corollary 5.3. *If A is extensional and well founded and A itself is transitive, then the Mostowski-Shepherdson map on A is the identity map.*

We also have the following.

Proposition 5.4. *Suppose A is extensional and well founded. If the Mostowski Shepherdson map F on A is the identity map, then A must be transitive.*

Proof The hypothesis implies that $F''(A) = A$, and since $F''(A)$ is transitive, then A is transitive. ∎

§6 Isomorphisms, Mostowski-Shepherdson, well orderings

Suppose now that an extensional, well founded class A is $\in$-isomorphic to a class B under a mapping F. Is F necessarily a Mostowski-Shepherdson mapping?—that is, is it necessarily the case that for all x in A, $F(x) = F''(x \cap A)$? No, it is not—an obvious counterexample being the identity map on a class that is not transitive. However, if B is transitive, then F must be a Mostowski-Shepherdson map. More generally we have the following.

Theorem 6.1. *Suppose Γ is a proper, extensional, well founded relational system (A, R), and F is an isomorphism from Γ to (B, G), where B is transitive. F is the Mostowski-Shepherdson map for Γ.*

Proof Assume the hypothesis. We will show that for any $x \in A$, $F(x) = F''(x^*)$ by showing that $F''(x^*) \subseteq F(x)$ and $F(x) \subseteq F''(x^*)$.

To show that $F''(x^*) \subseteq F(x)$ we do not even need the hypothesis that B is transitive, for suppose $y \in F''(x^*)$. Then $y = F(z)$ for some $z \in x^*$. Since $z \in x^*$ then zRx, hence $F(z) \in F(x)$ (since F is an isomorphism) and thus $y \in F(x)$. This proves that $F''(x^*) \subseteq F(x)$.

To show that $F(x) \subseteq F''(x^*)$, suppose $y \in F(x)$. Since $F(x) \in B$ and B is transitive then $y \in B$, and so $y = F(z)$ for some $z \in A$. Then $F(z) \in F(x)$, hence zRx (since F is an isomorphism), hence $z \in x^*$, and so $F(z) \in F''(x^*)$, and thus $y \in F''(x^*)$. This proves that $F(x) \subseteq F''(x^*)$ and so $F(x) = F''(x^*)$. ∎

Let us note that if A is properly well ordered under $\leq$ then there is a Mostowski-Shepherdson map F for $(A, <)$, by Theorems 1.5 and 4.1. Now we have the following.

Theorem 6.2. *Suppose F is the Mostowski-Shepherdson map for the relational system $(A, <)$, where $\leq$ is a proper well ordering of A.*

(1) If A is a set, then $F''(A)$ is an ordinal number.

(2) If A is not a set, then $F''(A) = On$.

Proof 1 of Theorem 6.2 The class $F''(A)$ is transitive. In addition, $(A, <)$ is isomorphic to $(F''(A), \in)$ under F. But since $(A, <)$ is well founded and is isomorphic to $(F''(A), \in)$, $(F''(A), \in)$ is well founded, and so $F''(A)$ is a well founded class. Now suppose y_1 and y_2 are distinct elements of $F''(A)$. Let x_1 and x_2 be the respective elements of A such that $F(x_1) = y_1$ and $F(x_2) = y_2$. Then $x_1 \neq x_2$, hence either $x_1 < x_2$ or $x_2 < x_1$, and therefore $y_1 \in y_2$ or $y_2 \in y_1$ (since F is an isomorphism). Thus the class $F''(A)$ is $\in$-connected (as defined in Chapter 8, §5). And so $F''(A)$ is well founded, transitive, and $\in$-connected, hence by Theorems 5.3 and 5.4 of Chapter 8, if A is a set, $F''(A)$ is an ordinal (since $F''(A)$ is then also a set), and if not, $F''(A) = On$. ∎

We remark that the theorem above provides an alternative and elegant proof of the counting theorem (Theorem 4.1, Chapter 6), which we recall is that every proper well ordered set is isomorphic to an ordinal and every proper well ordered class is isomorphic to On. Let $x \in' y$ be the relation "$x \in y$ or $x = y$". When we say that a class A that is well ordered under $\leq$ is isomorphic to a class B of ordinals, what is meant is that the relational system $(A, \leq)$ is isomorphic, not to $(B, \in)$, but to $(B, \in')$. But this is obviously equivalent to $(A, <)$ being isomorphic to $(B, \in)$.

Now, suppose that A is well ordered under $\leq$. Using Theorem 6.2, if F is the Mostowski-Shepherdson mapping for $(A, <)$, then $F''(A)$ is an ordinal, if A is a set, and is On, if A is a proper class. Also F is an isomorphism from $(A, <)$ to $(F''(A), \in)$, and therefore $(A, <)$ is isomorphic to $(F''(A), \in)$, hence $(A, \leq)$ is isomorphic to $(F''(A), \in')$, and so we have the counting theorem.

Also, if we use the counting theorem, we have the following alternative proof of Theorem 6.2.

Proof 2 of Theorem 6.2 Suppose A is properly well ordered under $\leq$. By the counting theorem, $(A, \leq)$ is isomorphic to $(B, \in')$ for some B, which is an ordinal if A is a set, or is On otherwise. Then $(A, <)$ is isomorphic to $(B, \in)$ under some mapping F. But B is transitive, hence by Theorem 6.1, F is the Mostowski-Shepherdson map for $(A, <)$. ∎

Theorem 6.3. *Suppose A is properly well ordered under $\leq$, Γ is the relational system $(A, <)$, and F is the Mostowski-Shepherdson map for Γ. Then for each $x \in A$, $F(x)$ is the Γ-rank of x.*

Proof Assume the hypothesis. We show by generalized induction that for each x in A, $F(x)$ is the Γ-rank of x. So suppose that x is such that for every $y \in x^*$, $F(y)$ is the Γ-rank of y. Then $F''(x^*)$ is the set of Γ-ranks of all the elements of x^*. Now, the Γ-rank α of x is the least ordinal that is greater than the Γ-ranks of all the elements of x^* (by P_2 of §2), and so α is the least ordinal greater than all ordinals in $F''(x^*)$, hence α is the least ordinal greater than all the ordinals in $F(x)$, but $F(x)$ is an ordinal (since $F(x) \in F''(A)$ and $F''(A)$ is a class of ordinals, by Theorem 6.2), so the least ordinal greater than all the ordinals in the ordinal $F(x)$ is $F(x)$ itself. Thus $\alpha = F(x)$, and so $F(x)$ is the Γ-rank of x. This completes the induction. ∎

A class A of ordinals is properly well ordered under $\leq$, hence there is a Mostowski-Shepherdson map F for $(A, <)$, but $(A, <)$ is simply $(A, \in)$ (since for ordinals α and β, $\alpha < \beta$ iff $\alpha \in \beta$), and so F is a Mostowski-Shepherdson map on A, which by Theorem 6.3 maps each ordinal α in A to its A-rank—i.e., $F(\alpha) = A$-rank of α. But A-rank of $\alpha \leq \alpha$, by Theorem 2.8, and so we have the following.

Corollary 6.4. *For any class A of ordinals, the Mostowski-Shepherdson map F for A does exist, and for each ordinal $\alpha \in A$, $F(\alpha) \leq \alpha$.*

We might add that if A is a class of ordinals and F is the Mostowski-Shepherdson map on A and if A is not transitive, then there must be at least one ordinal α in A such that $F(\alpha) < \alpha$, because if $F(\alpha) = \alpha$ held for every ordinal α in A, then F would be the identity map, hence A would be transitive by Proposition 5.4.

Corollary 6.5. *If S is a set of ordinals, each member of S is less than the ordinal β, and F is the Mostowski-Shepherdson map on S, then $F''(S)$ is an ordinal $\leq \beta$.*

§6. ISOMORPHISMS, MOSTOWSKI-SHEPHERDSON, WELL ORDERINGS

Remark Since the Mostowski-Shepherdson map on a set S of ordinals is the same thing as the order isomorphism of S onto an ordinal, then Corollary 6.5 is also a consequence of Lemma 7.2 of Chapter 9.

We next prove that for extensional, well founded A, and for F the Mostowski-Shepherdson map for A, for each ordinal α in A, $F(\alpha)$ is an ordinal and $F(\alpha) \leq \alpha$. We first show the following.

Lemma 6.6. *Suppose that A is well founded and extensional, A_0 is the class of ordinals in A, F is the Mostowski-Shepherdson map for A, and F_0 is the Mostowski-Shepherdson map for A_0. Then for any ordinal α in A_0, $F(\alpha) = F_0(\alpha)$.*

Proof For any $\alpha \in A_0$, since only ordinals are elements of α, then $\alpha \cap A = \alpha \cap A_0$, hence $F''(\alpha \cap A) = F''(\alpha \cap A_0)$, and so $F(\alpha) = F''(\alpha \cap A_0)$ (since $F(\alpha) = F''(\alpha \cap A)$). We now show by generalized induction on A_0 that for all $\alpha \in A_0$, $F(\alpha) = F_0(\alpha)$. So suppose that for each β in $\alpha \cap A_0$, $F(\beta) = F_0(\beta)$. Then $F''(\alpha \cap A_0) = F_0''(\alpha \cap A_0)$, but $F''(\alpha \cap A_0) = F(\alpha)$ (as we have shown) and $F_0''(\alpha \cap A_0) = F_0(\alpha)$, and so $F(\alpha) = F_0(\alpha)$. This completes the induction. ∎

Theorem 6.7. *Suppose A is well founded and extensional, and F is the Mostowski-Shepherdson map for A. Then for any ordinal α in A, $F(\alpha)$ is an ordinal $\leq \alpha$.*

Proof Assume the hypothesis. Let A be the class of ordinals in A and let F_0 be the Mostowski-Shepherdson map on A_0. For any ordinal α in A, $\alpha \in A_0$, hence by Corollary 6.4, $F_0(\alpha)$ is an ordinal $\leq \alpha$, but $F_0(\alpha) = F(\alpha)$ by the lemma above, hence $F(\alpha) \leq \alpha$. ∎

We conclude this chapter with the following result.

Theorem 6.8. *Suppose Γ is a proper, extensional, and well founded system (A, R), and F is its Mostowski-Shepherdson map. Then for any $x \in A$, the Γ-rank of x is the V-rank of $F(x)$.*

Proof Assume the hypothesis. To begin with, every element of $F''(A)$ does have V-rank, because $(F''(A), \in)$ is well founded, being isomorphic to (A, R), which is well founded, hence $F''(A)$ is a well founded class and is also transitive, hence every element $F(x)$ of $F''(A)$ has V-rank by Theorem 3.3 of Chapter 8.

We now show by generalized induction that for every x in A, the Γ-rank of x is the V-rank of $F(x)$. So suppose that for every y in x^*, the Γ-rank of y is the V-rank of $F(y)$. Let α be the Γ-rank of x and β be the V-rank of $F(x)$. We are to show that $\alpha = \beta$. Let S be the set of Γ-ranks of all the elements of x^*. By P_8, α is the least ordinal greater than all the ordinals in S. But also, S is the set of V-ranks of all the elements of $F''(x^*)$ (since for each y of x^*, the Γ-rank of y is the V-rank of $F(y)$), and so S is the set of V-ranks of all the elements of $F(x)$ (since $F(x) = F''(x^*)$). Then by P_9, β is the least ordinal greater than all the ordinals in S, hence $\alpha = \beta$. This completes the induction. ∎

CHAPTER 11

REFLECTION PRINCIPLES

We now turn to a beautiful group of theorems called *Reflection principles* which are sophisticated outgrowths of earlier results of Skolem and Löwenheim, which we will explain. One of these results—the *Tarski-Vaught theorem*, combined with the Mostowski-Shepherdson mapping theorem—will play a key role in the proof of the relative consistency of the continuum hypothesis.

Part I—The Tarski-Vaught theorem

§0 Preliminaries

We shall base our formulas of first-order logic on the primitive connectives $\neg$ (not), $\wedge$ (and), and $\exists$ (there exists). We shall consider the set of all formulas of first-order logic involving only one binary predicate P (which in applications to set theory will be interpreted as the $\in$-relation). We also employ an infinite sequence $x_1, x_2, \ldots, x_n, \ldots$ of symbols called *variables* and the parentheses "(" and ")" for punctuation. The notions of *formula, free occurrences, bound occurrences* of variables are defined inductively by the following conditions.

(1) By an *atomic formula* is meant an expression Pxy, where x and y are any individual variables. The occurrences of x and y in Pxy are both free. Every atomic formula is also a formula. (For set theory, the symbol "$\in$" is used instead of "P", and an atomic formula is then usually written $x \in y$ instead of $\in xy$.)

(2) For any formula φ, the expression $\neg\varphi$ is a formula, and the free occurrences of a variable x in $\neg\varphi$ are those of φ.

(3) For any formulas φ and ψ, the expression $(\varphi \wedge \psi)$ is a formula and the free occurrences of a variable x in $(\varphi \wedge \psi)$ are those of φ together with those of ψ.

(4) For any formula φ and any variable x, the expression $(\exists x)\varphi$ is a formula. All occurrences of x in $(\exists x)\varphi$ are *bound* occurrences, and for any variable y other than x, the free occurrences of y in $(\exists x)\varphi$ are those of φ.

Subformulas The *proper* subformulas of a formula are defined by the following inductive scheme.

(1) Atomic formulas have no proper subformulas.

(2) The proper subformulas of $\neg\varphi$ are φ and the proper subformulas of φ.

(3) The proper subformulas of $(\varphi \wedge \psi)$ are φ, ψ, the proper subformulas of φ, and the proper subformulas of ψ.

(4) The proper subformulas of $(\exists x)\varphi$ are φ and the proper subformulas of φ.

By a *subformula* of φ we mean either φ or a proper subformula of φ.

Abbreviations We introduce various other connectives and quantifiers via the usual definitions.

(1) We use $(\varphi \vee \psi)$ as an abbreviation of $\neg(\neg\varphi \wedge \neg\psi)$.

(2) We use $(\varphi \supset \psi)$ as an abbreviation of $\neg(\varphi \wedge \neg\psi)$.

(3) We use $(\varphi \equiv \psi)$ as an abbreviation of $((\varphi \supset \psi) \wedge (\psi \supset \varphi))$.

(4) We use $(\forall x)\varphi$ as an abbreviation of $\neg(\exists x)\neg\varphi$.

We generally delete outer parentheses from an entire formula (and sometimes other parentheses as well) when no ambiguity can result.

Formulas with constants Given a class A, we wish to define the notion of a formula with constants in A—sometimes called an *A-formula*. This is usually done by introducing a new group of symbols to serve as *names* of elements of A. Actually, there is no reason why we cannot use the elements of A as names for themselves. After all, we can look at a formula as simply a finite sequence of things called *symbols*, and there is no reason why we can't have a finite sequence of elements, some of which are symbols, and others of which are elements from a given class A. This is what we shall do, and so by an *atomic A-formula* we shall simply mean a triple (P, α, β) where each of α and β is either a variable or an element of A. We also write (P, α, β) as $P\alpha\beta$, for convenience. Thus an atomic A-formula is one of four types: Pxy, Pxa, Pax, Pab, where x and y are variables and a and b are elements of A. One then defines an A-formula, starting with atomic A-formulas, by the same inductive scheme we used to define "formula," only replacing "formula" by "A-formula." Thus for any A-formulas φ and ψ, the entities $\neg\varphi$, $(\varphi \wedge \psi)$, and $(\exists x)\varphi$ are also A-formulas. A-formulas are also called *formulas with constants in A*. A-formulas involving no constants will be called *pure formulas*. An A-formula with no free occurrences of variables will be called an *A-sentence*, or a *sentence over A*, or a *sentence whose constants are in A*.

Substitution For any variable x, and any element a of a class A, and any α which is either a variable or an element of A, by α_a^x is meant a, if α is the variable x, or α itself, if α is not the variable x. α_a^x is called the result of substituting a for x in α.

For any formula φ with constants in A (and this includes formulas with no constants at all), variable x and element a of A, we define φ_a^x (called *the result of substituting a for all free occurrences of x in φ*) by the following inductive scheme.

(1) For an atomic A-formula $P\alpha\beta$, by $(P\alpha\beta)_a^x$ is meant $P\alpha_a^x\beta_a^x$.

(2) $(\neg\varphi)_a^x = \neg(\varphi_a^x)$.

(3) $(\varphi \wedge \psi)_a^x = (\varphi_a^x \wedge \psi_a^x)$.

(4) If y is a variable distinct from x, then $((\exists y)\varphi)_a^x = (\exists y)(\varphi_a^x)$, but $((\exists x)\varphi)_a^x = (\exists x)\varphi$.

If φ has only one free variable, x, then φ_a^x is also written $\varphi(a)$, and no ambiguity can result. Also, for any formula $\varphi(x_1,\ldots,x_n)$ whose free variables are all included in the set $\{x_1,\ldots,x_n\}$, and for any elements $a_1,\ldots,a_n$ of A, by $\varphi(a_1,\ldots,a_n)$ is meant the result of substituting a_1 for all free occurrences of x_1 in φ, a_2 for all free occurrences of x_2, $\ldots$, a_n for all free occurrences of x_n in φ. $\varphi(a_1,\ldots,a_n)$ can also be written $\varphi_{a_1,\ldots,a_n}^{x_1,\ldots,x_n}$.

Degrees and induction By the *degree* of a formula is meant the number of occurrences of the symbols $\neg$, $\wedge$, and $\exists$. Thus:

(1) Atomic formulas are of degree 0.

(2) If φ is of degree n then $\neg\varphi$ is of degree $n+1$.

(3) If φ is of degree n and ψ is of degree m, then $(\varphi \wedge \psi)$ is of degree $n+m+1$.

(4) If φ is of degree n then $(\exists x)\varphi$ is of degree $n+1$.

We shall use the following induction principle. To show that a given property holds for all pure formulas (A-formulas) it suffices to show the following four things:

(1) The property holds for all *atomic* pure formulas (A-formulas).

(2) If the property holds for φ then it holds for $\neg\varphi$.

(3) If the property holds for both φ and ψ, then it holds for $(\varphi \wedge \psi)$.

(4) If the property holds for φ, then for every variable x, the property holds for $(\exists x)\varphi$.

The four conditions above imply that for every natural number n, if the property holds for all pure formulas (A-formulas) of degree $< n$, then the property holds for all pure formulas (A-formulas) of degree n. Then by the principle of complete mathematical induction (Exercise 3.1, Chapter 3) it follows that for every n, the property holds for all pure formulas (A-formulas) of degree n—and thus all pure formulas (A-formulas) have the property.

Truth In relational systems Consider now a relational system (A,R) where A is a set. By an A-sentence we mean a sentence whose constants are all in A. We now wish to define what it means for an A-sentence X to be *true* in the relational system (A,R). Intuitively this means that the sentence is true when the predicate symbol P is interpreted as a name of the relation R, but a more formal definition is necessary. The definition is by induction on the degree of X—that is, truth of sentences of degree n is defined in terms of truth of sentences of degree $< n$, for n positive, and truth of atomic sentences is defined outright. Here are the induction conditions.

(1) An atomic sentence Pab (a and b are elements of A) is called *true* (in the system (A, R)) if and only if aRb holds.

(2) For any A-sentence φ, the sentence $\neg\varphi$ is true in (A, R) if and only if φ is not true in (A, R).

(3) For any sentences φ and ψ with constants in A, the conjunction $(\varphi \wedge \psi)$ is true in (A, R) if and only if the sentences φ and ψ are both true in (A, R).

(4) An existential quantification $(\exists x)\varphi(x)$, with constants in A, is true in (A, R) if and only if there is at least one element a of A such that the sentence $\varphi(a)$ is true in (A, R).

A pure formula $\varphi(x_1, \ldots, x_n)$ is said to be *satisfiable* in a relational system (A, R) if there are elements $a_1, \ldots, a_n$ of A such that the sentence $\varphi(a_1, \ldots, a_n)$ is true in (A, R). A relational system (A, R) is called *denumerable* if A is a denumerable set.

One of the early major results in model theory is Löwenheim's theorem: Given a formula φ with no constants, if φ is satisfiable in any relational system at all, then it is satisfiable in a *denumerable* relational system.[1] Then Skolem extended this result to the famous Skolem-Löwenheim theorem: Given any denumerable set S of pure formulas, if there is any relational system at all in which all formulas of S are satisfiable, then there is a *denumerable* such relational system.[1] The Tarski-Vaught theorem carries this even further (as we shall see).

Consider now two relational systems (A_1, R_1) and (A_2, R_2). They are called *elementarily equivalent* ('first-order' equivalent might be a more apt phrase) if and only if for every *pure* sentence X, X is true in (A_1, R_1) if and only if X is true in (A_2, R_2). For example, if (A_1, R_1) is *isomorphic* to (A_2, R_2), then the two systems are certainly elementarily equivalent (see Exercise 0.1), but the converse is far from true! As we will see, it is possible for a non-denumerable relational system to be elementarily equivalent to a denumerable one — indeed, it will follow from the Tarski-Vaught theorem that *every* non-denumerable relational system is elementarily equivalent to a denumerable one!

A stronger notion than elementary equivalence is this: A relational system (A_1, R_1) is called an *elementary subsystem* of a system (A_2, R_2) if $A_1 \subseteq A_2$ and $R_2 \upharpoonright A_1 = R_1$ and for every sentence X with constants in the (possibly smaller) set A_1, X is true in (A_1, R_1) if and only if X is true in (A_2, R_2). (In particular, this implies that any *pure* sentence is true in (A_1, R_1) if and only if X is true in (A_2, R_2), and hence that (A_1, R_1) is elementarily equivalent to (A_2, R_2).)

EXERCISES

Exercise 0.1. Suppose π is an isomorphism from (A_1, R_1) to (A_2, R_2). Prove that not only are the systems (A_1, R_1) and (A_2, R_2) elementarily equivalent, but also that for any pure formula $\varphi(x_1, \ldots, x_n)$ and any elements $a_1, \ldots, a_n$ of A_1, sentence $\varphi(a_1, \ldots, a_n)$ is true in (A_1, R_1) if and only if $\varphi(\pi(a_1), \ldots, \pi(a_n))$ is true in (A_2, R_2). (Prove this by induction on degrees of formulas.)

---

[1] Actually these theorems were proved for formulas with predicates of any degrees ≥ 1.

§1 The Tarski-Vaught theorem

The relation R that is of most interest to us is the $\in$-relation, and so in this chapter, we will prove various results that *could* be proved for an arbitrary R for just the special case of the $\in$-relation. (In the exercises we will indicate which of our theorems can be generalized to an arbitrary R.)

We shall say that a sentence X with constants in A is *true over A* iff it is true in the relational system $(A, \in)$. Given a set A and a subset B, we will say that B *reflects A* if for every sentence X whose constants are all in B, X is true over B if and only if X is true over A. (This is the same thing as saying that the relational system $(B, \in \restriction B)$ is an elementary subsystem of $(A, \in \restriction A)$.)

Now, given a set A, a subset B, and a formula $\varphi(x_1, \ldots, x_n)$ whose constants are in B, we say that B reflects A *with respect to the formula φ* if for all constants $b_1, \ldots, b_n$ of B, the sentence $\varphi(b_1, \ldots, b_n)$ is true over B if and only if it is true over A. (Thus to say that B reflects A is to say that B reflects A with respect to every formula whose constants are all in B.) We shall also say that B *completely* reflects A with respect to φ if B reflects A with respect to all subformulas of φ (which includes φ itself).

We now aim to prove the following theorem.

Theorem 1.1 (Tarski-Vaught). *For any set A that can be well ordered, any infinite subset A_0 of A can be extended to a subset B of A of no higher cardinality than A_0 such that B reflects A. (In particular, any denumerable subset A_0 of A can be extended to a denumerable subset B of A such that B reflects A.)*

We first introduce a key notion (due to Leon Henkin, and possibly others) and establish some lemmas that will have other applications as well.

The Henkin closure condition Given a subset B of A and a formula φ whose constants are all in B, we shall say that B is *A-closed with respect to φ* if for every subformula of φ of the form $(\exists x)\psi(x, y_1, \ldots, y_n)$ (where the free variables of ψ are $x, y_1, \ldots, y_n$) and for all elements $b_1, \ldots, b_n$ of B, if there is some element a of A such that the sentence $\psi(a, b_1, \ldots, b_n)$ is true over A, then there is some b in B such that $\psi(b, b_1, \ldots, b_n)$ is true over A. (Yes we mean true over A!) And we shall say that B is *A-closed* (or *Henkin-closed* with respect to A) if B is A-closed with respect to every formula φ whose constants are all in B. (This is equivalent to saying that for every formula $\varphi(x)$ with constants in B and x as the only free variable, if $(\exists x)\varphi(x)$ is true over A, then for some b in B, the sentence $\varphi(b)$ is true over A.)

Lemma 1.2. *If B is A-closed with respect to φ, then B completely reflects A with respect to φ.*

Proof Assume the hypothesis. Call a formula (with constants in B) *good* if B reflects A with respect to it. We will show by induction on the degrees of subformulas of φ that every subformula of φ (including φ) is good.

(1) It is obvious that any atomic formula is good.

(2) It is obvious that the negation of any good formula is good.

(3) It is obvious that the conjunction of any two good formulas is good.

(4) Now consider a subformula of φ of the form $(\exists x)\psi(x, y_1, \ldots, y_n)$ $(x, y_1, \ldots, y_n$ are the free variables of ψ) such that $\psi(x, y_1, \ldots, y_n)$ is good. We must show that the formula $(\exists x)\psi(x, y_1, \ldots, y_n)$ is good, i.e., that for any elements $b_1, \ldots, b_n$ of B, the sentence $(\exists x)\psi(x, b_1, \ldots, b_n)$ is true over B iff it is true over A.

    (a) Suppose it is true over B, so for some $b \in B$, the sentence $\psi(b, b_1, \ldots, b_n)$ is true over B. Then this sentence is also true over A (by the inductive assumption that the formula $\psi(x, y_1, \ldots, y_n)$ is good). Also b is an element of A (since $B \subseteq A$), and so the sentence $(\exists x)\psi(x, b_1, \ldots, b_n)$ is true over A.

    (b) Conversely, suppose the sentence $(\exists x)\psi(x, b_1, \ldots, b_n)$ is true over A. Now we use the hypothesis of the lemma for the first time! By this hypothesis, there is some element b of B such that $\psi(b, b_1, \ldots, b_n)$ is true *over* A. But it is also true over B (since the formula $\psi(x, y_1, \ldots, y_n)$ is assumed good), and so the sentence $(\exists x)\psi(x, b_1, \ldots, b_n)$ is true over B. ∎

As an immediate corollary we have:

Lemma 1.3. *If B is Henkin closed with respect to A, then B reflects A.*

Now we can prove the Tarski-Vaught theorem. Given a well orderable set A and a subset A_0 of cardinality c (c infinite), we will extend A_0 to a subset B of A of cardinality c such that B is Henkin closed with respect to A.

For any subset S of A, we define the set S' as follows: We take all formulas with constants in S of the form $(\exists x)\varphi(x)$ that are true over A, and for each such formula we take the first element a of A (in the well ordering) such that $\varphi(a)$ is true over A, and we adjoin a to S. Now, if S is infinite and of cardinality c, there are only c formulas with constants in S, hence S' has no higher cardinality than S.

We now consider the denumerable sequence $A_0, A_1, \ldots, A_n, A_{n+1}, \ldots$, where for each n, we take A_{n+1} to be A'_n. We let B be the union of the set of A_n's. Since A_0 has cardinality c, so does A_1 (since $A_1 = A'_0$); hence A_2 (being A'_1) has cardinality c, and so forth. Thus by an obvious induction argument each A_n has cardinality c. And so their union B has cardinality c.

Now, B is Henkin-closed with respect to A, because for any formula $\varphi(x)$ with constants in B, since there are only finitely many constants in $\varphi(x)$, all these constants lie in A_n for some n. If $(\exists x)\varphi(x)$ happens to be true over A, then there is some element a in A_{n+1} such that $\varphi(a)$ is true over A, and a is therefore in B. Thus B is Henkin-closed with respect to A, hence B reflects A by Lemma 1.3.

Remark We shall sometimes refer to the set B constructed out of A_0 in the manner above as the *Henkin-closure* of A_0 with respect to A.

EXERCISES

Exercise 1.1. Prove the Tarski-Vaught theorem for an arbitrary relational system (A, R) rather than just the case where R is the $\in$-relation.

Exercise 1.2. Assuming the axiom of choice, how does the exercise above yield us the Skolem-Löwenheim theorem? (We remark that the Skolem-Löwenheim theorem can be proved without the axiom of choice.)

§2 We add extensionality considerations

We will have particular need of the following variant of Theorem 1.1.

Theorem 2.1. *Let A be a well-orderable infinite set that is* extensional *(with respect to $\in$). Then any infinite subset A_0 of A can be extended to an* extensional *subset B of A of no higher cardinality than A_0, such that B reflects A.*

Proof In the proof of the Tarski-Vaught theorem we modify the construction of the sequence $A_0, A_1, \ldots, A_n, A_{n+1}, \ldots$ as follows. Assuming A_n defined, we throw into A_{n+1} the same items as before, but in addition, we throw in the following: Since A is extensional, then for each x, y in A_n that are distinct, A must contain some element z such that z is in one of x, y but not in the other; throw one such z (say the first in the well ordering of A) into A_{n+1} (if A_n doesn't happen to contain such an element z already). If c is the cardinality of A_n, there are only c pairs $\{x, y\}$ of elements of A_n and so again A_{n+1} has just cardinality c. Again we take B to be the union of the A_n's and B reflects A for the same reasons as before. Also, for any x, y in B, they are both in some A_n, and if $x \neq y$, then A_{n+1} must contain some z such that z is in one of x, y but not in the other; such an element z is therefore in B, and so B is extensional. ∎

First-order definability For any class K, a formula $\varphi(x)$ whose constants are all in K is said to *define* over K the class of all elements $k \in K$ such that $\varphi(k)$ is true over K. And a subclass K_1 of K is called *definable* over K (more completely, *first-order definable* over K) if it is defined over K by some formula $\varphi(x)$ whose constants are all in K. For any set A we let $\mathcal{F}(A)$ be the set of all definable subsets of A.

The notion of first-order definability will play a key role in topics to come. For the time being, we suggest you try the exercises at the end of this section.

Discussion We have superficially stepped outside axiomatic set theory in stating (let alone proving) the Tarski-Vaught theorem, since the theorem says something about *formulas*, whereas the language of set theory talks only about sets. We said "superficially" because the problem is easily remediable. Later, in Chapter 14, we will show how we can assign to every formula φ a certain *set* $\ulcorner \varphi \urcorner$ (called the *code* of φ), and we will do this in such a way that the relation "x is the code of a formula that is true over y" is not only expressible in set theory, but is first-order definable over V! And thus the relation "x is first-order definable over y" turns out to be definable over V.

EXERCISES

Exercise 2.1. If A is denumerable, how many definable subsets of A are there?

Exercise 2.2. If A is finite and extensional and n is the number of elements of A, how many definable subsets of A are there?

Exercise 2.3. Let B be a subset of A. A subset of A is called *B-definable over A* if it is defined over A by a formula $\varphi(x)$ all of whose constants are in B. Now suppose that every non-empty subset of A which is B-definable over A contains at least one element of B. Does it necessarily follow that B reflects A?

Exercise 2.4. Suppose that B is an elementary subset of A. Let C be any subset of A that is B-definable over A. Prove that $C \cap B$ is definable over B.

§3 The class version of the Tarski-Vaught theorems

For any class K, let $\mathcal{T}(K)$ be the class of all sentences with constants in K which are true over K. In particular, $\mathcal{T}(V)$ is the class of all sentences that are true over V. Is the class $\mathcal{T}(V)$ definable over V? That is, is there a formula $\varphi(x)$ such that for every set a, the sentence $\varphi(a)$ is true over V if and only if a is (the code of) a sentence which is true over V? By a celebrated theorem of Alfred Tarski, the answer is *no*. And so in general, given a class K that is not a set, even if K is itself definable over V, the set $\mathcal{T}(K)$ is not.

However, the following can be done: For every natural number n, we let $\mathcal{T}_n$ be the class of all sentences *of degree less than or equal to n* that are true over V. Although the class $\mathcal{T}$ (which is the union of the classes $\mathcal{T}_0, \mathcal{T}_1, \ldots, \mathcal{T}_n, \ldots$) is not definable over V (Tarski's theorem), each of the classes $\mathcal{T}_n$ *is* definable over V! (This can be proved by induction on n. The class $\mathcal{T}_0$ is simply the class of all sentences $a \in b$ such that a is a member of b. Assuming $\mathcal{T}_n$ defined, the class $\mathcal{T}_{n+1}$ is the class of all elements of $\mathcal{T}_n$ together with all sentences $\neg X$ of degree $n+1$ such that X is not in $\mathcal{T}_n$, together with all sentences of degree $n+1$ of the form $X \wedge Y$ such that X and Y are both in $\mathcal{T}_n$, together with all sentences of degree $n+1$ of the form $(\exists x)\varphi(x)$ such that for at least one set a, the sentence $\varphi(a)$ is in $\mathcal{T}_n$.)

More generally, for any class K, let $\mathcal{T}_n^K$ be the class of all sentences of degree $\leq n$ with constants in K that are true over K. If K itself is definable over V, then for each particular n, the class $\mathcal{T}_n^K$ is definable over V (though the union $\mathcal{T}^K$ of these classes is in general not). And therefore for each particular definable class K and each particular formula φ with constants in K, the following properties of a set B are all definable:

(1) B reflects K with respect to φ and all its subformulas;

(2) B is K-closed with respect to φ.

And then, for each particular definable class K and each particular formula φ, the following lemmas and theorems can be formally proved in axiomatic set theory.

Lemma 3.1. *If $B \subseteq K$ and B is K-closed with respect to φ, then B completely reflects K with respect to φ.*

Theorem 3.2. *Suppose K is well orderable and A_0 is an infinite subset of K of cardinality c. Then for each formula φ:*

(1) A_0 can be extended to a subset B of K of cardinality c such that B completely reflects K with respect to φ;

§3. THE CLASS VERSION OF THE TARSKI-VAUGHT THEOREMS   149

(2) *if also K is extensional, then A_0 can be extended to an* extensional *subset B of K of cardinality c such that B completely reflects K with respect to φ.*

Lemma 3.1 is proved in the same manner as Lemma 1.2. Theorem 3.2 can be proved in the same manner as Theorems 1.1 and 2.1, except that in the constructions of the sequence $A_0, A_1, \ldots, A_n, A_{n+1}, \ldots$, to get from A_n to A_{n+1} we don't consider *all* formulas with constants in A_n, but only formulas that are of the form $(\exists x)\psi(x, a_1, \ldots, a_k)$, where $\psi(x, y_1, \ldots, y_k)$ is a proper subformula of φ (and $a_1, \ldots, a_k$ are constants in A_n).

As an obvious corollary we have the following.

Theorem 3.3. *Suppose K is well orderable and A_0 is an infinite subset of K of cardinality c. If X is a pure sentence that is true over K then we have the following.*

(1) *A_0 can be extended to a subset B of K of cardinality c such that X is true over B.*

(2) *If also K is extensional, then A_0 can be extended to an* extensional *subset B of K of cardinality c such that X is true over B.*

Addendum to part I (optional) There is a pretty theorem (due I believe to Richard Montague) that is closely related to the Tarski-Vaught theorem (Theorem 1.1—the "set" form) and yields an alternative proof of Theorem 1.1, provided a bit of extra hypothesis is added. Aside, however, from this alternative proof, the theorem is neat in its own right.

Given a subset B of A, an element a of A is called *B-describable* over A if there is a formula $\varphi(x)$ whose constants are all in B such that $\varphi(x)$ is true over A for the element a, but for no other element of A—in other words, $\varphi(a)$ is true over A, but for every a' in A such that $a' \neq a$, the sentence $\varphi(a')$ is not true over A. (Equivalently stated, the set defined by $\varphi(x)$ over A is $\{a\}$.) Now here is the theorem.

Theorem 3.4. *Suppose that A can be well ordered, and moreover some well ordering $W(x, y)$ of A is definable over A. For any subset B of A, let B^* be the set of all elements of A that are B-describable over A. Then B^* reflects A!*

To aid in the proof of this, we first establish two facts.

Fact 1 Suppose A has a well ordering W that is definable over A. Let $B \subseteq A$ and let $\varphi(x)$ be a formula whose constants are all in B. Then if $(\exists x)\varphi(x)$ is true over A and if a_0 is the first element of A (in the well ordering W) such that $\varphi(a_0)$ is true over A, then a_0 is B-describable over A.

Proof of fact 1 Assume the hypothesis. Let $w(x, y)$ be a formula that defines the well ordering $W(x, y)$ over A. Then a_0 is described by the following formula:

$$\varphi(x) \wedge (\forall y)(\varphi(y) \supset w(x, y)).$$

∎

Fact 2 Let B be a subset of A and let B^* be the set of all B-describable elements of A. Then any element of A that is B^*-describable over A is B-describable over A.

Proof of fact 2 Let a be B^*-describable over A by $\varphi(x, b_1, \ldots, b_n)$, where $b_1, \ldots, b_n$ are all constants in B^*. Each b_i is described over A by some formula $\varphi_i(x)$ whose constants are all in B. Then the following formula (whose constants are all in B) describes a over A:

$$(\exists x_1) \cdots (\exists x_n)(\varphi(x_1) \wedge \ldots \wedge \varphi_n(x_n) \wedge \varphi(x, x_1, \ldots, x_n)).$$

∎

Proof of theorem 3.4 Assume the hypothesis. We show that B^* is Henkin closed with respect to A (and hence by Lemma 1.3 B^* reflects A). Let $\varphi(x)$ be a formula with constants in B^* such that $(\exists x)\varphi(x)$ is true over A. We are to show that B^* contains at least one element b such that $\varphi(b)$ is true over A.

Let a_0 be the first element of A (in the well ordering W) such that $\varphi(a_0)$ is true over A. Then by fact 1 (reading B^* for B) a_0 is B^*-describable over A. Then by fact 2, a_0 is B-describable over A, hence $a_0 \in B^*$. Thus $\varphi(b)$ is true over A where b is the element a_0 (and $b \in B^*$). ∎

EXERCISES

Exercise 3.1.

(a) Suppose that the relation $x = y$ is definable over A. Show that for any subset B of A, $B \subseteq B^*$.

(b) Show that if A is extensional, then for any subset B of A, $B \subseteq B^*$.

(c) Now show the Montague theorem provides an alternative proof that if some well ordering of A is definable over A and if the relation $x = y$ is definable over A, then any infinite subset A_0 of A can be extended to a subset B of A of no higher cardinality than A_0 such that B reflects A.

Part II—The M.S.T.V. Theorem

§4 Mostowski, Shepherdson, Tarski, and Vaught

We now combine the Mostowski-Shepherdson mapping theorem, of the last chapter, with the Tarski-Vaught theorem (class version) to obtain the main result of this chapter, which will be seen to play a key role in Gödel's proof of the consistency of the continuum hypothesis. We shall refer to this important result as the "Mostowski-Shepherdson-Tarski-Vaught theorem"—abbreviated "M.S.T.V."

Theorem 4.1 (M.S.T.V.). *Let K be an extensional, well founded class that can be well ordered, and let X be a pure sentence that is true over K. Then any infinite transitive subset A of K is a subset of some transitive set T of the same cardinality as A, such that X is true over T.*

§5. THE MONTAGUE-LEVY REFLECTION THEOREM

Proof Assume the hypothesis. Suppose that A is an infinite transitive subset of K of cardinality c. By part 2 of Theorem 3.2, A is a subset of an *extensional* subset B of K of cardinality c such that X is true over B. Since K is well founded, so is its subset B, and so B is extensional and well founded, and so by Theorem 4.2 of Chapter 10 (Mostowski-Shepherdson) B is $\in$-isomorphic to a *transitive* set T. Obviously T also has cardinality c, and since T is $\in$-isomorphic to B and X is a pure sentence that is true over B, then X is also true over T. Finally, since A is transitive, then by Theorem 5.1 of Chapter 10, the Mostowski-Shepherdson mapping from B onto T carries every element of A to itself, and so A is a subset of T. ∎

Part III—More Reflection Principles

We now turn to some related principles, which will have applications in Chapter 13, and are also of interest in their own right.

§5 The Montague-Levy reflection theorem

We let $\mathcal{W}$ be an On-sequence $w_0, w_1, \ldots, w_\alpha, \ldots$, of sets satisfying the following two conditions:

(1) for $\alpha \leq \beta$, $w_\alpha \subseteq w_\beta$;

(2) for any limit ordinal λ, $w_\lambda = \cup_{\alpha < \lambda} w_\alpha$.

As an example, any ordinal hierarchy, as defined in Chapter 7, satisfies these two conditions. (Condition (2) is satisfied by definition, and condition (1) was proved in Chapter 7, Theorem 2.1). Actually, in the presence of (2), condition (1) is *equivalent* to the condition that for each ordinal α, $w_\alpha \subseteq w_{\alpha+1}$. Thus our present setup is like that of an ordinal hierarchy, except we do not require that $w_0 = \emptyset$.

Condition (1) is usually paraphrased as saying that the sequence is *monotone increasing*, and condition (2) is sometimes paraphrased as saying that the sequence is *continuous*.

We let W be the union of all the w_α's. (In general, each w_α can be a *proper* subset of $w_{\alpha+1}$, and so W can be a proper class.) As with ordinal hierarchies, for any $x \in W$, we define its $\mathcal{W}$-rank to be the first ordinal α such that $x \in w_{\alpha+1}$. And, as with ordinal hierarchies, any *subset* S of W must be a subset of some w_α (because by the axiom of substitution, the class of $\mathcal{W}$-ranks of the elements of S must be a set, and so we can take α to be the first ordinal not in that set).

Theorem 5.1. *If W can be well ordered, than for any pure formula φ and any ordinal α there exists a limit ordinal $\beta > \alpha$ such that w_β completely reflects W with respect to φ.*

Remark The hypothesis that W can be well ordered is actually not necessary for this theorem. However, the proof using the well orderability of W is considerably simpler and the only applications we have will involve a W that *is* well orderable. But we will give the other proof too. Both proofs utilize the following lemma.

Lemma 5.2. *Let $A_0, A_1, \ldots, A_n, \ldots$ be a monotone increasing denumerable sequence of subsets of a proper class W (each $A_n \subseteq A_{n+1}$) and let A be their union. Let φ be any pure formula. Then if each A_i completely reflects W with respect to φ, so does A.*

Proof Assume the hypothesis. By virtue of Lemma 1.2 it suffices to show that A is W-closed (in the Henkin sense) with respect to φ. So let $\psi(x, y_1, \ldots, y_n)$ be any proper subformula of φ and let $a_1, \ldots, a_n$ be constants of A such that $(\exists x)\psi(x, a_1, \ldots, a_n)$ is true over W. We must show that for some $a \in A$, $\psi(a, a_1, \ldots, a_n)$ is true over W.

All the constants $a_1, \ldots, a_n$ lie in some A_i. Since A_i completely reflects W with respect to φ, and $(\exists x)\psi(x, a_1, \ldots, a_n)$ is true over W, then it is also true over A_i. Hence for some $a \in A$, $\psi(a, a_1, \ldots, a_n)$ is true over A, and hence true over W (again, because A_i completely reflects W with respect to φ). Since $a \in A$, then for some $a \in A$, $\psi(a, a_1, \ldots, a_n)$ is true over W. ∎

Proof of theorem 5.1 This proof uses Theorem 1.1 which has already been proved. Using the fact that W can be well ordered, we know by the proof of Theorem 3.2 that any subset S of W can be extended in a canonical manner to a subset—call it S^*—of W that completely reflects W with respect to φ. (The fact that if S is infinite, S^* has no higher cardinality than S, though true, is not relevant for our present purposes.) Now, given an ordinal α, we construct a denumerable sequence $\beta_0, \beta_1, \ldots, \beta_n, \ldots$ of ordinals as follows: we let β_0 be any ordinal $> \alpha$. Now assuming β_n already defined, we let β_{n+1} be the smallest ordinal $> \beta_n$ such that $w^*_{\beta_n} \subseteq w_{\beta_{n+1}}$. ($w^*_{\beta_n}$ is an extension of w_{β_n} within W that completely reflects W with respect to φ. It exists by Theorem 3.2. Since it is a set, and a subset of W, it is a subset of some w_k, and so we take β_{n+1} to be the first such w_k that is greater than β_n.)

We thus have $w_{\beta_0} \subseteq w^*_{\beta_0} \subseteq w_{\beta_1} \subseteq w^*_{\beta_1} \subseteq \ldots \subseteq w_{\beta_n} \subseteq w^*_{\beta_n} \subseteq \ldots$. We let β be the smallest ordinal greater than all the β_n, and so w_β is the union of the sets $w_{\beta_0}, w_{\beta_1}, \ldots, w_{\beta_n}, \ldots$. It is also the union of the sets $w^*_{\beta_0}, w^*_{\beta_1}, \ldots, w^*_{\beta_n}, \ldots$ (because any element of W is an element of some w_{β_n}, and hence of $w^*_{\beta_n}$, and any element of any $w^*_{\beta_n}$ is an element of $w_{\beta+1}$, hence of W).

Since for each n, $w^*_{\beta_n}$ completely reflects W with respect to φ, by Lemma 5.2 so does w_β. This concludes the proof. ∎

Theorem 5.3 (Montague-Levy). *Same as Theorem 5.1, without the assumption that W can be well ordered.*

Proof This proof does *not* use Theorem 3.2. It more or less proceeds from scratch (although it does use Lemma 5.2).

We prove the theorem by mathematical induction on degrees of formulas. Call a pure formula φ *good* if for every α there exists a limit ordinal $\beta > \alpha$ such that w_β completely reflects W with respect to φ. We show by induction that every pure formula φ is good.

To save much locution, let us say that an ordinal α *takes care of* φ if w_α completely reflects W with respect to φ. Now for the inductive proof:

(1) Any atomic formula $x \in y$ is obviously good. (Given any α, let β be any limit ordinal $> \alpha$ and w_β will reflect W with respect to $x \in y$—indeed, any subset of W does this.)

§5. THE MONTAGUE-LEVY REFLECTION THEOREM

(2) If φ is good, then obviously $\neg\varphi$ is good (why?).

(3) Now suppose φ_1 and φ_2 are both good. It is *not* obvious that $\varphi_1 \wedge \varphi_2$ is good; nevertheless, $\varphi_1 \wedge \varphi_2$ *is* good, as we can see by using the following trick.

Given α, let β_1 be the first limit ordinal $> \alpha$ that takes care of φ_1 (such a β_1 exists by the inductive assumption that φ_1 is good), then let β_2 be the first limit ordinal $> \beta_1$ such that β_2 takes care of φ_2; then let β_3 be the next limit ordinal that takes care of φ_1; and so forth. Thus for all odd n, β_n takes care of φ_1, and for all even n, β_n takes care of φ_2. We let β be the first ordinal greater than all the β_n's (β is clearly a limit ordinal). Then w_β is the union of all the w_{β_n}'s. It is also the union of the sets $w_{\beta_1}, w_{\beta_3}, \ldots, w_{\beta_{2n+1}}, \ldots$, hence by Lemma 5.2, w_β completely reflects W with respect to φ_1 (β takes care of φ_1). But also β takes care of φ_2 (since w_β is also the union of the $w_{\beta_{2n}}$'s), and so β clearly takes care of $\varphi_1 \wedge \varphi_2$.

(4) Now comes the most ingenious part of the proof! Suppose that $\psi(x, y_1, \ldots, y_n)$ is good; we show that $(\exists x)\psi(x, y_1, \ldots, y_n)$ is also good.

We use a clever idea due to Dana Scott. For elements $a_1, \ldots, a_n$ of W we define the set $\Gamma(a_1, \ldots, a_n)$ as follows: If there is some element $a \in w_n$ such that $\psi(a, a_1, \ldots, a_n)$ is true over W, we let $\Gamma(a_1, \ldots, a_n)$ be the set of all a's of *minimal W-rank* such that $\psi(a, a_1, \ldots, a_n)$ is true over W. (The important thing is that $\Gamma(a_1, \ldots, a_n)$ is then a *set*, since it is a subclass of some w_α.) If there is no $a \in W$ such that $\psi(a, a_1, \ldots, a_n)$ is true over W, we take $\Gamma(a_1, \ldots, a_n)$ to be the empty set. For any class K, we let K^n be the class of all n-tuples $\langle k_1, \ldots, k_n \rangle$ of K. Then the operation $\Gamma(x_1, \ldots, x_n)$ has W^n for its domain. Let us note that for any subset S of W, S^n is a set, hence $\Gamma''(S^n)$ is a set (by the axiom of substitution), hence $\cup \Gamma''(S^n)$ is a set. Let us also note that for any elements $a_1, \ldots, a_n$ of S, $\Gamma(a_1, \ldots, a_n)$ is a *subset* of $\cup \Gamma''(S^n)$ (because $\Gamma(a_1, \ldots, a_n)$ is an *element* of $\Gamma''(S^n)$, hence a *subset* of $\cup \Gamma''(S^n)$). Now, given an ordinal α, we define a denumerable sequence $\beta_1, \beta_2, \ldots, \beta_n, \ldots$ of ordinals as follows: we let β_1 be the first ordinal $> \alpha$ such that β_1 takes care of the formula $\psi(x, y_1, \ldots, y_n)$ (which exists, since ψ is good). We take β_2 to be the first ordinal after β_1 such that $\cup \Gamma''(w_{\beta_1}) \subseteq w_{\beta_2}$. Then we take β_3 to be the first ordinal after β_2 that takes care of ψ. And so forth, alternating in this manner. And so for each odd number n, w_{β_n} takes care of ψ and $\cup \Gamma''(w_{\beta_n}) \subseteq w_{\beta_{n+1}}$. We let β be the least ordinal greater than all the β_n's. Now $w_\beta = \cup\{w_{\beta_1}, w_{\beta_3}, \ldots, w_{\beta_{2n+1}}, \ldots\}$ and so by Lemma 5.2, β takes care of $\psi(x, y_1, \ldots, y_n)$ (since each β_{2n+1} does).

It remains to show that w_β reflects W with respect to $(\exists x)\psi(x, y_1, \ldots, y_n)$. So let $a_1, \ldots, a_n$ be elements of w_β.

(a) Suppose $(\exists x)\psi(x, a_1, \ldots, a_n)$ is true over W. Then $\psi(a, a_1, \ldots, a_n)$ is true over W for some $a \in W$, hence there is some a *of lowest W-rank* such that $\psi(a, a_1, \ldots, a_n)$ is true over W—such an a is an element of $\Gamma(a_1, \ldots, a_n)$. Now, the constants $a_1, \ldots, a_n$ are all in some w_{β_j}, for j an odd number. Then $\Gamma(a_1, \ldots, a_n) \subseteq w_{\beta_{j+1}}$ (since $\cup \Gamma''(w_{\beta_j}) \subseteq w_{\beta_{j+1}}$), and so a lies in $w_{\beta_{j+1}}$, hence lies in w_β. Thus there is some a in w_β such that $\psi(a, a_1, \ldots, a_n)$ is true over W. Then $\psi(a, a_1, \ldots, a_n)$ is also true over w_β (since w_β reflects W with respect to $\psi(x, y_1, \ldots, y_n)$). Thus $(\exists x)\psi(x, a_1, \ldots, a_n)$ is true over w_β.

(b) Conversely, suppose $(\exists x)\psi(x, a_1, \ldots, a_n)$ is true over w_β. Then for some $a \in w_\beta$, $\psi(a, a_1, \ldots, a_n)$ is true over W. Then, since $a \in W$, the sentence $(\exists x)\psi(x, a_1, \ldots, a_n)$ is true over W. ∎

CHAPTER 12

CONSTRUCTIBLE SETS

Now we come to the central notion involved in Gödel's proof of the relative consistency of the continuum hypothesis and the axiom of choice. We are going to define a certain subclass L of V whose elements are called the *constructible* sets and show (in this and the next several chapters) that if we reinterpret "set" to mean *constructible set*, then under this reinterpretation, all the formal axioms of Zermelo-Fraenkel set theory (which we state precisely in the next chapter) are true, and moreover so are the continuum hypothesis and the axiom of choice. From this the relative consistency of the axiom of choice and the continuum hypothesis will be seen to follow. In this and subsequent chapters we assume V is a Zermelo-Fraenkel universe in which Axiom D, the axiom of well foundedness, holds.

§0 More on first-order definability

We are denoting the set of all subsets of A that are first-order definable over A (allowing formulas with constants in A) by $\mathcal{F}(A)$.

Let us observe a few properties of first-order definability—or just "definability" for short.

P$_1$ A is definable over A. So is $\emptyset$.

P$_2$ If A is transitive (or even extensional) then every finite subset of A is definable over A.

P$_3$ If A is finite and extensional, then $\mathcal{F}(A) = \mathcal{P}(A)$.

P$_4$ If $a \in A$ and $a \subseteq A$, then a is definable over A.

P$_5$ If A is transitive then $A \subseteq \mathcal{F}(A)$. (Every element of A is definable over A.)

P$_6$ If A is transitive, so is $\mathcal{F}(A)$.

P$_7$ If x, y are both in $\mathcal{F}(A) - A$, then $x \notin y$.

Proofs

P$_1$ A is defined over A by the formula $x = x$. $\emptyset$ is defined over A by the formula $\neg(x = x)$.

P$_2$ If B is a finite set $\{a_1, a_2, \ldots, a_n\}$, of A, then B is defined over A by the formula $x = a_1 \lor x = a_2 \lor \ldots \lor x = a_n$.

P₃ Immediate from **P₂**.

P₄ Suppose $a \in A$ and $a \subseteq A$. Then a is defined, not by the formula $x = a$ (which defines $\{a\}$ rather than a), but by the formula $x \in a$. [Note: if we did not assume that $a \subseteq A$, then $x \in a$ would define $a \cap A$. But since we are assuming $a \subseteq A$, then $a \cap A = a$.]

P₅ Immediate from **P₄**.

P₆ Assume A is transitive. Suppose $x \in \mathcal{F}(A)$; we must show $x \subseteq \mathcal{F}(A)$. Take any $y \in x$. Since $x \in \mathcal{F}(A)$, then of course $x \subseteq A$, and so $y \in A$. Since A is transitive, $A \subseteq \mathcal{F}(A)$ (by **P₅**), hence $y \in \mathcal{F}(A)$ (since $y \in A$).

P₇ Suppose x, y are both in $\mathcal{F}(A) - A$. Suppose it were true that $x \in y$. Then we would have $x \in y \subseteq A$ (since $y \in \mathcal{F}(A)$), and hence $x \in A$, contrary to supposition.

§1 The class L of constructible sets

For each ordinal α we define the set L_α by the following transfinite recursion[1]

(1) $L_0 = \emptyset$

(2) $L_{\alpha+1} = \mathcal{F}(L_\alpha)$

(3) $L_\lambda = \cup_{\alpha < \lambda} L_\alpha$ (for λ a limit ordinal).

We let L be the union of all the L_α—the elements of L are called the *constructible sets*. For any constructible set x, by its *order* is meant the first α such that $x \in L_{\alpha+1}$.

We aim to show (in the next few chapters) that this class L is a model of the Zermelo-Fraenkel axioms, in which the axiom of choice (in the strong form of Axiom E) and the continuum hypothesis are both true.

Now we list some key properties of the constructible sets which we wish to establish.

C₀ Each L_α is transitive.

C₁ $L_\alpha \in L_{\alpha+1}$ and $L_\alpha \subseteq L_{\alpha+1}$.

C₂ The L_α-sequence is an ordinal hierarchy (see Chapter 7, §2).

C₃ If $\alpha \leq \beta$ then $L_\alpha \subseteq L_\beta$.

C₄ For each α, $L_\alpha \subseteq R_\alpha$.

C₅ L is well founded.

C₆ $L_\alpha \neq L_{\alpha+1}$.

C₇ The set of all elements of order α is $L_{\alpha+1} - L_\alpha$.

---

[1] Gödel used the notation M_α for the sequence that approximates to L as a "limit." More recently L_α has become standard notation, and it is what we use.

§1. THE CLASS L OF CONSTRUCTIBLE SETS

$\mathbf{C_8}$ Each L_α is constructible and of order α.

$\mathbf{C_9}$ L_α is the set of all constructible sets of order $< \alpha$.

$\mathbf{C_{10}}$ If x is constructible, then every element of x is constructible and has order less than the order of x.

$\mathbf{C_{11}}$ Every set of constructible sets is a *subset* of some constructible set.

$\mathbf{C_{12}}$ For any constructible sets a and b, the following sets are constructible:

  (1) $a \cap b$,

  (2) $a \cup b$,

  (3) $a - b$,

  (4) $\{a, b\}$,

  (5) $\langle a, b \rangle$.

$\mathbf{C_{13}}$ For any constructible sets $x_1, \ldots, x_n$, the set $\langle x_1, \ldots, x_n \rangle$ is constructible ($n \geq 2$).

Proofs

$\mathbf{C_0}$ Proof is by transfinite induction. (Note that L_α transitive implies $L_{\alpha+1}$ transitive, by $\mathbf{P_6}$). The rest of the proof is obvious.

$\mathbf{C_1}$ That $L_\alpha \in L_{\alpha+1}$ follows from $\mathbf{P_1}$. Then by transitivity of $L_{\alpha+1}$, $L_\alpha \subseteq L_{\alpha+1}$.

$\mathbf{C_2}$ Since $L_0 = \emptyset$ and $L_\alpha \subseteq L_{\alpha+1}$, and $L_\lambda = \bigcup_{\alpha < \lambda} L_\alpha$ for limit ordinals λ, then the L_α sequence is by definition an ordinal hierarchy.

$\mathbf{C_3}$ Since the L_α-sequence is an ordinal hierarchy, the result follows from condition $\mathbf{O_1}$, Theorem 2.1, Chapter 7.

$\mathbf{C_4}$ Proof is by transfinite induction on α. Since $L_0 = R_0$, then $L_0 \subseteq R_0$. Now suppose $L_\alpha \subseteq R_\alpha$. Then $\mathcal{P}(L_\alpha) \subseteq \mathcal{P}(R_\alpha)$. Also $\mathcal{F}(L_\alpha) \subseteq \mathcal{P}(L_\alpha)$. Hence $\mathcal{F}(L_\alpha) \subseteq \mathcal{P}(R_\alpha)$, i.e., $L_{\alpha+1} \subseteq R_{\alpha+1}$. For a limit ordinal λ, if for each $\alpha < \lambda$, $L_\alpha \subseteq R_\alpha$, then obviously $\bigcup_{\alpha < \lambda} L_\alpha \subseteq \bigcup_{\alpha < \lambda} R_\alpha$, i.e., $L_\lambda \subseteq R_\lambda$.

$\mathbf{C_5}$ Immediate from $\mathbf{C_4}$. Since for each α, $L_\alpha \subseteq R_\alpha$, then $L \subseteq R_\Omega$; and so every constructible set has rank. Therefore L is well founded.

$\mathbf{C_6}$ Since L is well founded, then no constructible set can be a member of itself. But for each α, $L_{\alpha+1}$ is constructible (it is a member of $L_{\alpha+2}$), so $L_{\alpha+1} \notin L_{\alpha+1}$. But $L_\alpha \in L_{\alpha+1}$, hence $L_\alpha \neq L_{\alpha+1}$.

$\mathbf{C_7}$ Let us note that if we take $\mathcal{O}$ to be the L_α-sequence, the *order* of any constructible set x is simply the $\mathcal{O}$-*rank* of x, as defined in §2 of Chapter 7, and so the result is but a special case of condition $\mathbf{O_5}$, Theorem 2.1, Chapter 7.

$\mathbf{C_8}$ Since $L_\alpha \in L_{\alpha+1}$, and $L_\alpha \notin L_\alpha$, then $L_\alpha \in L_{\alpha+1} - L_\alpha$, and so L_α has order α by $\mathbf{C_7}$.

C$_9$ This follows by condition **O$_4$**, Theorem 2.1, Chapter 7.

C$_{10}$ Suppose x is constructible of order α. Then $x \in L_{\alpha+1}$, so $x \subseteq L_\alpha$. Then any element y of x is a member of L_α, hence is constructible of order $< \alpha$ (by **C$_9$**).

C$_{11}$ This follows from condition **O$_6$**, Theorem 2.1, Chapter 7.

C$_{12}$ Given constructible sets a and b, let α be an ordinal greater than the orders of a and b. Then a, b are both in L_α. The sets $a \cap b$, $a \cup b$, and $a - b$ are respectively defined over L_α by the formulas $x \in a \wedge x \in b$, $x \in a \vee x \in b$, and $x \in a \wedge \neg(x \in b)$, hence are in $L_{\alpha+1}$, hence are constructible. As for $\{a, b\}$, since L_α is transitive, then $\{a, b\}$ is defined over L_α by the formula $x = a \vee x = b$ (by **P$_2$**) and hence is constructible. Also $\{a\}$ is constructible (since $\{a\} = \{a, a\}$), hence using (4), $\{\{a\}, \{a, b\}\}$ is constructible, and so $\langle a, b \rangle$ is constructible.

C$_{13}$ This follows from (5) of **C$_{12}$** by an obvious mathematical induction. (We recall that $\langle x_1, \ldots, x_n, x_{n+1} \rangle =_{df} \langle\langle x_1, \ldots, x_n\rangle, x_{n+1}\rangle$.)

EXERCISES

Exercise 1.1. Prove that for *finite* α, $L_\alpha = R_\alpha$.

Exercise 1.2. What is the smallest ordinal α such that $L_\alpha \neq R_\alpha$? For this α, does R_α have higher cardinality than L_α?

§2 Absoluteness

We now turn to a crucial notion due to Gödel that will play a major role in this and subsequent chapters.

Consider a class K and a sentence X whose constants (if there are any) are all in K. To say that X is true *over* K means that X is true if we interpret "$(\forall x)$" to mean "for all x in K" and "$(\exists x)$" as "for some x in K" (x is any bound variable of X). To say that X is true over V (or that X is *really* true) is to say that X is true if we interpret "$(\forall x)$" as "for all x in V" and "$(\exists x)$" as "for some x in V." Now, it may or may not be that the truth value of X over K is the same as its real truth value (its truth value over V); if it is, then we say that X is *absolute* over K.

Definition 2.1. Consider a formula $\varphi(x_1, \ldots, x_n)$ with no constants. We say that the formula φ is absolute over K if for any elements $a_1, \ldots, a_n$ of K, the sentence $\varphi(a_1, \ldots, a_n)$ is absolute over K. We shall say that a class (or a property, or a relation) is absolute over K if it is defined over K by at least one formula that is absolute over K. Finally, we shall say that a formula or class is *absolute* if it is absolute over every *transitive* class K.

We will eventually show that the very property of *constructibility* is absolute over L. To do this we first need to establish, of many classes and relations, that they are absolute. Here are some of them.

§2. ABSOLUTENESS

(1) $x \in y$

(2) $x \subseteq y$

(3) $x = y$

(4) $x \in (y \cup z)$

(5) $x = (y \cup z)$

(6) $x = (y \cap z)$

(7) x is empty

(8) $x \cap y$ is empty

(9) $z = \{x, y\}$

(10) $y = \{x\}$

(11) $y = x^+$

(12) $\emptyset \in x$

(13) $y^+ \in x$

(14) $y = \cup x$

(15) x is transitive

(16) x is $\in$-connected

(17) x is an ordinal

(18) x is a successor ordinal

(19) x is a limit ordinal

(20) x is a natural number, $\mathsf{Num}(x)$

(21) For each natural number n, the conditions $x \in n$, $x = n$, $n \in x$

(22) $\{x, y\} \in z$

(23) $y \in \{x\}$, $y = \{x\}$, $\{x\} \in y$

(24) $z \in \langle x, y \rangle$, $z = \langle x, y \rangle$, $\langle x, y \rangle \in z$

(25) For each $n \geq 2$, the relations: $z \in \langle x_1, \ldots, x_n \rangle$, $z = \langle x_1, \ldots, x_n \rangle$, $\langle x_1, \ldots, x_n \rangle \in z$

(26) x is an unordered pair

(27) x is an ordered pair

(28) x is a binary relation

(29) x is a function, $\mathsf{Fun}(x)$

(30) y is the domain of x $(y = \mathsf{Dom}(x))^2$

(31) y is the range of x $(y = \mathsf{Ran}(x))^2$

(32) f is a function from x onto y

(33) f is a 1-1 function from x onto y

(34) f is a function from x *into* y

(35) $z \subseteq x \times y$, $x \times y \subseteq z$, $z = x \times y$

Now, one can directly verify that the items above are all absolute, but much labor is saved by the following considerations.

Δ_0 **formulas** For a formula φ and variables x, y, we abbreviate $(\exists x)(x \in y \wedge \varphi)$ by $(\exists x \in y)\varphi$; we also use $(\forall x \in y)\varphi$ as an abbreviation of $(\forall x)(x \in y \supset \varphi)$.

By a Δ_0-*formula*[3] is meant any formula formed by the rules:

(1) Any atomic formula $x \in y$ is Δ_0.

---

[2] For any class X, whether a relation or not, it is technically convenient to define $\mathsf{Dom}(X)$ as the class of all x such that $\langle x, y \rangle \in X$ for some y, and $\mathsf{Ran}(X)$ as the class of all y such that $\langle x, y \rangle \in X$ for some x. It may be that X contains no ordered pairs at all, in which case $\mathsf{Dom}(X)$ and $\mathsf{Ran}(X)$ are of course both empty.

[3] Δ_0 formulas are also commonly called Σ_0 formulas.

(2) If φ, ψ are Δ_0, so are $\neg\varphi$, $\varphi \wedge \psi$ (and hence also $\varphi \vee \psi, \varphi \supset \psi, \varphi \equiv \psi$).

(3) If φ is Δ_0, then for any distinct variables x, y, $(\exists x \in y)\varphi$ and $(\forall x \in y)\varphi$ are Δ_0.

(Roughly speaking, in a Δ_0-formula, we do not have any unrestricted quantifiers $(\exists x)$, $(\forall x)$, but only restricted quantifiers $(\exists x \in y)$, $(\forall x \in y)$.)

A class or relation is called Δ_0 if it is definable over V by a Δ_0-formula.

We will first show that every Δ_0-class and relation is absolute over any transitive class, and we will then show that each of the thirty-five items above is in fact Δ_0. (The point is that although they are not normally defined by Δ_0-formulas, they can be *equivalently* defined by Δ_0-formulas, and this is due to the transitivity of V!)

Theorem 2.2. *Every Δ_0-class and relation is absolute (over any transitive class K).*

Proof We are to show that any Δ_0-formula is absolute over any transitive class K. We do this by induction on the degree of the formula.

Obviously any atomic formula $x \in y$ is absolute over any class K (whether transitive or not). And if φ, ψ are formulas that are absolute over K, then $\neg\varphi$, $\varphi \wedge \psi$ are also absolute over K (whether K is transitive or not).

Now consider a formula $\varphi(x, y, x_1, \ldots, x_n)$ (whose free variables are all in the set $\{x, y, x_1, \ldots, x_n\}$) that is absolute over K. We must show that if K is transitive, then the formula $(\exists x \in y)\varphi(x, y, x_1, \ldots, x_n)$ is absolute over K. Well, take any elements $a, b_1, \ldots, b_n$ in K; we must show that the sentence $(\exists x \in a)\varphi(x, a, b_1, \ldots, b_n)$ is true over K if and only if it is true over V. Now, suppose the sentence $(\exists x)(x \in a \wedge \varphi(x, a, b_1, \ldots, b_n))$ (which is the sentence above) is true over K. Then for some $k \in K$, the sentence $k \in a \wedge \varphi(k, a, b_1, \ldots, b_n)$ is true over K. Then k is an element of a, and also $\varphi(k, a, b_1, \ldots, b_n)$ is true over V (since $\varphi(x, y, x_1, \ldots, x_n)$ is absolute over K). Thus the sentence $k \in a \wedge \varphi(k, a, b_1, \ldots, b_n)$ is true over V, and since k is of course in V, the sentence $(\exists x)(x \in a \wedge \varphi(x, a, b_1, \ldots, b_n))$ is true over V.

Conversely, suppose the sentence $(\exists x)(x \in a \wedge \varphi(x, a, b_1, \ldots, b_n))$ is true over V. Then for some element b of V, the sentence $b \in a \wedge \varphi(b, a, b_1, \ldots, b_n)$ is true over V. Then $b \in a$ is true over V, and so b is an element of a. But since $a \in K$ and K is transitive, b is in K. Thus for some b in K, the sentence $b \in a \wedge \varphi(b, a, b_1, \ldots, b_n)$ is true over V, hence also true over K (by the absoluteness of $\varphi(x, y, x_1, \ldots, x_n)$ over K, and the absoluteness of $x \in y$), and hence the sentence $(\exists x)(x \in a \wedge \varphi(x, a, b_1, \ldots, b_n))$ is true over K. This concludes the proof. ∎

Now we shall verify that each of the thirty-five conditions above is Δ_0 (and hence absolute by Theorem 2.2).

(1) The formula $x \in y$ is already Δ_0.

(2) The formula $x \subseteq y$ is $(\forall z)(z \in x \supset z \in y)$, which is $(\forall z \in x)(z \in y)$, hence is Δ_0.

(3) $x = y$ iff $(x \subseteq y) \wedge (y \subseteq x)$.

(4) $x \in (y \cup z)$ iff $x \in y \vee x \in z$.

§2. ABSOLUTENESS 161

(5) $x = y \cup z$ is normally defined as $(\forall w)(w \in x \equiv (w \in y \vee w \in z))$, and this formula is not Δ_0, but it can be replaced by the formula: $(\forall w \in x)(w \in y \vee w \in z) \wedge (\forall w \in y)(w \in x) \wedge (\forall w \in z)(w \in x)$, and this formula *is* Δ_0.

(6) $x = y \cap z$ iff $(\forall w \in x)(w \in y \wedge w \in z) \wedge (\forall w \in y)(w \in z \supset w \in x)$.

(7) $x = \emptyset$ iff $\neg(\exists w \in x)(w = w)$.

(8) $x \cap y = \emptyset$ iff $\neg(\exists w \in x)(w \in y)$.

(9) $z = \{x, y\}$ iff $x \in z \wedge y \in z \wedge (\forall w \in z)(w = x \vee w = y)$. (This replaces the usual definition: $(\forall w)(w \in z \equiv (w = x \vee w = y))$, which is not a Δ_0-formula.)

(10) $y = \{x\}$ iff $y = \{x, x\}$.

(11) $y = x^+$ iff $x \in y \wedge x \subseteq y \wedge (\forall z \in y)(z \in x \vee z = x)$.

(12) $\emptyset \in x$ iff $(\exists w \in x)(w = \emptyset)$.

(13) $y^+ \in x$ iff $(\exists w \in x)(w = y^+)$.

(14) $y = \cup x$ iff $(\forall z \in y)(\exists w \in x)(z \in w) \wedge (\forall z \in x)(\forall w \in z)(w \in y)$.

(15) We already know that the condition "x is transitive" is absolute over any transitive class, but it is even Δ_0: x is transitive iff $(\forall y \in x)(\forall z \in y)(z \in x)$.

(16) x is $\in$-connected iff $(\forall y \in x)(\forall z \in x)(y \in z \vee z \in y \vee z = y)$.

(17) We recall from Chapter 5 that a set is an ordinal if and only if it is both transitive and $\in$-connected (this is Raphael Robinson's theorem—we assume in this and subsequent chapters that V is well founded). And so as a Δ_0-definition of the property of being an ordinal, simply let $\mathsf{Ordinal}(x)$ be the conjunction of the Δ_0 formulas saying x is transitive and x is $\in$-connected.

We now pause in our list and show the fact that the class of ordinals is absolute has the following consequence:

C$_{14}$ Every ordinal α is constructible, and of order α.

Proof To begin with, it is not possible that $\alpha \in L_\alpha$, because $L_\alpha \subseteq R_\alpha$, and we already know that $\alpha \notin R_\alpha$. So it remains to show that $\alpha \in L_{\alpha+1}$. We do this by transfinite induction on the class of ordinals. We assume that for every $\beta < \alpha$, $\beta \in L_{\beta+1}$ and we show it then follows that $\alpha \in L_{\alpha+1}$.

So suppose for every $\beta < \alpha$, $\beta \in L_{\beta+1}$. Then for every $\beta < \alpha$, $\beta \in L_\alpha$ (because if $\beta < \alpha$, then $\beta + 1 \leq \alpha$, hence $L_{\beta+1} \subseteq L_\alpha$, and so $\beta \in L_\alpha$). Thus $\alpha \subseteq L_\alpha$. Also $\alpha \notin L_\alpha$, and since L_α is transitive, no ordinal greater than α can be in L_α. So all ordinals in L_α are less than α, and also all ordinals $< \alpha$ are members of α, hence members of L_α (since $\alpha \subseteq L_\alpha$). Therefore α *is* the set of all ordinals in L_α.

Now, L_α is transitive and the formula $\mathsf{Ordinal}(x)$ is absolute, hence it defines over L_α the set of all ordinals in L_α (because for any element a of L_α, $\mathsf{Ordinal}(a)$ is true over L_α if and only if Ordinal(a) is true over V—in other words, if and only if a is really an ordinal!). But this set *is* the ordinal α, and so α is a *definable* subset of L_α, hence is in $L_{\alpha+1}$. This completes the proof. ∎

Now let us return to our list.

(18) x is a successor ordinal iff
Ordinal(x) $\wedge$ ($\exists y \in x$)($x = y^+$).

(19) x is a limit ordinal iff Ordinal(x) $\wedge$ $\neg(x = 0)$ $\wedge$ $\neg(x$ is a successor ordinal).

(20) x is a natural number, Num(x), iff
Ordinal(x) $\wedge$ ($\forall y \in x$)$\neg(y$ is a limit ordinal) $\wedge$ $\neg(x$ is a limit ordinal).

(21) We show by induction on n that the conditions $x \in n$ and $x = n$ are Δ_0. Obviously $x \in 0$ iff $\neg(x = x)$, and $x = 0$ iff $\neg(\exists y \in x)(y = y)$. Now suppose n is such that the properties $x \in n$ and $x = n$ are Δ_0. Then $x \in n^+$ iff $x \in n \vee x = n$, and $x = n^+$ iff ($\exists y \in x$)($y = n \wedge x = y^+$). This completes the induction. As for the condition $n \in x$, it is equivalent to ($\exists y \in x$)($y = n$).

(22) $\{x, y\} \in z$ iff ($\exists w \in z$)($w = \{x, y\}$).

(23) $y \in \{x\}$ iff $y = x$. $y = \{x\}$ iff $y = \{x, x\}$. $\{x\} \in y$ iff ($\exists w \in y$)($w = \{x\}$).

(24) $z \in \langle x, y \rangle$ iff $z = \{x\} \vee z = \{x, y\}$. $z = \langle x, y \rangle$ iff ($\forall w \in z$)($w \in \langle x, y \rangle$) $\wedge$ $\{x\} \in z \wedge \{x, y\} \in z$. $\langle x, y \rangle \in z$ iff ($\exists w \in z$)($w = \langle x, y \rangle$).

(25) We defined $\langle x, y, z \rangle$ to be $\langle \langle x, y \rangle, z \rangle$, $\langle x, y, z, w \rangle$ to be $\langle \langle \langle x, y \rangle, z \rangle, w \rangle$, etc. Thus for each n, $\langle x_1, \ldots, x_n, x_{n+1} \rangle = \langle \langle x_1, \ldots, x_n \rangle, x_{n+1} \rangle$, and so $\langle x_1, \ldots, x_n, x_{n+1} \rangle = \{\{\langle x_1, \ldots, x_n \rangle\}, \{\langle x_1, \ldots, x_n \rangle, x_{n+1}\}\}$.

The proof is by induction on $n \geq 2$. By the previous item, the relations $y \in \langle x_1, x_2 \rangle$ and $y = \langle x_1, x_2 \rangle$ are Δ_0. Now suppose $n \geq 2$ and all the relations $y \in \langle x_1, \ldots, x_n \rangle$ and $y = \langle x_1, \ldots, x_n \rangle$ are Δ_0; we show the same for $n + 1$.

Since the relation $y = \langle x_1, \ldots x_n \rangle$ is Δ_0, by the induction hypothesis, it follows that $y = \{\langle x_1, \ldots, x_n \rangle\}$ and $y = \{\langle x_1, \ldots, x_n \rangle, x_{n+1}\}$ are Δ_0, for the former holds iff ($\exists z \in y$)($z = \langle x_1, \ldots, x_n \rangle \wedge y = \{z\}$), and the latter holds iff ($\exists z \in y$)($z = \langle x_1, \ldots, x_n \rangle \wedge y = \{z, x_{n+1}\}$).

The relation $y \in \langle x_1, \ldots, x_n, x_{n+1} \rangle$ is Δ_0 because it holds iff $y = \{\langle x_1, \ldots, x_n \rangle\} \vee y = \{\langle x_1, \ldots, x_n \rangle, x_{n+1}\}$.

Next, $y = \langle x_1, \ldots, x_n, x_{n+1} \rangle$ iff ($\exists z_1 \in y$)($\exists z_2 \in y$)($z_1 = \{\langle x_1, \ldots, x_n \rangle\} \wedge z_2 = \{\langle x_1, \ldots, x_n \rangle, x_{n+1}\} \wedge y = \{z_1, z_2\}$), and this is Δ_0. This completes the induction.

As for the relation $\langle x_1, \ldots, x_n \rangle \in y$, it holds iff ($\exists z \in y$)($z = \langle x_1, \ldots, x_n \rangle$).

(26) x is an unordered pair iff ($\exists y \in x$)($\exists z \in x$)($x = \{y, z\}$).

(27) x is an ordered pair iff ($\exists y \in x$)($\exists x_1 \in y$)($\exists x_2 \in y$)($x = \langle x_1, x_2 \rangle$).

(28) x is a relation iff $(\forall y \in x)(y$ is an ordered pair).

(29) x is a function ($\mathsf{Fun}(x)$) iff x is a relation $\wedge\ [(\forall z \in x)(\forall w \in z)(\forall x_1 \in w)(\forall x_2 \in w)(\forall x_3 \in w)((\langle x_1, x_2 \rangle \in x \wedge \langle x_1, x_3 \rangle \in x) \supset x_2 = x_3)]$.

(30) Let us first verify that the relation $y \in \mathsf{Dom}(x)$ (y is in the domain of x) is Δ_0. Well, $y \in \mathsf{Dom}(x)$ iff $(\exists z \in x)(\exists w \in z)(\exists v \in w)(z = \langle y, v \rangle)$. Then $y = \mathsf{Dom}(x)$ iff $(\forall z \in y)(z \in \mathsf{Dom}(x)) \wedge (\forall z_1 \subset x)(\forall z_2 \subset z_1)(\forall z_3 \in z_2)(z_3 \in \mathsf{Dom}(x) \supset z_3 \in y)$.

(31) Range is treated similarly to domain, and is left to the reader.

(32) f is a function from x onto y iff $\mathsf{Fun}(f) \wedge \mathsf{Dom}(f) = x \wedge \mathsf{Ran}(f) = y$.

(33) f is a 1-1 function from x onto y iff the following two Δ_0 conditions hold:

  (a) f is a function from x onto y;

  (b) $(\forall x_1 \in x)(\forall x_2 \in x)(\forall y_1 \in y)[(\langle x_1, y_1 \rangle \in f \wedge \langle x_2, y_1 \rangle \in f) \supset x_1 = x_2]$.

(34) f is a function from x *into* y iff the following two Δ_0 conditions hold:

  (a) $\mathsf{Fun}(f) \wedge x = \mathsf{Dom}(f)$;

  (b) $(\forall z \in f)(\forall z_1 \in z)(\forall z_2 \in z_1)(z_2 \in \mathsf{Ran}(f) \supset z_2 \in y)$.

(35)

  (a) $z \subseteq x \times y$ iff $(\forall z_1 \in z)(\exists x_1 \in x)(\exists y_1 \in y)(z_1 = \langle x_1, y_1 \rangle)$.

  (b) $(x \times y) \subseteq z$ iff $(\forall x_1 \in x)(\forall y_1 \in y)(\langle x_1, y_1 \rangle \in z)$.

  (c) $z = x \times y$ iff $z \subseteq x \times y \wedge x \times y \subseteq z$.

§3 Constructible classes

Definition 3.1. Following Gödel, we will say that a *class A* is a *constructible class* if the following two conditions are satisfied:

(1) Every element of A is a constructible set.

(2) The intersection of A with any constructible set is a constructible set.

Constructible classes will play a significant role in the next chapter, and so we will now establish some of their basic properties.

Proposition 3.2.

(1) Any constructible set is also a constructible class.

(2) Any constructible class which is a set is a constructible set.

Thus for any set a, a is a constructible set iff a is a constructible class. (This is *not* double-talk, as it might superficially appear to be!)

Proof

(1) Suppose a is a constructible set. Then every element of a is a constructible set (by $\mathbf{C}_{10}$) and also the intersection of a with any constructible set is a constructible set (by (1) of $\mathbf{C}_{12}$). Hence a satisfies the definition of being a constructible class.

(2) Suppose a satisfies the definition of being a constructible class and also a is a set. Then all elements of a are constructible sets (by condition (1) of Definition 3.1), hence a is a subset of a constructible set b (by $\mathbf{C}_{11}$). Then by condition (2) of Definition 3.1, $a \cap b$ is a constructible set. But $a \cap b = a$ (since $a \subseteq b$), and so a is a constructible set. ∎

The next proposition will have a neat application later on.

Proposition 3.3. *Let A be a class of constructible sets. Then the following two conditions are equivalent:*

(1) A is a constructible class.

(2) Every subset b of A is a subset of some constructible subset b' of A.

Proof Assume all elements of A are constructible sets.

(a) Suppose that A is a constructible class. Let b be any subset of A. By $\mathbf{C}_{11}$, b is a subset of some constructible set d. Since A is a constructible class, then $A \cap d$ is a constructible set. Since $b \subseteq d$ and $b \subseteq A$, then $b \subseteq A \cap d$. And so we can take b' to be $A \cap d$. Thus b is a subset of b', and b' is a constructible subset of A.

(b) Suppose, conversely, that every subset b of A is a subset of some constructible subset b' of A. We show that A is a constructible class. Since all elements of A are constructible (by hypothesis) it remains to show that the intersection of A with any constructible set is a constructible set. Well, let c be any constructible set and let $b = A \cap c$. We are to show that b is constructible. By the present hypothesis, b is a subset of some constructible subset b' of A. Since b' is constructible, then $b' \cap c$ is constructible. But $b' \cap c = b$ (because $b' \cap c \subseteq A \cap c$, since $b' \subseteq A$, hence $b' \cap c \subseteq b$. Also $b \subseteq b' \cap c$, since b is a subset of both b' and c. Thus $b' \cap c = b$). Hence b is constructible. ∎

We next need the following proposition.

Proposition 3.4. *If A, B are constructible classes, so are $A \cap B$ and $A - B$.*

Proof Suppose A, B are constructible classes.

(a) To show that $A \cap B$ is a constructible class, let c be any constructible set. We must show that $(A \cap B) \cap c$ is a constructible set. Well, $(A \cap B) \cap c = A \cap (B \cap c)$. Since B is a constructible class, $B \cap c$ is a constructible set, and therefore $A \cap (B \cap c)$ is a constructible set (since A is a constructible class).

(b) To show that $A - B$ is a constructible class, we must show that for any constructible set c, $(A - B) \cap c$ is a constructible set, so let c be a constructible set. Now

$(A - B) \cap c = (A \cap c) - (B \cap c)$. The sets $A \cap c$ and $B \cap c$ are constructible sets (since A, B are constructible classes), hence $(A \cap c) - (B \cap c)$ is a constructible set (by (3) of C_{12}). ∎

Range and domain We recall that for any class A (whether a binary relation or not) by the domain of A is meant the class of all x such that for at least one y, the ordered pair $\langle x, y \rangle$ is in A, and by the range of A we mean the class of all y such that for some x, $\langle x, y \rangle \in A$.

We now wish to show that for any constructible class A, the domain of A and the range of A are constructible classes. This is a bit tricky; the proof is based on the following lemma (which generalizes a clever idea of Gödel, used in his monograph).

Lemma 3.5. *Let Q be a function from constructible sets to constructible sets such that for any constructible set x, the set $Q''(x)$ is constructible. Then for any constructible class A, the class $Q''(A)$ is a constructible class.*

Proof We shall assume the axiom of choice for this proof (though it is not actually needed, since L can be well-ordered, as we shall later see).

Suppose Q satisfies the hypothesis of the theorem. Let A be any constructible class. We are to show that $Q''(A)$ is a constructible class. By Proposition 3.3 it suffices to show that any subset b of $Q''(A)$ is a subset of some constructible subset b' of $Q''(A)$. So let b be any subset of $Q''(A)$.

Each element y of b is $Q(y')$ for at least one element y' of A. Well, for each $y \in b$, "pick" one such element y' and let a be the set of these elements y' (just one y' for each y). (a is certainly a set, since it is in 1-1 correspondence with b.) By Proposition 3.3, a is a subset of some constructible subset a' of A. By the hypothesis concerning Q, $Q''(a')$ is a constructible set. Since $a \subseteq a'$, then $Q''(a) \subseteq Q''(a')$, and so b is a subset of $Q''(a')$. And, of course, $Q''(a')$ is a subset of $Q''(A)$, and so b is a subset of the constructible subset $Q''(a')$ of $Q''(A)$. Then by Proposition 3.3, $Q''(A)$ is a constructible class. ∎

A generalization of lemma 3.5 This lemma has an interesting generalization which was proved in Gödel's monograph (Gödel 1940): Consider a binary relation R and a class A. By $R''(A)$ is meant the class of all y such that xRy holds for at least one $x \in A$. (If R happens to be a function Q, then $R''(A)$ is the same class as $Q''(A)$.) Here is the generalization of Lemma 3.5.

Lemma 3.5* Suppose that R is a relation between constructible sets ($R \subseteq L \times L$) such that for any constructible set x, the class $R''(x)$ is a constructible class. (Even though x is a set, $R''(x)$ might be a proper class!) Then for any constructible class A, $R''(A)$ is a constructible class.

Proof Suppose R satisfies the hypothesis. Let A be a constructible class and let b be a subset of $R''(A)$. Again we must show that b is a subset of some constructible subset b' of $R''(A)$.

For each y in b there is at least one $y' \in A$ such that $y'Ry$; again we pick one such y' for each $y \in b$ and let a be the set of these elements y'. Again, a is a set, because although a is not necessarily in 1-1 correspondence with b, if we let $\varphi(y) = y'$, for each $y \in b$, then φ maps b (possibly many-one) onto a. Also, a is a subset of some constructible subset a' of A (by Proposition 3.3), but now we do not know that $R''(a')$ is a set! However, $R''(a')$ is a constructible class (by the hypothesis concerning R), and so $R''(a')$ is a constructible subclass of $R''(A)$—and, of course, b is a subset of $R''(a')$ (since $b = R''(a)$ and $a \subseteq a'$). We now apply Proposition 3.3 to the constructible class $R''(a')$ and so b is a subset of some constructible subset b' of $R''(a')$. Of course, b' is also a subset of $R''(A)$, which completes the proof. ∎

It is Lemma 3.5 that now concerns us (rather than the more general Lemma 3.5*). To apply this to our domain and range problem, it will be useful to first consider the following lemma, which is of interest in its own right.

Lemma 3.6. *Let Q be a function from constructible sets to constructible sets such that the relation $Q(x) = y$ is absolute. Then for any constructible set c, the set $Q''(c)$ is constructible.*

Proof Assume the hypothesis. Suppose c is a constructible set. Let α be an ordinal such that $c \in L_\alpha$. Since the relation $Q(x) = y$ is absolute, then it is defined by some absolute formula $\varphi(x, y)$. Now, $Q''(c)$ is the set of all elements a such that for some element $b \in c$, $a = Q(b)$, hence $Q''(c)$ is the set of all a such that the sentence $(\exists y)(y \in c \land \varphi(y, a))$ is true over V. Thus $Q''(c)$ is defined over V by the formula $(\exists y)(y \in c \land \varphi(y, x))$. This same formula defines $Q''(c)$ over L_α (as is easily seen, using the facts that L_α is transitive, hence $\varphi(x, y)$ is absolute over L_α, and $c \in L_\alpha$, hence all elements of c are in L_α).

Since $Q''(c)$ is definable over L_α, then $Q''(c) \in L_{\alpha+1}$, and so $Q''(c)$ is a constructible set. ∎

Now we can prove the following.

Proposition 3.7. *The range and domain of any constructible class A are constructible classes.*

Proof For any ordered pair $\langle a, b \rangle$ of constructible sets, let $Q(\langle a, b \rangle) = a$. (If x is not an ordered pair, we let $Q(x)$ be undefined.) The relation $Q(x) = y$ is Δ_0 (because $Q(x) = y$ iff $(\exists c \in x)(\exists a \in c)(\exists b \in c)(x = \langle a, b \rangle \land y = a))$, hence is absolute. By the lemma above, for any constructible set c the set $Q''(c)$ is constructible. Then by Lemma 3.5, for any constructible class A, the class $Q''(A)$ is a constructible class, but $Q''(A)$ is the domain of A. Thus for any constructible class A, its domain is a constructible class.

The proof for "range" is the same, only taking $Q(\langle a, b \rangle) = b$, instead of a. ∎

More on constructible classes In preparation for the next chapter we need to establish some further properties of constructible classes.

We recall that we are taking $\langle x_1, x_2, x_3 \rangle$ to be $\langle \langle x_1, x_2 \rangle, x_3 \rangle$, and $\langle x_1, x_2, x_3, x_4 \rangle$ to be $\langle \langle x_1, x_2, x_3 \rangle, x_4 \rangle$, ..., $\langle x_1, \ldots, x_n, x_{n+1} \rangle$ to be $\langle \langle x_1, \ldots, x_n \rangle, x_{n+1} \rangle$. By an n-ary relation R on a class K is meant a class of n-tuples of elements of A. We let K^n be the class of all n-tuples of elements of K. (Thus $K^1 = K$, $K^2 = K \times K$, $K^3 = (K \times K) \times K$, and so forth.) A unary (1-ary) relation on K is simply a subclass of K.

§3. CONSTRUCTIBLE CLASSES

Let us note that if A is a relation on K of $(n + 1)$ arguments (i.e., an $(n + 1)$-ary relation on K), the *domain* of A is simply the class of all n-tuples $\langle a_1, \ldots, a_n \rangle$ such that for at least one element $a \in K$, the $(n + 1)$-tuple $\langle a_1, \ldots, a_n, a \rangle$ is in A. If R is an n-ary relation, we often write $R(a_1, \ldots, a_n)$ for $\langle a_1, \ldots, a_n \rangle \in R$.

We now introduce the following definitions: Let K be a class fixed for the discussion. For any positive integer n and any positive integers i, j both less than or equal to n, by $K_{i,j}^n$ we shall mean the class of all n-tuples $\langle a_1, \ldots, a_n \rangle$ of elements of K such that $a_i \in a_j$. Also, for any element a of K, by $K_{i,\bar{a}}^n$ we shall mean the class of all n-tuples $\langle a_1, \ldots, a_n \rangle$ of elements of K such that $a_i \in a$. And by $K_{\bar{a},i}^n$ we shall mean the class of all n-tuples $\langle a_1, \ldots, a_n \rangle$ of elements of K such that $a \in a_i$. Finally, for any elements a, b of K, by $K_{\bar{a},\bar{b}}^n$ we shall mean the class of all n-tuples $\langle a_1, \ldots, a_n \rangle$ of elements of K such that $a \in b$. Note that if $a \in b$ is true, then $K_{\bar{a},\bar{b}}^n$ is the class K^n, whereas if $a \in b$ is false, then $K_{\bar{a},\bar{b}}^n$ is simply the empty set $\emptyset$. Thus $K_{\bar{a},\bar{b}}^n$ is either K^n or $\emptyset$.

In the next chapter we will need the following proposition.

Proposition 3.8. *For any positive integer n, any $i, j \leq n$, and any constructible set a, the classes L^n, $L_{i,j}^n$, $L_{i,\bar{a}}^n$ are constructible classes.*

Proof

(1) We show that L^n is a constructible class (i.e., that for any constructible set c, the set $L^n \cap c$ is constructible). (We already know that any n-tuple of constructible sets is constructible, hence L^n is certainly a class of constructible sets.)

Well, let c be any constructible set. Let α be an ordinal such that $c \in L_\alpha$. Then the formula $x \in c \wedge (\exists x_1 \in c)(\exists x_2 \in c) \ldots (\exists x_n \in c)(x = \langle x_1, \ldots, x_n \rangle)$ defines $L^n \cap c$ over L_α (as can be seen from the fact that if $x = \langle x_1, \ldots, x_n \rangle \in c$, then $x_1, \ldots, x_n$ must have lower orders than c, and hence must all be in L_α).

(2) For $L_{i,j}^n$ the proof is the same, but using the formula
$x \in c \wedge (\exists x_1 \in c)(\exists x_2 \in c) \ldots (\exists x_n \in c)(x = \langle x_1, \ldots, x_n \rangle \wedge x_i \in x_j)$.

(3) For $L_{i,\bar{a}}^n$, almost the same proof, but we must now take an α such that c, a are both in L_α, and then use the formula above, replacing the part "$x_i \in x_j$" by "$x_i \in a$." ∎

EXERCISES

Exercise 3.1. Prove that $L_{\bar{a},i}^n$ and $L_{\bar{a},\bar{b}}^n$ are constructible classes, where a and b are constructible sets.

CHAPTER 13

L IS A WELL FOUNDED FIRST-ORDER UNIVERSE

§1 First-order universes

We have defined a class K to be *swelled* if K contains with each element x, all subsets of x as well. We shall now define K to be *first-order swelled* if K contains with every element x, all subsets of x which are *first-order definable over K*.

We now define K to be a *first-order* Zermelo-Fraenkel universe—more briefly a *first-order universe*—if the following conditions hold:

F_1 K is transitive;

* F_2 K is first-order swelled;

F_3 $\emptyset \in K$;

F_4 For every x, y in K, $\{x, y\} \in K$;

F_5 For every x in K, $\cup x \in K$;

* F_6 For every x in K, $\mathcal{P}(x) \cap K \in K$;

F_7 $\omega \in K$;

* F_8 For every function F from elements of K to elements of K, if F is *first-order definable over K*, then for every x in K, $F''(x)$ is in K.

In addition, we say K is a *well founded* first-order Zermelo-Fraenkel universe or a *well founded first-order universe* if, in addition to the conditions above, we also have a version of what we earlier called Axiom D.

F_9 Every non-empty member of K contains an initial element.

We have put an asterisk before those conditions which differ from the corresponding conditions defining a "genuine" Zermelo-Fraenkel universe. Let us now discuss some of those differences.

We have already discussed F_2; let us turn to F_6: Condition A_6, of the definition of a Zermelo-Fraenkel universe (the power set axiom) says that for every $x \in V$, the set $\mathcal{P}(x)$ of all subsets of x is also in V. But condition F_6 says merely that for any $x \in K$, the set of all subsets of x which happen to be in K is an element of K. This difference is quite drastic. A genuine Zermelo-Fraenkel universe (or even a genuine Zermelo universe) contains the non-denumerable set $\mathcal{P}(\omega)$. But if all we know about

K is that it is a first-order universe, then it may be that K contains only denumerably many subsets of ω, and hence $\mathcal{P}(\omega) \cap K$ may be only a denumerable set. Indeed, as we will see, it is possible that a first-order universe may have only denumerably many elements.

We might note that if K is genuinely swelled, then F_6 is equivalent to the genuine power set axiom A_6, because then $\mathcal{P}(x) \cap K = \mathcal{P}(x)$.

Condition F_8 is the axiom of substitution restricted to those functions F that are first-order definable over K.

Why are the first-order universes of any importance? The answer is that given any transitive class K, a necessary and sufficient condition that K be a first-order universe is that all the formal axioms of Zermelo-Fraenkel set theory are true over K. From now on we often refer to these as the ZF axioms—here they are (they are *formulas*):

Ax 1 [Extensionality] $(\forall z)(z \in x \equiv z \in y) \supset x = y$.[1]

Ax 2 [Separation] This is an axiom schema—an infinite collection of axioms—one for each formula $\varphi(x, y_1, \ldots, y_n)$ (n may be 0). The separation (or Aussonderungs) axioms are all formulas of the following form: $(\forall x)(\exists z)(\forall y)(y \in z \equiv (y \in x \wedge \varphi(x, y_1, \ldots, y_n)))$.[2]

Ax 3 [Axiom of the empty set] $(\exists x)\neg(\exists y)(y \in x)$.

Ax 4 [Axiom of unordered pairs] $(\forall x_1)(\forall x_2)(\exists y)(\forall z)(z \in y \equiv (z = x_1 \vee z = x_2))$.

Ax 5 [Union axiom] $(\forall x)(\exists y)(\forall z)(z \in y \equiv (\exists w)(z \in w \wedge w \in x))$.

Ax 6 [Power axiom] $(\forall x)(\exists y)(\forall z)(z \in y \equiv z \subseteq x)$.

Ax 7 [Axiom of infinity] (In the form, "there exists an inductive set")[3]

$$(\exists w)(\emptyset \in w \wedge$$
$$(\forall x)(x \in w \supset (\exists z)(z \in w \wedge (\forall v)(v \in z \equiv v \in x \vee v = x)))).$$

Ax 8 [Axioms of substitution] All formulas of the form:

$$(\forall x)(\forall y)(\forall z)(\varphi(x, y) \wedge \varphi(x, z) \supset y = z) \supset$$
$$(\exists v)(\forall y)(y \in v \equiv (\exists x)(x \in w \wedge \varphi(x, y))).$$

In addition to the ZF axioms, we generally will also want to consider the following.

---

[1] We are taking "$x = y$" as an abbreviation of "$(\forall z)(z \in x \equiv z \in y)$." Thus $x = y$ means that x and y are members of the same sets, and so the axiom of extensionality asserts that if x and y contain the same members, then they are members of the same sets. (Some formalizations of Zermelo-Fraenkel set theory take identity as undefined, and then provide axioms which imply that $x = y$ is *equivalent* to $(\forall z)(x \in z \equiv y \in z)$.)

[2] Actually it is not necessary to assume these axioms since they can be derived from the axioms A_8, as explained in §3, Chapter 6.

[3] "$\emptyset \in w$" is simply an abbreviation of "$(\exists y)(y \in w \wedge \neg(\exists z)(z \in y))$".

§1. FIRST-ORDER UNIVERSES

Ax 9 [Axiom of well foundedness]

$$(\forall a)\{(\exists y)(y \in a) \supset (\exists y)[y \in a \land (\forall z)(z \in a \supset \neg(z \in y))]\}.$$

It is a routine matter to verify that for a transitive class K, K is a first-order universe iff the ZF axioms (the Zermelo-Fraenkel axioms) are all true over K. Likewise K is a well founded first-order universe iff the ZF axioms and the well foundedness axiom are all true over K. (For an open formula $\varphi(x_1, \ldots, x_n)$, with free variables $x_1, \ldots, x_n$, when we say that φ is true over K we mean that the sentence $(\forall x_1) \cdots (\forall x_n)\varphi(x_1, \ldots, x_n)$ is true over K.) In particular, the separation axioms **Ax 2** are all true over K iff K is first-order swelled. This is a consequence of the lemma below.

The condition F_2 is the condition of being first-order swelled. Let F'_2 be the following condition: the intersection of any first-order definable subclass A of K with any element b of K is an element of K.

Lemma 1.1. *If K is transitive, the conditions F_2 and F'_2 are equivalent.*

Proof

(1) Suppose K is transitive and first-order swelled. Let A be any definable subclass of K and let b be any element of K. Since K is transitive, then b is definable over K (by the formula $x \in b$), and hence $A \cap b$ is definable over K (it is defined by $\varphi(x) \land x \in b$, where $\varphi(x)$ is any formula that defines A over K). Then, by the assumption that K is swelled, $A \cap b$ is an element of K.

(2) The reverse direction is quite trivial (and doesn't even require the transitivity of K). Suppose F'_2 holds. Now suppose $b \in K$ and A is any subset of b that is definable over K. Then by F'_2, $A \cap b$ is an element of K, but $A \cap b = A$. ∎

Now, what the separation axioms (**Ax 2**) collectively say (when interpreted over K) is not that K is first-order swelled, but that K has property F'_2 (Exercise 1.1). But, by the lemma above, this is equivalent to K being first-order swelled.

Let us remark that although the separation axioms force K to be first-order swelled, they do not force K to be genuinely swelled. Indeed no set of first-order axioms can possibly force K to be genuinely swelled since the very notion of being swelled involves talking about sets which might be outside K and asserting that they are in fact inside K, but first-order formulas (interpreted over K) can only talk about elements of K, since the quantifier "$\forall x$" means "for all x in K" and "$\exists x$" means "for some x in K."

We aim in this chapter to show that the class L of constructible sets is a well founded first-order universe. Whether it is a genuine Zermelo-Fraenkel universe, no one knows. If all sets are constructible, then of course it is (since then $L = V$). But no one knows whether or not all sets are constructible. We will later prove that it is *consistent* to assume all sets are constructible (provided the ZF axioms are themselves consistent). Further, if all sets are constructible, then the axiom of choice and the continuum hypothesis are both true (and this fact can be proved from the ZF axioms). From this it will follow that if the ZF axioms are consistent, then they remain consistent under the adjunction of the axiom of choice and the continuum hypothesis.

The consistency of the statement that all sets are constructible will be established by showing that the property of constructibility is itself absolute, and hence (since L is transitive) that for any constructible set a, the statement that a is constructible is not only true over V, but true over L, and therefore the statement that all sets are constructible is true *in* L. But this is jumping ahead—let us now turn to the project of showing that L is a first-order universe.

EXERCISES

Exercise 1.1.

(a) Prove the assertion made above—that K has property F'_2 iff all the separation axioms are true over K (this does not require transitivity of K).

(b) Prove that if K is a transitive class, then K is a first-order universe iff all the axioms of ZF (Zermelo-Fraenkel set theory) are true over K.

Exercise 1.2. Establish the following relative consistency result: If the ZF axioms are consistent, they remain consistent if the axiom of well foundedness is added, and consequently the axiom of well foundedness cannot be *disproved* in ZF. Hint: see Theorem 4.3 of Chapter 7, and Theorem 3.5 of Chapter 8.

§2 Some preliminary theorems about first-order universes

Let us momentarily forget about the particular class L and establish some general sufficient conditions for a transitive class K to be a first-order universe. We will then show that the conditions do hold for L.

The following theorem may be a bit surprising (but of course it must be understood that K is a subclass of the universe V already given).

Theorem 2.1. *A sufficient condition for a transitive class K to be a first-order universe is that every subset of K that is definable over K is an element of K.*

Proof Assume the hypothesis. We will verify that each of the conditions F_1–F_8 hold.

F_1 We are given that K is transitive.

F_2 Since every definable subset of K is an element of K, then of course for any $x \in K$, all subsets of x that are definable over K are elements of K, and so K is first-order swelled.

F_3 Obviously $\emptyset$ is definable over K, hence $\emptyset \in K$.

F_4 For any a, b in K, the set $\{a, b\}$ is defined over K by the formula $x = a \vee x = b$ (which is absolute over K, since K is transitive), hence $\{a, b\} \in K$.

F_5 For any $a \in K$, the set $\cup a$ is defined over K by the Δ_0 formula $(\exists y \in a)(x \in y)$.

F_6 For any $a \in K$, the set $\mathcal{P}(a) \cap K$ is defined over K by the Δ_0 formula $(\forall y \in x)(y \in a)$.

§2. SOME PRELIMINARY THEOREMS ABOUT FIRST-ORDER UNIVERSES

F$_7$ We already know that $\emptyset \in K$, that is, $0 \in K$. For any $a \in K$, the element a^+ is in K, because a^+ is defined over K by the formula $x \in a \lor x = a$. It then follows by mathematical induction that all the natural numbers are in K. Now, the class ω is defined over V by the Δ_0 formula Num(x), which is absolute (see item (20) of Chapter 12, §2). Since K is transitive and all the natural numbers are in K, then Num(x) defines ω over K. Since ω is definable over K and is a set, $\omega \in K$.

F$_8$ Suppose F is a function from elements of K to elements of K such that F is definable over K; let $\varphi(x, y)$ be a formula that defines over K the relation $F(x) = y$. We assert that for any element $a \in K$, the set $F''(a)$ is definable over K by the formula $(\exists y \in a)\varphi(y, x)$. Thus we must show that for any element b of K, $b \in F''(a)$ iff the sentence $(\exists y)(y \in a \land \varphi(y, b))$ is true over K.

If the sentence $(\exists y)(y \in a \land \varphi(y, b))$ is true over K, then it is obvious that $b \in F''(a)$ (the proof of this, which we leave to the reader, doesn't require that K be transitive). Now, conversely, suppose that $b \in F''(a)$. Then for some $c \in a$, $F(c) = b$. But since $a \in K$ and $c \in a$ and K is transitive, c must be in K. Thus for some c in K the sentence $c \in a \land \varphi(c, b)$ is true over K, which means that the sentence $(\exists y)(y \in a \land \varphi(y, b))$ is true over K. This completes the proof. ∎

In application to the class L, Theorem 2.1 means that to show L is a first-order universe, it suffices to show that every set of constructible sets that is definable over L is a constructible set. We shall prove this by two entirely different methods—the first (which is both more elementary and more elaborate) will establish the stronger result that every definable subclass of L is a constructible class. But first we turn to some more general facts about first-order universes. Theorem 2.1 has the following simple corollary.

Corollary 2.2. *Suppose that K is a transitive class satisfying the following two conditions: (1) K is first-order swelled; (2) every subset of K is a subset of some element of K. Then K is a first-order universe.*

Proof Assume the hypothesis. Suppose that a is a subset of K that is first-order definable over K; we wish to show that $a \in K$. Well, a is a subset of some element b of K, by (2). Thus a is a subset of b and is first-order definable over K, so by (1), $a \in K$. Thus the hypothesis of Theorem 2.1 is satisfied, and so K is a first-order universe. ∎

Distinguished subclasses By a distinguished subclass of a class K we shall mean a subclass of K such that its intersection with any element of K is an element of K. (This generalizes the notion of a constructible class, as it was defined in the last chapter. A constructible class is simply a distinguished subclass of L.) Lemma 1.1, restated is this: If K is transitive, then K is first-order swelled if and only if every first-order definable subclass of K is distinguished. This, together with the corollary above, at once yields the following.

Theorem 2.3. *The following two conditions are jointly sufficient for a transitive class K to be a first-order universe.*

(1) Every definable subclass of K is distinguished.

(2) Every subset of K is a subset of some element of K.

§3 More on first-order universes

We now turn to a deeper theorem about first-order universes (Theorem 3.1 below) which, with certain propositions proved in the last chapter, will yield our desired result that L is a first-order universe. The interesting thing about the theorem is that it makes no reference to the notion of first-order definability.

The reader should recall the definitions of K^n, $K^n_{i,j}$, $K^n_{i,\bar{a}}$, $K^n_{\bar{a},j}$, $K^n_{\bar{a},\bar{b}}$ from the previous Chapter 12, §3. Now, here is the theorem which, we note again, is stated without reference to the notions of "formula" or "first-order definability."

Theorem 3.1. *The following conditions are jointly sufficient for a transitive class K to be a first-order universe:*

(1) For any positive integer n, any positive integers $i, j \leq n$, and any element $a \in K$, the classes K^n, $K^n_{i,j}$, $K^n_{i,\bar{a}}$ are distinguished subclasses of K.

(2) For any distinguished subclasses A, B of K, the classes $A - B$ and $\text{Dom}(A)$ are distinguished.

(3) Every subset of K is a subset of an element of K.

To prove Theorem 3.1 we need the following basic lemma.

Lemma 3.2. *From the first two hypotheses of Theorem 3.1 (and without using the fact that K is transitive) it follows that all definable subclasses of K are distinguished.*

We pause to remark on a curious fact about the proof of the lemma: it nowhere utilizes the definition of "distinguished." Indeed, we could take a perfectly arbitrary property of classes and call those classes having the property *special* classes, and if the two hypotheses hold (reading "special" for "distinguished") it would follow that all first-order definable subclasses of L are special.

To prove Lemma 3.2 we must first say more about formulas. Given a formula φ, a sequence $(x_1, \ldots, x_n)$ of distinct variables will be said to *cover* φ if all free variables of φ (and possibly others) are in the set $\{x_1, \ldots, x_n\}$. Now, given a sequence $(x_1, \ldots, x_n)$ of distinct variables that covers φ, we shall say that $\langle \varphi, x_1, \ldots, x_n \rangle$ *defines* (over K) the class of all n-tuples $\langle a_1, \ldots, a_n \rangle$ of elements of K such that the sentence $\varphi(a_1, \ldots, a_n)$ is true over K. (By $\varphi(a_1, \ldots, a_n)$ is meant the result of substituting each a_i for all free occurrences of x_i in φ.)

What we now aim to show is that for any formula φ and any sequence $(x_1, \ldots, x_n)$ that covers φ, the class defined by $\langle \varphi, x_1, \ldots, x_n \rangle$ is a distinguished subclass of K (under the first two hypotheses of Theorem 3.1 of course). Actually, we need not consider *all* formulas, but only those built up from atomic formulas of one of the three forms $x_i \in x_j$, $x_i \in a$, $a \in b$ (a, b being constants of K), because an atomic formula of the form $a \in x_i$ can be rewritten as $(\exists y \in x_i)(y = a)$. And so we will consider only those formulas having the property that no atomic subformula of it is of the form $a \in x_i$—such formulas as we now consider will temporarily be called *relevant* formulas.

Proof of Lemma 3.2 Assume hypotheses (1) and (2) hold. Let us call a (relevant) formula φ *good* if for every sequence $(x_1, \ldots, x_n)$ that covers φ, the relation defined by $\langle \varphi, x_1, \ldots, x_n \rangle$ is a distinguished subclass of K. We are to show that every (relevant) formula φ is good. We do this by mathematical induction on the structure of φ.

Atomic case Suppose φ is a (relevant) atomic formula. Then it is of one of the forms $x_i \in x_j$, $x_i \in a$, $a \in b$. Let $(x_1, \ldots, x_n)$ be any sequence that covers φ. If φ is of the first of the three forms, then $\langle \varphi, x_1, \ldots, x_n \rangle$ defines the class $K^n_{i,j}$, which is distinguished, by hypothesis (1). If φ is of the second form, then $\langle \varphi, x_1, \ldots, x_n \rangle$ defines $K^n_{i,a}$, which is also distinguished by (1). Now suppose φ is of the third form $a \in b$. Then $\langle \varphi, x_1, \ldots, x_n \rangle$ defines either K^n or $\emptyset$, depending on whether $a \in b$ is true or not. If the former, K^n is distinguished by (1). If the latter, $\emptyset$ is distinguished because $\emptyset = K - K$ (and K, which is K^1, is distinguished by (1), hence $K - K$ is distinguished by (2)). This proves that every (relevant) atomic formula is good.

Negation If $\langle \varphi, x_1, \ldots, x_n \rangle$ defines R, then $\langle \neg \varphi, x_1, \ldots, x_n \rangle$ defines $K^n - R$, and if R is distinguished, so is $K^n - R$ (by (1) and (2)). Thus it easily follows that if φ is good, so is $\neg \varphi$.

Conjunction If $\langle \varphi_1, x_1, \ldots, x_n \rangle$ defines R_1 and $\langle \varphi_2, x_1, \ldots, x_n \rangle$ defines R_2, then $\langle \varphi_1 \wedge \varphi_2, x_1, \ldots, x_n \rangle$ defines $R_1 \cap R_2$, and if R_1 and R_2 are both distinguished, so is $R_1 \cap R_2$ (by (2), since $R_1 \cap R_2 = R_1 - (R_1 - R_2)$). Thus it follows that the conjunction of two good formulas is good.

Existential quantification Suppose $\varphi = (\exists y)\psi$ and ψ is good; we must show that φ is good. We assume that y is actually a free variable of ψ, because if it isn't, then φ is logically equivalent to ψ, and there is nothing to prove.

Let $(x_1, \ldots, x_n)$ be a sequence that covers φ. We can assume without loss of generality that no bound variable of φ is in the set $\{x_1, \ldots, x_n\}$ (for otherwise we can simply rewrite bound variables and work with a logically equivalent formula φ', none of whose bound variables occur in the set $\{x_1, \ldots, x_n\}$). In particular, we assume that y (which is a bound variable of φ) doesn't occur in the set $\{x_1, \ldots, x_n\}$. The sequence $(x_1, \ldots, x_n, y)$ obviously covers ψ, and since ψ is good (by the inductive assumption) then $\langle \psi, x_1, \ldots, x_n, y \rangle$ defines a distinguished relation $R(x_1, \ldots, x_n, y)$. Then $\langle (\exists y)\psi, x_1, \ldots, x_n \rangle$ defines the domain of R, which is distinguished by (2). Thus $\langle \varphi, x_1, \ldots, x_n \rangle$ defines a distinguished relation, hence φ is good.

This completes the induction and proves that for any formula φ and any sequence $(x_1, \ldots, x_n)$ that covers φ, $\langle \varphi, x_1, \ldots, x_n \rangle$ defines a distinguished subclass of K. In particular, if φ is a formula with just one free variable x, then $\langle \varphi, x \rangle$ defines a distinguished subclass A of K, but this A is the class defined by φ. Thus all definable subclasses of K are distinguished. ∎

Theorem 3.1 is now immediate from Lemma 3.2 and Theorem 2.3 (because, according to Lemma 3.2, the hypotheses (1) and (2) of Theorem 3.1 imply that all definable subclasses of K are distinguished, which is the first hypothesis of Theorem 2.3, and hypothesis (3) of Theorem 3.1 is the second hypothesis of Theorem 2.3). Now we can prove the following.

Theorem 3.3. *L is a first-order universe.*

Proof We know that L is transitive. We know by Proposition 3.8 of Chapter 12 that the classes L^n, $L^n_{i,j}$, $L^n_{i,\bar{a}}$ ($i, j \le n$) are constructible classes (distinguished subclasses of L), and so hypothesis (1) of Theorem 3.1 is satisfied. Hypothesis (2) is satisfied by virtue of Propositions 3.4 and 3.7 of Chapter 12. And hypothesis (3) is satisfied by C_{11}, §1 of Chapter 12. Then by Theorem 3.1, L is a first-order universe. ∎

Corollary 3.4. *L is a well founded first-order universe.*

Proof All that remains is to show the axiom of well foundedness is true over L. That is, we must show any non-empty member of L has an initial element, one having no members in common with a. But this is easy. If a is a non-empty member of L, let y be a member of a of lowest order. If z were a member of $y \cap a$, z would be a member of y, hence of lower order than y, but z would also be a member of a and this is impossible, since y was a member of a of lowest order. Consequently $y \cap a$ is empty, so y is an initial member of a. ∎

A second proof of Theorem 3.3 By using the Montague-Levy reflection principle (Theorem 5.3, Chapter 11), together with Theorem 2.1, we have a far simpler proof that L is a first-order universe. This proof doesn't involve the notion of constructible classes, but only that of constructible sets. Here is the proof.

By Theorem 2.1, we have only to verify that every definable subset of L is a member of L. Well, suppose a is a definable subset of L; let $\varphi(x, a_1, \ldots, a_n)$ be a formula, whose constants $a_1, \ldots, a_n$ are in L, that defines a over L. Let α_0 be an ordinal greater than the orders of all elements of a, so that $a \subseteq L_{\alpha_0}$. Let $\alpha_1, \ldots, \alpha_n$ be the respective orders of $a_1, \ldots, a_n$. Let α be the largest of the ordinals $\alpha_0, \alpha_1, \ldots, \alpha_n$.

We can now use the Montague-Levy reflection principle for the sequence $L_0, L_1, \ldots, L_\alpha, \ldots$. By this principle there is a limit ordinal $\beta > \alpha$ (hence greater than the ordinals $\alpha_0, \alpha_1, \ldots, \alpha_n$) such that L_β reflects L with respect to the formula $\varphi(x, y_1, \ldots, y_n)$. Now, all of the constants $a_1, \ldots, a_n$ are in L_β and the set a is a subset of L_β. Then, since L_β reflects L with respect to $\varphi(x, y_1, \ldots, y_n)$ it follows that $\varphi(x, a_1, \ldots, a_n)$ defines a over L_β. Reason: a is the set of all elements b of L such that $\varphi(b, a_1, \ldots, a_n)$ is true over L. But for any $b \in L_\beta$, $\varphi(b, a_1, \ldots, a_n)$ is true over L iff it is true over L_β (by reflection), and so a is the set of all $b \in L_\beta$ such that $\varphi(b, a_1, \ldots, a_n)$ is true over L_β—and thus a is the set defined by $\varphi(x, a_1, \ldots, a_n)$ over L_β.

Since a is definable over L_β, then $a \in L_{\beta+1}$, and so a is constructible, i.e., $a \in L$. ∎

§4. ANOTHER RESULT

Discussion We have now given two very different proofs that L is a first-order universe. The first (using constructible classes) is a variant of the proof in Gödel's monograph (Gödel 1940). I do not know who is mainly responsible for the second proof, using the Montague-Levy reflection principle. Actually this reflection principle is necessary only to establish F_7 (the first-order substitution axioms), because F_1–F_5 all follow easily from results of the last chapter, and F_6 (the relativized power set axiom) can be established by the following simple argument.

For K being the class L, F_6 says that for any constructible set a, the set b of all *constructible* subsets of a is a constructible set. Well, let α be any ordinal greater than the orders of all the elements of b and greater than the order of a. Then $a \in L_\alpha$ and b is a subset of L_α. Then b is defined over L_α by the (absolute) formula $x \subseteq a$, thus $b \in L_{\alpha+1}$ so b is constructible.

Now a word concerning the hypotheses of Theorem 3.1 (and also of Theorem 2.3) that says any subset of K is a subset of some element of K. (This implies, by the way, that K must be a proper class (assuming V well founded), because if K were a set, we would then have $K \subseteq x$ for some element x in K, hence we would have $x \in x$.) Now, this condition automatically holds if K (like V and L) is the union of an ordinal hierarchy $K_0, K_1, \ldots, K_\alpha, \ldots$ of *elements* of K, for then by the axiom of substitution, any subset of K is a subset of some K_α (see condition O_4, Theorem 2.1, Chapter 7).

§4 Another result

We shall have later use for the following theorem.

Theorem 4.1. *Let K (a subclass of V) be a (transitive) first-order universe such that every subset of K is a subset of an element of K. Then K contains all ordinals (of V).*

Proof Assume the hypothesis. We will first show that no element of K can contain all the ordinals in K.

We let Ordinal(x) be the Δ_0 formula of Chapter 12 that defines the class of all ordinals. This formula is absolute (over all transitive classes). We have proved that no set can contain all ordinals (since the class of ordinals is not a set), and this fact can be formally proved from the axioms of Zermelo-Fraenkel set theory, i.e., the sentence $(\forall x)(\exists y)(\text{Ordinal}(y) \wedge \neg(y \in x))$ is a theorem of ZF, hence is true over all first-order universes, and so is true over K. Now let k be any element of K. It follows that the sentence $(\exists y)(\text{Ordinal}(y) \wedge \neg(y \in k))$ is true over K, hence there exists an element a in K such that the sentence Ordinal(a) $\wedge \neg(a \in k)$ is true over K. Since Ordinal(a) is true over K and the formula Ordinal(x) is absolute, then a is really an ordinal. And since $\neg(a \in k)$ is true over K, then a is not an element of k. Thus some ordinal a in K is not an element of k. Thus no element k of K contains all ordinals of K.

Let K_0 be the class of ordinals of K. We have just proved that K_0 is not a subclass of any element of K, but every sub*set* of K *is* a subset of some element of K (by hypothesis), from which it follows that K_0 is not a set; it is a proper class. Also K_0 is transitive (because for any ordinal α in K, every element of α is both an ordinal and a member of K, hence is in K_0). Thus K_0 is a transitive class of ordinals, and K_0 is not a set, hence K_0 is the class On of all the ordinals of V. ∎

CHAPTER 14

CONSTRUCTIBILITY IS ABSOLUTE OVER L

In this chapter we prove, among other things, that the property of being a constructible set is absolute over L—i.e., that there is a formula "Const(x)" such that: (1) For any set a, Const(a) is true over V iff a is constructible; (2) For any $a \in L$, the sentence Const(a) is true, not only over V, but over L!

(1) can be more simply stated that Const(x) defines L over V. (2) can be equivalently stated that Const(x) defines L over L.

From this it will follow that the sentence $(\forall x)$Const(x) (which may or may not be true over V) *is* true over L. This sentence is called *the axiom of constructibility*. From the fact that constructibility is absolute over L will follow that it is impossible to prove the existence of a non-constructible set from the axioms of ZF, even with the axiom of well foundedness (unless ZF is inconsistent). In the next chapter we will prove that the axiom of constructibility implies the continuum hypothesis! Moreover, this fact is provable in ZF. From this, and the relative consistency of the axiom of constructibility, will follow the relative consistency of the continuum hypothesis.

Along the way to proving constructibility is absolute over L, we will prove that the relation $y = L_\alpha$ (as a relation between y and α) is absolute (over *all* transitive classes!). This fact will have a key application in the next chapter. The most involved part of this is the proof that the relation $y = \mathcal{F}(x)$ is absolute (over all transitive classes). This relation, incidentally, is not Δ_0, but it is nevertheless absolute, as we will see. To prove this, we first introduce the notions of Σ-formulas, *upward absoluteness*, and *downward absoluteness* and prove several results concerning them in Part A of this chapter. In Part B we prove the absoluteness of the relation $y = \mathcal{F}(x)$, and in Part C we prove the absoluteness of constructibility over L. We conclude this chapter (Part D) with a proof that L can be well ordered (without using the axiom of choice, of course).

Part A—Upward absoluteness and Σ-formulas

§1 Σ-formulas and upward absoluteness

We will say that a formula $\varphi(x_1, \ldots, x_n)$ with no constants is *absolute upwards* over a class K if for any elements $a_1, \ldots, a_n$ of K, if $\varphi(a_1, \ldots, a_n)$ is true over K, then it is true over V. We will also say that φ is absolute *downwards* over K if, for any $a_1, \ldots, a_n$ in K, if $\varphi(a_1, \ldots, a_n)$ is true over V, then it is true over K. Thus φ is absolute over K iff it is both upwards and downwards absolute over K. We shall say that a *relation* $R(x_1, \ldots, x_n)$ is absolute upwards (downwards) over K if it is defined by at least one formula that is absolute upwards (downwards) over K. And we shall say that a formula or relation is *absolute upwards* if it is absolute upwards over all *transitive* classes (and similarly with *downwards*). We will now be largely concerned with upward absoluteness. (Upward absolute formulas and relations are sometimes called *stable*.)

Proposition 1.1.

(1) Every Δ_0-formula is absolute upwards.

(2) If φ, ψ are absolute upwards, so are $\varphi \wedge \psi$ and $\varphi \vee \psi$.

(3) If φ is absolute upwards, so is $(\exists x)\varphi$.

(4) If φ is absolute upwards, so are $(\forall x \in y)\varphi$ and $(\exists x \in y)\varphi$.

Proof

(1) Since Δ_0-formulas are absolute, then they are of course absolute upwards.

(2) Obvious.

(3) Easy exercise.

(4) Suppose $\varphi(x, y, z_1, \ldots, z_n)$ is absolute upwards; we wish to show the same for $(\forall x \in y)\varphi(x, y, z_1, \ldots, z_n)$. So take any transitive class K and any elements $a, b_1, \ldots, b_n$ of K such that $(\forall x \in a)\varphi(x, a, b_1, \ldots, b_n)$ is true over K. Then for any element c of K, the sentence $c \in a \supset \varphi(c, a, b_1, \ldots, b_n)$ is true over K. Also for any element c in a, c is also in K (since K is transitive), hence $c \in a$ is a true sentence of K, hence $\varphi(c, a, b_1, \ldots, b_n)$ is true over K, and hence true over V (since $\varphi(x, y, z_1, \ldots, z_n)$ is absolute upwards). This proves that for every c in a, $\varphi(c, a, b_1, \ldots, b_n)$ is true over V, hence $(\forall x \in a)\varphi(x, a, b_1, \ldots, b_n)$ is true over V. We have shown that the formula with a bounded quantifier, $(\forall x \in y)\varphi(x, y, z_1, \ldots, z_n)$, is absolute upwards.

As to the formula $(\exists x \in y)\varphi$, this is $(\exists x)(x \in y \wedge \varphi)$. Of course $x \in y$ is absolute upwards (it is Δ_0), then if φ is absolute upwards, so is $x \in y \wedge \varphi$ (by (2)), hence also $(\exists x)(x \in y \wedge \varphi)$ (by (3)). ∎

Σ-formulas We inductively define the class of Σ-formulas by the following rules.

(1) Every Δ_0 formula is a Σ-formula.

(2) If φ and ψ are Σ-formulas, so are $\varphi \wedge \psi$ and $\varphi \vee \psi$.

(3) If φ is a Σ-formula, so is $(\exists x)\varphi$.

(4) If φ is a Σ-formula, so are $(\forall x \in y)\varphi$ and $(\exists x \in y)\varphi$.

By a Σ-*relation* we mean a relation that is defined over V by a Σ-formula.

By induction on the structure of Σ-formulas, the proposition above at once yields the following.

Theorem 1.2. *All Σ-formulas (and hence all Σ-relations) are absolute upwards.*

Proposition 1.3.

(1) φ is absolute upwards iff $\neg\varphi$ is absolute downwards; φ is absolute downwards iff $\neg\varphi$ is absolute upwards.

(2) If φ and $\neg\varphi$ are both absolute upwards, then φ is absolute.

Proof (1) and (2) are obvious. ∎

Definition 1.4. Let us say a formula $\varphi(x, y)$ is *function-like* over a transitive class K if $(\forall x)(\exists y)(\forall z)[\varphi(x, z) \equiv (z = y)]$ is true over K. Equivalently we could use the formula $(\forall x)(\exists y)[\varphi(x, y) \land (\forall z)(\varphi(x, z) \supset (z = y))]$.

Theorem 1.5. *Suppose $\varphi(x, y)$ is a Σ formula that is function-like over both the transitive class K and over V. Then $\varphi(x, y)$ is absolute over K.*

Proof Let $a, b \in K$. If $\varphi(a, b)$ is true over K, since φ is Σ, $\varphi(a, b)$ is true over V because Σ formulas are absolute upwards. Now suppose $\varphi(a, b)$ is true over V. Since $a \in K$ and φ is function-like over K, there must be some $c \in K$ such that $\varphi(a, c)$ is true over K. Since φ is absolute upward, $\varphi(a, c)$ is true over V, but so is $\varphi(a, b)$, so since φ is function-like over V, $b = c$ is true over V. Thus b and c are the same set, hence $\varphi(a, b)$ is true over K. ∎

EXERCISES

Exercise 1.1. Suppose K and K' are classes, such that K is transitive, $K \subseteq K'$, $a_1, \ldots, a_n$ are elements of K, and $\varphi(x_1, \ldots, x_n)$ is a formula whose constants are all in K.

(a) Show that if $\varphi(x_1, \ldots, x_n)$ is Δ_0, then $\varphi(a_1, \ldots, a_n)$ is true over K if and only if it is true over K'. (This generalizes the fact that Δ_0-formulas are absolute, which is the special case $K' = V$.)

(b) Show that if $\varphi(x_1, \ldots, x_n)$ is Σ, then if $\varphi(a_1, \ldots, a_n)$ is true over K, it is true over K'. (This generalizes the fact that Σ-formulas are absolute upwards. This generalization has an application in the next chapter.)

§2 More on Σ definability

Given a function $f(x)$ and a relation $R(x, y, z_1, \ldots, z_n)$ we write $(\forall x \in f(y))R(x, y, z_1, \ldots, z_n)$ to mean $(\forall x)(x \in f(y) \supset R(x, y, z_1, \ldots, z_n))$.

Proposition 2.1. *If the relations $f(x) = y$ and $R(x, y, z_1, \ldots, z_n)$ are both Σ, so is $(\forall x \in f(y))R(x, y, z_1, \ldots, z_n)$.*

Proof The relation $(\forall x \in f(y))R(x, y, z_1, \ldots, z_n)$ is the relation $(\exists w)(f(y) = w \land (\forall x \in w)R(x, y, z_1, \ldots, z_n))$. ∎

Clearly related to this is the following.

Proposition 2.2. *Suppose $R(x, y_1, \ldots, y_n)$ is Σ and a is a set such that the property $x = a$ is Σ (i.e., the set $\{a\}$ is Σ). Then the following relations are Σ:*

(1) $R(a, y_1, \ldots, y_n)$ (as a relation among $y_1, \ldots, y_n$);

(2) $(\forall x \in a)R(x, y_1, \ldots, y_n)$.

Proof
(1) This can be written: $(\exists x)(x = a \land R(x, y_1, \ldots, y_n))$.
(2) This can be written: $(\exists z)(z = a \land (\forall x \in z)R(x, y_1, \ldots, y_n))$. ∎

Proposition 2.3. *Composition of Σ-functions is Σ—i.e., if relations $g_1(x_1, \ldots, x_n) = y$, $\ldots$, $g_k(x_1, \ldots, x_n) = y$ are Σ and if $f(x_1, \ldots, x_k) = y$ is Σ, then the following relation is also Σ: $f(g_1(x_1, \ldots, x_n), \ldots, g_k(x_1, \ldots, x_n)) = y$.*

Proof This relation can be written:

$$(\exists z_1) \cdots (\exists z_k)(g_1(x_1, \ldots, x_n) = z_1 \land \cdots \land g_k(x_1, \ldots, x_n) = z_k \land f(z_1, \ldots, z_k) = y)$$

∎

Proposition 2.4. *The relation $w = (x \times y) \times z$ is Σ.*

Proof Since the function $x \times y$ is Σ (in fact, Δ_0—see item 35, §2, Chapter 12) and since composition of Σ-functions is Σ, then the function $(x \times y) \times z$ is Σ. (Alternatively, the relation $w = (x \times y) \times z$ can be written $(\exists v)(x \times y = v \land v \times z = w)$. But this is simply a special case of the previous proposition.) ∎

Proposition 2.5. *For any sets a and b, if the properties $x = a$, $x = b$ are Σ, so is the property $x = a \cup b$. (This is not to be confused with the statement that the union of two Σ-sets is Σ!)*

Proof $x = (a \cup b)$ can be written $(\exists y)(\exists z)(y = a \land z = b \land x = y \cup z)$. ∎

Proposition 2.6. *The relation $y = f''(x)$ (as a relation among y, f, x, where f is a set) is Σ (in fact, Δ_0).*

Proof $y = f''(x)$ is the conjunction of the following two Δ_0 conditions:

(1) $(\forall z \in x)(f(z) \in y)$. (Note: $f(z) \in y$ can be written as $(\exists w \in y)(f(z) = w)$.)

(2) $(\forall z \in y)(\exists w \in x)(z = f(w))$. □

Proposition 2.7. *The relation $y = \cup f''(x)$ (as a relation among y, f, x) is Σ.*

Proof This can be written $(\exists z)(z = f''(x) \land y = \cup z)$. ∎

The two propositions above were stated mainly as lemmas for the following important result.

Theorem 2.8. *Let F be a function defined on V and let F^* be the function on On defined from F by the following transfinite recursion:*

(1) $F^*(0) = \emptyset$

(2) $F^*(\alpha + 1) = F(F^*(\alpha))$

(3) $F^*(\lambda) = \cup_{\alpha<\lambda} F^*(\alpha)$ (λ a limit ordinal).

Then if F is Σ, F^* is also Σ.

Proof Suppose F is Σ. Then the relation $F^*(\alpha) = x$ is also Σ, because it can be written:

$$\alpha \in On \land (\exists f)[\text{Fun}(f) \land \text{Dom}(f) = \alpha + 1 \land f(\alpha) = x$$
$$\land f(0) = \emptyset \land (\forall \beta \in \alpha + 1)(f(\beta + 1) = F(f \upharpoonright \beta))$$
$$\land (\forall \beta \in \alpha + 1)(\beta \text{ is a limit ordinal} \supset f(\beta) = \cup f''(\beta)]$$

It is a routine matter to verify that the condition above is Σ, though a few remarks may help.

(1) Writing ($\forall \beta \in \alpha + 1$) is O.K. by Proposition 2.1, since the function assigning $\alpha + 1$ to α is Σ.

(2) The relation $y = \cup f''(\beta)$ is Σ, by Proposition 2.7, and $f(\beta) = \cup f''(\beta)$ can be written $(\exists y)(y = f(\beta) \land y = \cup f''(\beta))$, hence it is Σ.

(3) (β is a limit ordinal) $\supset (\ldots)$ can also be written: (β is not a limit ordinal) $\lor (\ldots)$. The property (x is a limit ordinal) is Δ_0, so the property (x is not a limit ordinal) is also Δ_0, hence Σ. Then if $(\ldots)$ is Σ, so is the condition (β is a limit ordinal) $\supset (\ldots)$. ∎

Let us remark right now that our main application of Theorem 2.8 is the case where F is the function $\mathcal{F}$ (that assigns to each set x the set of all subsets of x that are first-order definable over x). The reader should note that for any ordinal α, $\mathcal{F}^*(\alpha)$ is the set L_α. In the section that follows, we will show that the relation $y = \mathcal{F}(x)$ is Σ, hence it will follow by Theorem 2.8 that the relation $y = L_\alpha$ (as a relation between y and α) is also Σ (and hence absolute).

Part B—The Σ-definability of $y = \mathcal{F}(x)$

§3 The relation $y = \mathcal{F}(x)$

We now turn to the project of showing that the relation $y = \mathcal{F}(x)$ is Σ. To give all the details of the proof would be quite tedious, but we shall give the main outline. We must first turn to the coding of all formulas (with or without constants) into sets.

Coding of formulas We arrange our individual variables in some fixed, but arbitrary, infinite sequence $v_0, v_1, \ldots, v_n, \ldots$. By an *individual* we shall mean either a variable v_i or a set a. By the *code* $\ulcorner v_i \urcorner$ of v_i, we shall mean the ordered pair $\langle 0, i \rangle$. By the code $\ulcorner a \urcorner$ of a set a, we shall mean the ordered pair $\langle 1, a \rangle$.

We now assign to every formula φ a set $\ulcorner \varphi \urcorner$ called the *code* of φ. We do this by the following inductive scheme:

(1) If φ is an atomic formula $m_1 \in m_2$ (m_1, m_2 are both individuals) we take $\ulcorner \varphi \urcorner$ to be the ordered triple $\langle \ulcorner m_1 \urcorner, \ulcorner m_2 \urcorner, 0 \rangle$. (As examples, $\ulcorner v_i \in v_j \urcorner = \langle \langle 0, i \rangle, \langle 0, j \rangle, 0 \rangle$; $\ulcorner v_i \in a \urcorner = \langle \langle 0, i \rangle, \langle 1, a \rangle, 0 \rangle$.)

(2) $\ulcorner \neg \varphi \urcorner = \langle \ulcorner \varphi \urcorner, 1 \rangle$.

(3) $\ulcorner \varphi \land \psi \urcorner = \langle \ulcorner \varphi \urcorner, \ulcorner \psi \urcorner, 2 \rangle$.

(4) $\ulcorner(\exists v_i)\varphi\urcorner = \langle i, \ulcorner\varphi\urcorner, 3\rangle$.

We will sometimes write $\mathsf{Neg}(x)$ for $\langle x, 1\rangle$; $\mathsf{Con}(x, y)$ for $\langle x, y, 2\rangle$; $\mathsf{E}_i(x)$ for $\langle i, x, 3\rangle$. Thus if x, y are respective codes of φ, ψ, then $\mathsf{Neg}(x)$ is the code of $\neg\varphi$; $\mathsf{Con}(x, y)$ is the code of $\varphi \wedge \psi$; and $\mathsf{E}_i(x)$ is the code of $(\exists v_i)\varphi$.

For any set a, by $\Pi(a)$ we shall mean the set of all elements of a, together with all elements $\mathsf{Neg}(x)$ such that $x \in a$, together with all elements $\mathsf{Con}(x, y)$ such that x, y are both in a, together with all elements $\mathsf{E}_i(x)$ such that $x \in a$ and $i \in \omega$.

Lemma 3.1. *The function $\Pi(x)$ is Σ.*

Proof $\Pi(x) = x \cup (x \times \{1\}) \cup ((x \times x) \times \{2\}) \cup ((\omega \times x) \times \{3\})$. ∎

Definition 3.2. For any set a we let $c(a)$ be the set of codes of all the elements of a together with the codes of all variables.

For any set a and any $n \in \omega$ we define the set $\mathcal{E}_n^a$ by the following inductive scheme:

(1) $\mathcal{E}_0^a$ is the set of all (codes of) *atomic* formulas with constants in a.

(2) $\mathcal{E}_{n+1}^a = \Pi(\mathcal{E}_n^a)$.

We let $\mathcal{E}^a$ be the union of all the sets $\mathcal{E}_0^a$, $\mathcal{E}_1^a$, ..., $\mathcal{E}_n^a$, ... Thus $\mathcal{E}^a$ is the set of (codes of) *all* formulas with constants in a.

Proposition 3.3. *The following relations are all Σ:*

(1) $y = c(x)$

(2) $y = \mathcal{E}_0^a$ (as a relation between y and a)

(3) $y = \mathcal{E}_n^a$ (as a relation among y, a, n)

(4) $y \in \mathcal{E}^a$ (as a relation between y and a)

(5) $y = \mathcal{E}^a$ (as a relation between y and a).

Proof

(1) $c(a) = (\{0\} \times \omega) \cup (\{1\} \times a)$.

(2) $\mathcal{E}_0^a = ((c(a) \times c(a)) \times \{0\})$.

(3) We must now formalize the inductive scheme of defining $\mathcal{E}_{n+1}^a$ from $\mathcal{E}_n^a$. Well, $y = \mathcal{E}_n^a$ is equivalent to the following Σ-condition:

$$(\exists g)[\mathsf{Fun}(g) \wedge \mathsf{Dom}(g) = \omega \wedge g(n) = y \wedge g(0) = \mathcal{E}_0^a \wedge (\forall m \in \omega)(g(m+1) = \Pi(g(m)))].$$

(4) $y \in \mathcal{E}^a$ iff $(\exists n \in \omega)(y \in \mathcal{E}_n^a)$.

(5) $y = \mathcal{E}^a$ iff $(\forall z \in y)(z \in \mathcal{E}^a) \wedge (\forall n \in \omega)(\mathcal{E}_n^a \subseteq y)$. ∎

Types By the *type* $t(\varphi)$ of a formula φ (and also the type of its code) we shall mean the set of indices of its free variables. Thus:

(0)   (a) $t(x_i \in x_j) = \{i, j\}$
     (b) $t(x_i \in a) = \{i\}$
     (c) $t(a \in x_i) = \{i\}$
     (d) $t(a \in b) = \emptyset$

(1) $t(\neg \varphi) = t(\varphi)$

(2) $t(\varphi \wedge \psi) = t(\varphi) \cup t(\psi)$

(3) $t((\exists v_i)\varphi) = t(\varphi) - \{i\}$.

The four conditions of (0) can be collectively stated: $t(m_1 \in m_2) = \{\ulcorner m_1 \urcorner, \ulcorner m_2 \urcorner\} \cap \omega$, where m_1, m_2 are individuals.

Lemma 3.4. *The relation $z \in \mathcal{E}^a \wedge y = t(z)$ (as a relation among z, a, y) is Σ.*

Proof We let $R(f, a)$ be the conjunction of the following Σ-conditions:

(1) $\mathsf{Fun}(f) \wedge \mathsf{Dom}(f) = \mathcal{E}^a$

(2) $(\forall x \in c(a))(\forall y \in c(a))(f(\langle x, y, 0 \rangle) = \{x, y\} \cap \omega)$

(3) $(\forall x \in \mathcal{E}^a)(f(\mathsf{Neg}(x)) = f(x))$

(4) $(\forall x \in \mathcal{E}^a)(\forall y \in \mathcal{E}^a)(f(\mathsf{Con}(x, y)) = f(x) \cup f(y))$

(5) $(\forall x \in \mathcal{E}^a)(\forall i \in \omega)(f(\mathsf{E}_i(x)) = f(x) - \{i\})$.

(These conditions collectively say that f is a function on $\mathcal{E}^a$ that assigns to each $x \in \mathcal{E}^a$ its type.)

Then the relation $z \in \mathcal{E}^a \wedge y = t(z)$ can be written:

$$z \in \mathcal{E}^a \wedge (\exists f)(R(f, a) \wedge f(z) = y).$$

■

We shall use the term "*a*-formula" to mean formulas whose constants are all in a. For any set a and any finite set b of natural numbers, by $\mathcal{E}_b(a)$ we shall mean the set of codes of all *a*-formulas of type b. We let $\overline{\mathcal{E}}_b(a)$ be the set of codes of all *a* formulas *not* of type b.

Proposition 3.5. *The following relations (among y, a, b) are all Σ:*

(1) $y \in \mathcal{E}_b(a)$

(2) $y \in \overline{\mathcal{E}}_b(a)$

(3) $y = \mathcal{E}_b(a)$.

Proof
(1) $y \in \mathcal{E}_b(a)$ is equivalent to $y \in \mathcal{E}^a \wedge b = t(y)$. By the lemma above, this condition is Σ.
(2) $y \in \overline{\mathcal{E}}_b(a)$ is equivalent to $y \in \mathcal{E}^a \wedge (\exists z)(y \in \mathcal{E}_z(a) \wedge z \neq b)$.
(3) The relation $y \subseteq \mathcal{E}_b(a)$ is Σ—it can be written: $(\forall x \in y)(x \in \mathcal{E}_b(a))$. The relation $\mathcal{E}_b(a) \subseteq y$ is also Σ—it can be written: $(\forall x \in \mathcal{E}^a)(x \in \mathcal{E}_b(a) \supset x \in y)$. (Note: For $x \in \mathcal{E}^a$, the condition $x \in \mathcal{E}_b(a) \supset x \in y$ is equivalent to $x \in \overline{\mathcal{E}}_b(a) \vee x \in y$, which is Σ.) Then $y = \mathcal{E}_b(a)$ is Σ since it is equivalent to $y \subseteq \mathcal{E}_b(a) \wedge \mathcal{E}_b(a) \subseteq y$. ∎

A formula is called *monadic* if it has exactly one free variable — i.e., it is of type $\{i\}$ for some natural number i. A formula is called *closed*, or a *sentence* if it is of type $\emptyset$—i.e., has no free variables. By an *a-sentence* we mean a closed *a*-formula—i.e., a sentence whose constants are all in a.

We let $M(a)$ be the set of codes of monadic *a*-formulas and $S(a)$ be the set of codes of closed *a*-formulas (*a*-sentences).

Proposition 3.6. *The relations $y = M(a)$ and $y = S(a)$ (as relations between y, a) are* Σ.

Proof First, $y = M(a)$ iff $(\exists i \in \omega)(y = \mathcal{E}_{\{i\}}(a))$. Second, $y = S(a)$ iff $y = \mathcal{E}_\emptyset(a)$. ∎

Substitution For any individual m, variable v_i, and set b, by $Sub(\varphi, v_i, b)$ we mean the result of substituting b for all *free* occurrences of v_i in φ. This substitution operation is the one and only operation defined on all triples $\langle \varphi, v_i, b \rangle$ that satisfies the following conditions (for all individuals m_1, m_2, formulas φ, ψ, variables v_i, v_j, and sets b).

(0) $Sub(m_1 \in m_2, v_i, b) = Sub(m_1, v_i, b) \in Sub(m_2, v_i, b)$

(1) $Sub(\neg\varphi, v_i, b) = \neg Sub(\varphi, v_i, b)$

(2) $Sub(\varphi \wedge \psi, v_i, b) = Sub(\varphi, v_i, b) \wedge Sub(\psi, v_i, b)$

(3) (a) If $j \neq i$ then $Sub((\exists v_j)\varphi, v_i, b) = (\exists v_j)Sub(\varphi, v_i, b)$
    (b) But $Sub((\exists v_j)\varphi, v_j, b) = (\exists v_j)\varphi$.

We must now "codify" all this. We let Sub_a be the function with domain $(\mathcal{E}^a \times \omega) \times a$ such that for every $x \in \mathcal{E}^a$, $i \in \omega$, and $b \in a$, $Sub_a(x, i, b)$ is the code of the formula $Sub(\varphi, v_i, b)$, where φ is the formula whose code is x.

Proposition 3.7. *The relation* $Sub_a(x, i, y) = z$ *(understood as a 5-place relation among a, x, i, y, z) is* Σ.

Proof We first define $\Psi_a(x, i, y) = z$ to be the Σ-condition:

$$x \in c(a) \wedge i \in \omega \wedge y \in a \wedge [(\langle 0, i \rangle = x \wedge z = \langle 1, y \rangle) \vee (\langle 0, i \rangle \neq x \wedge z = x)]$$

(We note that for any individual m, either a variable or an element of a, and for any $y \in a$, and any $i \in \omega$, if x is the code of m, then $Sub_a(x, i, y)$ is the code of $Sub(m, v_i, y)$.)

§3. THE RELATION $y = \mathcal{F}(x)$

Now let us call f a *substitution function* for $\mathcal{E}^a$ if $\mathsf{Fun}(f) \wedge \mathsf{Dom}(f) = (\mathcal{E}^a \times \omega) \times a$ and for all m_1, m_2 in $c(a)$ and all x, y in $\mathcal{E}^a$ and all i, j in ω, and all $b \in a$, the following conditions hold:

(0) $f(\langle m_1, m_2, 0 \rangle, i, b) = \langle \Psi_a(m_1, i, b), \Psi_a(m_2, i, b), 0 \rangle$

(1) $f(\mathsf{Neg}(x), i, b) = \mathsf{Neg}(f(x, i, b))$

(2) $f(\mathsf{Con}(x, y), i, b) = \mathsf{Con}(f(x, i, b), f(y, i, b))$

(3) (a) If $j \neq i$ then $f(\mathsf{E}_j(x), i, b) = \mathsf{E}_j(f(x, i, b))$
　　(b) If $j = i$ then $f(\mathsf{E}_j(x), i, b) = \mathsf{E}_j(x)$.

It is a routine matter to verify that the relation "f is a substitution function for $\mathcal{E}^a$" is a Σ-relation between f and a.

Then $\mathsf{Sub}_a(x, i, y) = z$ iff the following holds:

$$x \in \mathcal{E}^a \wedge i \in \omega \wedge (\exists f)(f \text{ is a substitution function for } \mathcal{E}^a \wedge f(x, i, y) = z).$$

∎

We now define $\mathsf{Sub}(x, i, y) = z$ to mean $(\exists a)(x \in \mathcal{E}^a \wedge \mathsf{Sub}_a(x, i, y) = z)$. Thus if x is the code of the formula φ, then $\mathsf{Sub}(x, i, y)$ is the code of $\mathsf{Sub}(\varphi, v_i, y)$.

Corollary 3.8. *The relation* $\mathsf{Sub}(x, i, y) = z$ *is* Σ.

We write $\varphi(a) = b$ to mean that φ is a monadic formula and $\mathsf{Sub}(\varphi, v_i, a) = b$, where v_i is the free variable of φ.

We write $x(y) = z$ to mean that x is the code of a monadic formula φ and $\varphi(y) = z$.

Proposition 3.9. *The relation* $x(y) = z$ *is* Σ.

Proof It can be written:

$$(\exists a)(\exists i \in \omega)(x \in \mathcal{E}_{\{i\}}(a) \wedge \mathsf{Sub}(x, i, y) = z).$$

∎

Valuations and truth Consider a closed a-formula X. The definition of X being *true* over a is given by the following inductive scheme ($i, j \in \omega$, $a_1, a_2 \in a$).

(0) $a_1 \in a_2$ is true over $a \Leftrightarrow a_1$ is a member of a_2

(1) $\neg X$ is true over $a \Leftrightarrow X$ is not true over a

(2) $X \wedge Y$ is true over $a \Leftrightarrow X$ and Y are both true over a

(3) $(\exists v_i)\varphi(v_i)$ is true over $a \Leftrightarrow$ for some $a_1 \in a$, $\varphi(a_1)$ is true over a.

Now, given any x in $S(a)$—any x that is the code of some a-sentence X—we define $V_a(x)$ (called the *value* of x in a) to be 1 or 0, depending respectively on whether X is or is not true over a. The function V_a is thus the one and only function from $S(a)$ into $\{0, 1\}$ that satisfies the following conditions:

C_0 $(\forall a_1 \in a)(\forall a_2 \in a)[(a_1 \in a_2 \wedge V_a(\langle \ulcorner a_1 \urcorner, \ulcorner a_2 \urcorner, 0 \rangle) = 1) \vee$
$(a_1 \notin a_2 \wedge V_a(\langle \ulcorner a_1 \urcorner, \ulcorner a_2 \urcorner, 0 \rangle) = 0)].$

C_1 $(\forall x \in S(a))[V_a(\mathsf{Neg}(x)) = 1 - V_a(x)].$ (Note: $1 - 1 = 0$ and $1 - 0 = 1$.)

C_2 $(\forall x \in S(a))(\forall y \in S(a))[V_a(\mathsf{Con}(x, y)) = V_a(x) \cap V_a(y)].$ (Note: $1 \cap 1 = 1$; $1 \cap 0 = 0$; $0 \cap 1 = 0$; $0 \cap 0 = 0$—just like the truth-table rules: $t \wedge t = t$, $t \wedge f = f$; $f \wedge t = f$; $f \wedge f = f$.)

C_3 $(\forall i \in \omega)(\forall x \in \mathcal{E}_{\{i\}}(a))[(\exists b \in a)(V_a(x(b)) = 1 \wedge V_a(\mathsf{E}_i(x)) = 1) \vee$
$(\forall b \in a)(V_a(x(b)) = 0 \wedge V_a(\mathsf{E}_i(x)) = 0)].$

We shall call a function f a *valuation function* for $S(a)$ if it is from $S(a)$ to $\{0, 1\}$ and if conditions C_0–C_3 hold, replacing "V_a" by "f." It is a routine matter to verify that conditions C_0–C_3 (replacing "V_a" by "f") are all Σ-relations between f and a, and hence the relation "f is a valuation function for $S(a)$" is a Σ-relation between f and a. And so we have the following.

Proposition 3.10. *The relation* $V_a(x) = y$ *is a Σ-relation among* a, x, y.

Proof $V_a(x) = y$ is equivalent to

$[x \in S(a) \wedge y \in \{0, 1\} \wedge (\exists f)$ is a valuation function for $S(a) \wedge f(x) = y]$.

∎

A monadic formula φ with constants in a is said to *define* over a the set of all $b \in a$ such that $\varphi(b)$ is true over a.

We let $\mathsf{Def}(x, y, a)$ mean that x is the code of a formula φ whose constants are in a such that φ defines y over a.

Proposition 3.11. *The relation* $\mathsf{Def}(x, y, a)$ *is a Σ-relation among* x, y, a.

Proof This condition can be written as

$x \in M(a) \wedge (\forall b \in a)[(V_a(x(b)) = 1 \wedge b \in y) \vee (V_a(x(b)) = 0 \wedge b \notin y)] \wedge y \subseteq a.$

[We recall that $M(a)$ is the set of codes of monadic formulas with constants in a.] ∎

By $\mathcal{F}(a)$ is meant (as you know) the set of all subsets of a that are definable over a. Now (at last) we have the following.

Theorem 3.12. *The relation* $y = \mathcal{F}(x)$ *is* Σ.

Proof
(a) The relation $y \in \mathcal{F}(x)$ is Σ, for it can be written $(\exists z \in M(x))(\mathsf{Def}(z, y, x))$.
(b) The relation $y \subseteq \mathcal{F}(x)$ is Σ, for it can be written $(\forall z \in y)(z \in \mathcal{F}(x))$.
(c) The relation $\mathcal{F}(x) \subseteq y$ is Σ, for it can be written
$(\forall z \in M(x))(\exists w \in y)(\mathsf{Def}(z, w, x))$.
(d) $y = \mathcal{F}(x)$ is Σ since it is equivalent to $y \subseteq \mathcal{F}(x) \wedge \mathcal{F}(x) \subseteq y$. ∎

Part C—The *L*-absoluteness of constructibility

§4 Constructibility is absolute over *L*

By Theorem 3.12 and Theorem 2.8 it follows that the function $\mathcal{F}^*$—that assigns to each ordinal α the set L_α—is Σ. Suppose we extend this artificially to a function $\mathcal{A}$, on all sets, by mapping non-ordinals to 0.

We let $\mathcal{M}(x,y)$ be a Σ formula, fixed for the discussion, that defines the relation $L_x = y$; that is, $\mathcal{M}(x,y)$ defines the graph of the function $\mathcal{F}^*$. Also let $\mathcal{N}(x,y)$ be the formula [Ordinal(x) $\wedge$ $\mathcal{M}(x,y)$] $\vee$ [$\neg$Ordinal(x) $\wedge$ $y = 0$]. Then $\mathcal{N}(x,y)$ defines the function $\mathcal{A}$. We have given an informal proof of the fact that $\mathcal{N}(x,y)$ defines a function, but of course it could be formalized using the axioms of set theory. By Corollary 3.4 of Chapter 13, *L* is a well founded first-order universe, so it follows that $\mathcal{N}(x,y)$ is function-like over *L*. Of course $\mathcal{N}(x,y)$ is also function-like over *V*. Then by Theorem 1.5 $\mathcal{N}(x,y)$ is absolute. It follows that $\mathcal{M}(x,y)$ is also absolute, because it is equivalent to Ordinal(x) $\wedge$ $\mathcal{N}(x,y)$. We thus have established the following.

Theorem 4.1. *The formula $\mathcal{M}(x,y)$ is Σ and absolute over L.*

We now let $\mathcal{L}(x,y)$ be the Σ-formula $(\exists z)(\mathcal{M}(y,z) \wedge x \in z)$. This formula defines the relation $x \in L_y$. Since Σ-formulas are absolute upwards (Theorem 1.2) we have the following.

Theorem 4.2. *The formula $\mathcal{L}(x,y)$ (and hence also the relation $x \in L_y$ defined by it) is absolute upwards—in fact it is Σ.*

The formula $\mathcal{L}(x,y)$ is *not* absolute over all transitive classes, but we do have the following.

Theorem 4.3. *The formula $\mathcal{L}(x,y)$, and hence also the relation $x \in L_y$, is absolute over L.*

Proof By Theorem 4.2, $\mathcal{L}(x,y)$ is absolute upwards, hence absolute upwards over *L*. We must now show that it is absolute downwards over *L*.

Take any elements a, b of *L* such that $\mathcal{L}(a,b)$ is true over *V*. We must show that it is true over *L*. Since it is true over *V*, then b is really an ordinal and a is really a member of L_b. Let $c = L_b$. Then $a \in c$. Also c, of course, is in *L*. Then the sentence $a \in c$ is true over *L*. Also, the sentence $\mathcal{M}(b,c)$ is true over *V* (since $c = L_b$), hence it is also true over *L* (because the formula $\mathcal{M}(x,y)$ is absolute over *L*, both upwards and downwards). Therefore the sentence $\mathcal{M}(b,c) \wedge a \in c$ is true over *L*, and since c is in *L*, the sentence $(\exists y)(\mathcal{M}(b,y) \wedge a \in y)$ is true over *L*. This sentence is the sentence $\mathcal{L}(a,b)$. Thus $\mathcal{L}(a,b)$ is true over *L*. ∎

We now let Const(x) (sometimes written $L(x)$) be the formula $(\exists y)\mathcal{L}(x,y)$. It obviously defines over *V* the property of being a constructible set—that is, it defines the class *L* over *V*. Clearly, it is a Σ-formula. And so we have the following.

Theorem 4.4. *The property of being a constructible set is absolute upwards—indeed, it is Σ.*

Next we have the following vital result.

Theorem 4.5. *The property of being a constructible set is absolute over L—more specifically, the formula* $\text{Const}(x)$ *defines the whole of L over L.*

Proof Since the formula $\text{Const}(x)$ is Σ, then it is certainly absolute upwards over L. We must now show that for any $a \in L$, the sentence $\text{Const}(a)$ (which of course is true over V) is true over L.

Well, suppose $a \in L$. Then a is constructible, hence for some ordinal α, $a \in L_\alpha$, and therefore the sentence $\mathcal{L}(a, \alpha)$ is true over V. Also $\alpha \in L$ (since L contains all ordinals), hence $\mathcal{L}(a, \alpha)$ is true over L (by Theorem 4.3). Since $\alpha \in L$, then the sentence $(\exists y)\mathcal{L}(a, y)$ is true over L—and this is the sentence $\text{Const}(a)$. ∎

The Axiom of constructibility This is the sentence $(\forall x)\text{Const}(x)$.

For any a in L, a is constructible, hence $\text{Const}(a)$ is true over V, hence also true over L (by Theorem 4.5). Therefore the sentence $(\forall x)\text{Const}(x)$ is true over L. (Remember, if we say $(\forall x)\text{Const}(x)$ is true over L, this does not mean that for *every* set a, $\text{Const}(a)$ is true over L, but rather that for every a in L, the sentence $\text{Const}(a)$ is true over L, which it is, by Theorem 4.5.) And so we have the following.

Theorem 4.6. *The axiom of constructibility—the sentence* $(\forall x)\text{Const}(x)$ *— is true over the first-order universe L.*

§5 Further results

We now obtain some interesting and important consequences of Theorems 4.1 and 4.2.

Theorem 5.1. *For every first-order universe K that contains all ordinals of V*, $L \subseteq K$.

Proof Let $\text{Ordinal}(x)$ be a Δ_0-formula that defines over V the class of all ordinals. We have proved that for every ordinal α, the set L_α exists, and this proof can be formalized in ZF—i.e., the sentence $(\forall y)(\text{Ordinal}(y) \supset (\exists z)\mathcal{M}(y, z))$ is a theorem of ZF, hence is true in all first-order universes. Now suppose K is a first-order universe containing all ordinals of V. Then the sentence $(\forall y)(\text{Ordinal}(y) \supset (\exists z)\mathcal{M}(y, z))$ is true over K. Now take any ordinal α. Then $\alpha \in K$, so the sentence $\text{Ordinal}(\alpha) \supset (\exists z)\mathcal{M}(\alpha, z)$ has all its constants (the only one being α) in K and is true over K (since $(\forall y)(\text{Ordinal}(y) \supset (\exists z)\mathcal{M}(y, z))$ is true over K). Also the antecedent, $\text{Ordinal}(\alpha)$, of the sentence is true over K (because α is an ordinal and the formula $\text{Ordinal}(\alpha)$ is absolute over K), and hence the consequent $(\exists z)\mathcal{M}(\alpha, z)$ is true over K. Hence for some element $b \in K$, the sentence $\mathcal{M}(\alpha, b)$ is true over K. But the formula $\mathcal{M}(y, z)$ is absolute over K (Theorem 4.1) hence $b = L_\alpha$. Thus $L_\alpha \in K$.

We have now proved that for every ordinal α, the set L_α is in K, and since K is transitive, then every element of every L_α is in K—i.e., all constructible sets are in K, and so $L \subseteq K$. ∎

From Theorem 5.1 and Theorem 3.3 of Chapter 13, and the fact that L contains all ordinals (C_{14}, Chapter 12) we have the following.

Theorem 5.2. *A necessary and sufficient condition that a set x be constructible is that x belongs to all first-order universes that contain all ordinals.*

§6. A PROOF THAT L CAN BE WELL ORDERED   191

Proof By Theorem 5.1, every constructible set does belong to all first-order universes that contain all ordinals.

Conversely, suppose that x belongs to all first-order universes that contain all ordinals. Well, L is a first-order universe that contains all ordinals (by Theorem 3.3, Chapter 13 and C_{14}, Chapter 12) hence x belongs to L. And so x is constructible. ∎

Theorem 5.2, in conjunction with Theorem 4.1 of Chapter 13, yields the following.

Theorem 5.3. *If K is a first-order universe such that every subset of K is a subset of some element of K, then $L \subseteq K$.*

Here is another consequence of Theorem 4.2, which will be crucial for the next chapter.

Theorem 5.4. *Let T be a transitive class such that the axiom of constructibility is true over T. Then every element of T is constructible and its order is in T.*

Proof Suppose T is transitive and $(\forall x)\text{Const}(x)$ is true over T. Let a be any element of T. Then $\text{Const}(a)$ is true over T—i.e., the sentence $(\exists y)\mathcal{L}(a, y)$ is true over T. Hence for some $b \in T$, the sentence $\mathcal{L}(a, b)$ is true over T. But T is transitive and the formula $\mathcal{L}(x, y)$ is absolute upwards (Theorem 4.2), hence the sentence $\mathcal{L}(a, b)$ is true over V, which means that b is an ordinal and $a \in L_b$. Since $a \in L_b$, then the order of a is $\leq b$. And since $b \in T$ and T is transitive, T contains all ordinals $\leq b$, and so T contains the order of a. ∎

Part D—The well orderability of L

§6 A proof that L can be well ordered

Without the axiom of choice, the class L can be properly well ordered—a fact that we will need. There are many ways of well ordering L, and we shall now describe one.

For any set A, let A^* be the smallest set that includes A and is closed under the ordered pairing operation—i.e., for all x and y in A^*, the ordered pair $\langle x, y \rangle$ is in A^*. A^* is actually the union of the denumerable sequence $A_0, A_1, \ldots, A_n, \ldots$ of sets defined by the inductive scheme:

(1) $A_0 = A$

(2) $A_{n+1} = A_n \cup (A_n \times A_n)$.

A^* can be called the *closure* of A under the ordered pairing operation.

Our plan is the following: Consider a well ordering W of a set A. We will successively show how to extend W to:

(1) a well ordering W' of $A \cup (A \times A)$;

(2) a well ordering $W^\#$ of A^*;

(3) a well ordering W_F of $\mathcal{F}(A)$.

Then we will see how to well order L.

So, let W be a well ordering of A.

(1) We let W' be the well ordering of $A \cup (A \times A)$ defined as follows:

   (a) The elements of A are ordered as in W.
   (b) All elements of A are put before all elements of $(A \times A)$ that are not in A.
   (c) For any elements x and y of $(A \times A) - A$, $x = \langle x_1, x_2 \rangle$ for some x_1, x_2 in A and $y = \langle y_1, y_2 \rangle$ for some y_1, y_2 in A. Then we put x before y iff either x_1 is before y_1 in W, or $x_1 = y_1$ and x_2 is before y_2 in W.

   Thus W' is a well ordering of $A \cup (A \times A)$ for which $W \subseteq W'$.

(2) Now consider the sequence $A_0, A_1, \ldots, A_n, \ldots$ in which $A_0 = A$ and $A_{n+1} = A_n \cup (A_n \times A_n)$. We then consider the sequence $W_0, W_1, \ldots, W_n, \ldots$ in which $W_0 = W$ and for each n, $W_{n+1} = W'_n$. For each n, W_n is a well ordering of A_n, and also $W_n \subseteq W_{n+1}$, hence the set $\{W_0, W_1, \ldots, W_{n+1}, \ldots\}$ is a nest whose union is a well ordering of A^*. We let $W^\#$ be this well ordering.

(3)

Step 1 Let $C(A)$ be the set of codes of all formulas whose constants are in A. Let $B = A \cup \omega$. Let W_B be the well ordering of B in which all elements of A come before all elements of ω not in A, any elements x, y of A are ordered according to W, and any natural numbers n and m not in A are ordered according to magnitude. Thus W_B is a well ordering of B and $W \subseteq W_B$. Now, the set $C(A)$ is a subset of B^* (since every ordered triple $\langle a, b, c \rangle$ is the ordered pair $\langle \langle a, b \rangle, c \rangle$) and we now order the elements of $C(A)$ according to their order in B^*. We let W_C be this ordering. Thus W_C is a well ordering of the set of codes of all formulas whose constants are in A.

Step 2 Each x in $\mathcal{F}(A)$ is defined over A by at least one formula with constants in A. We let x_0 be the least element of $C(A)$ (in the well ordering W_C) that is the code of a formula that defines x over A. We now order $\mathcal{F}(A)$ as follows:

   (a) All elements of A are put before all elements of $\mathcal{F}(A) - A$.
   (b) For any x, y in A, we order them as in W.
   (c) For any distinct x, y in $\mathcal{F}(A) - A$, we put x before y iff x_0 precedes y_0 in the ordering W_C of $C(A)$.

Thus we have a well ordering W_F of $\mathcal{F}(A)$ in which $W \subseteq W_F$.

Now we well order L as follows. By transfinite recursion, we have the following On-sequence of well orderings:

(1) $W_0 = \emptyset$

(2) $W_{\alpha+1} = (W_\alpha)_F$

(3) $W_\lambda = \cup_{\alpha<\lambda} W_\alpha$ (λ a limit ordinal).

By transfinite induction, each W_α is a well ordering of L_α, and $W_\alpha \subseteq W_{\alpha+1}$, so the class $\mathcal{N}$ of the W_α's is a nest whose union is a proper well ordering of L.

CHAPTER 15

CONSTRUCTIBILITY AND THE CONTINUUM HYPOTHESIS

§0 What we will do

We now come to the most interesting part of the proof of the relative consistency of the continuum hypothesis. In this chapter, the Mostowski-Shepherdson mapping theorem, the Tarski-Vaught theorem, the absoluteness of constructibility, and the fact that L is a first-order universe—these four principles—all play their roles.

We will first show that if all sets are constructible, then Cantor's generalized continuum hypothesis must be true. From this—together with results of the last chapter—we will see that the generalized continuum hypothesis can never be disproved from the axioms of ZF, unless ZF itself is inconsistent.

For the remainder of this chapter, we let c stand for any *infinite* cardinal, and by c^* we shall mean the next cardinal after c. (If c is the cardinal $\aleph_\alpha$—the αth infinite cardinal—then c^* is the cardinal $\aleph_{\alpha+1}$.)

Cantor's generalized continuum hypothesis (GCH) is that the cardinality of $\mathcal{P}(c)$ is c^* (or equivalently, that for any set A of cardinality c, the number of subsets of A is c^*). The *continuum hypothesis* (CH) is the special case of this where c is the cardinal $\aleph_0(=\omega)$ — i.e., the statement that the number of sets of natural numbers is $\aleph_1(=\omega^*)$—the *next* cardinal after $\aleph_0$.

Now, by Cantor's *theorem*, the cardinality of $\mathcal{P}(c)$ is *at least* c^* (since it is greater than c and c^* is the *least* cardinal greater than c), and so by Cantor's theorem, GCH is equivalent to the statement that $\overline{\overline{\mathcal{P}(c)}} \leq c^*$—i.e., that there are *at most* c^* subsets of c.

In this chapter we will prove Gödel's remarkable result that there are at most c^* *constructible* subsets of c. From this, of course, it follows that if all sets are constructible, then the generalized continuum hypothesis must be true! Then we will show (using the absoluteness of constructibility proved in the last chapter) that whether or not all sets are really constructible, it is at least *consistent* with the axioms of ZF that they are (assuming, of course, that ZF itself is consistent) and hence that the generalized continuum hypothesis can never be disproved in ZF (unless ZF itself is inconsistent).

That there are at most c^* constructible subsets of c will be seen to be an easy consequence of Gödel's basic result (Theorem G, 1.1 below) that every constructible subset of L_c (and hence every constructible subset of c) is an element of L_{c^*}. This equivalently says that the order of every constructible subset of L_c is less than c^*—or what is the same thing, has cardinality $\leq c$. Now, for each infinite cardinal c, the set L_c has cardinality c (see Exercise 0.1 below) and so L_{c^*} has cardinality c^*, hence Theorem G, 1.1, implies that there are at most c^* constructible subsets of c (in fact at most c^* constructible subsets of L_c, since each of them is in L_{c^*}).

EXERCISES

Exercise 0.1.

(a) Show that for any infinite set A, the set of all finite sequences of elements of A has the same cardinality as A.

(b) Show, therefore, that for any infinite set A of cardinality c, the set of all formulas with constants in A has cardinality c.

(c) Show, therefore, that if A is infinite of cardinality c, and if every element of A is definable over A (which happens when A is transitive) then $\mathcal{F}(A)$ has the same cardinality as that of A. Show, therefore, that for every infinite ordinal α, $L_{\alpha+1}$ has the same cardinality as L_α.

(d) Then, using transfinite induction on all infinite ordinals, and the fact that L_ω is denumerable, show that for each infinite ordinal α, the set L_α has cardinality at most $\overline{\overline{\alpha}}$. Show therefore that $\overline{\overline{L_c}} \leq c$.

(e) Using the fact that each $L_{\alpha+1}$ contains at least one element not in L_α (namely L_α itself), show that for each infinite cardinal c, L_c has at least c elements. Then conclude that $\overline{\overline{L_c}} = c$.

§1 The key result

Theorem 1.1 (G). *For any infinite cardinal c, every constructible subset of L_c is an element of L_{c^*}.*

Proof We use Theorem 4.1, Chapter 11 (the M.S.T.V. theorem which, we recall, combines the Mostowski-Shepherdson theorem with the Tarski-Vaught theorem). In this theorem, we take K to be the class L and X to be the sentence $(\forall x)\text{Const}(x)$ (the axiom of constructibility). We proved in the last chapter that the axiom of constructibility is true over L, and also that L can be well ordered.

Now, let m be any constructible subset of L_c (c an infinite cardinal). The set $L_c \cup \{m\}$ is obviously of cardinality c and is transitive (since $m \subseteq L_c$). Then by the M.S.T.V. theorem, $L_c \cup \{m\}$ is a subset of a *transitive* set T of cardinality c such that $(\forall x)\text{Const}(x)$ is true over T. Of course m is an element of T (since $\{m\} \subseteq T$) and since $(\forall x)\text{Const}(x)$ is true over T, then by Theorem 5.4 of Chapter 14, the order α of m lies in T. But T is transitive, hence $\alpha \subseteq T$. Then, since T has cardinality c, α must have cardinality $\leq c$, hence $\alpha < c^*$. And so $m \in L_{c^*}$. ∎

Corollary 1.2. $V = L$ *implies* GCH. *(If all sets are constructible, then the generalized continuum hypothesis is true.)*

§1. THE KEY RESULT

Levy's basis theorem In a sense, Theorem G, 1.1, is a special case of the interesting theorem below, due to Lévy (1965).

By the *transitive closure* of a set x is meant the set consisting of x, the elements of x, the elements of the elements of x, and so forth. More precisely, let $\cup^n x$ be the set $\underbrace{\cup \cup \cdots \cup}_{n} x$. Then the transitive closure of x is the union of the sets $\{x\}$, x, $\cup x$, $\cup^2 x$, ..., $\cup^n x$, ... $(n \in \omega)$. It is easy to see that the transitive closure of x is transitive and that it is a subset of *every* transitive set containing x. Thus the transitive closure of x can be alternatively characterized as the smallest transitive set containing x (i.e., the intersection of all transitive sets which have x as a member).

By the *hereditary cardinality* of x (written $\mathsf{HC}(x)$) is meant the cardinality of the transitive closure of x. (For example, the set $\{\omega, \omega+1\}$ has cardinality 2, but hereditary cardinality ω.) Of course the cardinality of a set x is always less than or equal to its hereditary cardinality, and if x is transitive, the two are then the same (since x is then its own transitive closure). In particular, the hereditary cardinality of an ordinal is the same as its cardinality. Let us also note that if x is a member of a transitive set T of cardinality c, then x has hereditary cardinality $\leq c$ (because the transitive closure of x is then a subset of T).

Theorem 1.3 (Levy's Basis Theorem). *Let $R(x, y)$ be a Σ-relation on a well orderable class K. Then for any element $a \in K$, of hereditary cardinality c (c infinite), if $R(a, b)$ holds for some b in K, then $R(a, b')$ holds for some b' in K of hereditary cardinality $\leq c$.*

Proof Let $\varphi(x, y)$ be a Σ-formula that defines $R(x, y)$ over K. Now let a be any element of K of hereditary cardinality c such that $R(a, b)$ holds for at least one b in K. Let A be the transitive closure of a. Then A has cardinality c. By the class version of the Tarski-Vaught theorem, Theorem 3.2 of Chapter 11, A is a subset of some extensional set B of cardinality c such that B reflects K with respect to the formula $(\exists y)\varphi(x, y)$. Of course a is an element of B. Now, the sentence $(\exists y)\varphi(a, y)$ is true over K (since there is at least one y in K such that $R(a, y)$), hence it is true over B, and so there is some element $b \in B$ such that $\varphi(a, b)$ is true over B. (We might remark that B is not necessarily transitive, and so b might have cardinality far bigger than c!) Now we apply the Mostowski-Shepherdson map π to B; we let T be the transitive set $\pi''(B)$. Each element x of B is mapped to some element x' of T. Since $\varphi(a, b)$ is true over B and B is $\in$-isomorphic to T, then $\varphi(a', b')$ is true over T. But $a' = a$ (since a is an element of the transitive subset A of B) and so the sentence $\varphi(a, b')$ is true over T. Then, since $\varphi(x, y)$ is Σ and T is transitive, $\varphi(a, b')$ must be true over K (Exercise 1.1, Chapter 14), which means that $R(a, b')$ holds. Of course T also has cardinality c (since it is in 1-1 correspondence with B) and since $b' \in T$ and T is transitive, b' has hereditary cardinality $\leq c$. This concludes the proof. ∎

Remark This theorem holds in the following more general form:

If $R(x_1, \ldots, x_n, y)$ is a Σ-relation over K, where K is well orderable, then for any $a_1, \ldots, a_n$ in K, all of hereditary cardinality $\leq c$, if $R(a_1, \ldots, a_n, b)$ is true for some b, then there is some b' of hereditary cardinality $\leq c$ such that $R(a_1, \ldots, a_n, b')$.

Now we obtain Theorem G, 1.1, as a consequence of Levy's theorem and the facts that L itself is well orderable and the relation $x \in L_y$ is Σ. Suppose m is a constructible subset of L_c. Then the hereditary cardinality of m— $HC(m)$—is less than or equal to c. Since m is constructible, there is some b such that $m \in L_b$. Then by Levy's theorem (applied to the Σ-relation $x \in L_y$ and taking L for K) there is some b of hereditary cardinality $\leq HC(m)$ such that $m \in L_b$. Hence b is an ordinal of cardinality $\leq HC(m)$, hence of cardinality $\leq c$, hence $b < c^*$ and since $m \in L_b$, then $m \in L_{c^*}$.

§2 Gödel's isomorphism theorem (optional)

We have now seen two somewhat different proofs of Theorem G, 1.1. This theorem (called "Theorem 2" in Gödel's 1939 paper) was originally proved along quite different lines than the streamlined proofs we have given (which combine ideas of Scott, Levy, Karp and others). It is considerably more elaborate, but brings to light certain facts about constructibility and Mostowski-Shepherdson mappings that are of interest in their own right. An interesting thing about Gödel's original proof of 1939 is that he proved Theorem G *before* introducing the notion of absoluteness! Thus, neither the L-absoluteness of constructibility, nor the upward absoluteness of the relation $x \in L_y$ played any role.

The original proof reveals some additional interesting features not displayed in the more recent proofs. His proof is based on the following theorem—which we will call *Gödel's isomorphism theorem, Theorem GI* for short.

Theorem 2.1 (Gödel's isomorphism theorem, GI). *Any infinite set A whose members are constructible sets is a subset of some extensional set K of constructible sets such that K has the same cardinality as A and K is $\in$-isomorphic to some L_α.*

To make the proof easier to follow, let us define a set K to be *pseudo-constructible* if there is a limit ordinal γ and a γ-tuple $(K_0, K_1, \ldots, K_\alpha, \ldots)$ $(\alpha < \gamma)$ such that K is the union of the K_α's and the γ-tuple (which we will call a *pseudo-construction*) satisfies the following three conditions.

C_1 $K_0 \neq \emptyset$.

C_2 For any $\alpha < \gamma$, $\mathcal{F}(K_\alpha)$ is the set of all elements $a \cap K$, where $a \in K_{\alpha+1}$.

C_3 For any limit ordinal $\delta < \gamma$, $K_\delta = \cup \alpha < \delta K_\alpha$.

(We remark in passing that if K happens to be transitive, then $K = L_\gamma$.) The heart of the proof of Theorem GI is the following result, which we will call *Theorem K*. Though not explicit in Gödel's proof, Theorem K is certainly implicit.

Theorem 2.2 (K). *If K is pseudo-constructible and extensional, then K is Mostowski-Shepherdson mappable to some L_γ. (We are assuming well foundedness.)*

Finally, any infinite set A of constructible sets can be extended to an extensional pseudo-constructible set K of the same cardinality as A.

Further details of the proofs of these results are given in the exercises below.

EXERCISES

§2. GÖDEL'S ISOMORPHISM THEOREM (OPTIONAL)

Exercise 2.1. To prove Theorem K, let us suppose that $(K_0, K_1, \ldots, K_\alpha, \ldots)$ $(\alpha < \gamma)$ is a pseudo-construction and that $K = \cup_{\alpha < \gamma} K_\alpha$. Let π be the Mostowski-Shepherdson map on K.

(a) Show by transfinite induction that for each $\alpha < \gamma$, $\pi''(K_\alpha) = L_\alpha$.

(b) Then show that $\pi''(K) = L_\gamma$.

Exercise 2.2. Given two sets K and M, define K to be *bound up* with M if the following two conditions hold:

B_1 $K \cap M$ is Henkin closed with respect to M (see §1, Chapter 11);

B_2 for any subset a of M such that $a \in \mathcal{F}(M)$, $a \in K$ iff a is definable over M by a formula whose constants are all in $K \cap M$.

Show that if K is bound up with M, then $\mathcal{F}(K \cap M)$ is the set of all elements $a \cap K$ such that $a \in (K \cap \mathcal{F}(M))$.

Exercise 2.3. For any set K of constructible sets, define $o(K)$ to be the set of all ordinals $\alpha + n$, where α is the order of some element of K and n is a natural number. Let us say that K is *B-closed* if for every $\alpha \in o(K)$, K is bound up with L_α. We need the following theorem: If K is B-closed, then K is pseudo-constructible. Here is an outline of the proof (which uses the result of the preceding exercise).

The set $o(K)$, ordered according to magnitude, is isomorphic to a unique ordinal γ. Since $o(K)$ is closed under successor, then γ is a limit ordinal. For each $\alpha < \gamma$, we let $\overline{\alpha}$ be the ordinal in $o(K)$ that corresponds to α under the isomorphism—in other words, $\overline{\alpha}$ is the αth ordinal in $o(K)$ in order of magnitude. For each $\alpha < \gamma$, let $K_\alpha = K \cap L_{\overline{\alpha}}$. K is then the union of the K_α's.

Show that the γ-tuple $(K_0, \ldots, K_\alpha, \ldots)$ $(\alpha < \gamma)$ is a pseudo-construction. (The result of the preceding exercise is crucial for the verification of condition C_2 of the definition of a pseudo-construction.)

Exercise 2.4. It must next be shown that any infinite set A of constructible sets can be extended to a set K of the same cardinality as A such that K is extensional and B-closed. Show this by inductively defining the following denumerable sequence A_0, $A_1, \ldots, A_n, \ldots$ of sets and taking K to be their union:

Take $A_0 = A \cup \{0\}$.

Then, assuming A_n defined, take A_{n+1} to be the result of adding to A_n the following items.

(1) For each x, y in A_n such that $x \neq y$, add to A_n some element of $(x - y) \cup (y - x)$ (say the first, in the well ordering of L).

(2) For every $\alpha \in o(A_n)$, put α into A_{n+1}.

(3) For every $\alpha \in o(A_n)$, put the following items into A_{n+1}.

(a) For every sentence $(\exists x)\varphi(x)$ with constants in $A_n \cap L_\alpha$ that is true over L_α, take some $a \in L_\alpha$ such that $\varphi(a)$ is true over L_α, and put a into A_{n+1}.

(b_1) For each $a \in A_n$ that is definable over L_α, take a formula that defines a over L_α and put its constants into A_{n+1}.

(b_2) For every subset of L_α that is definable over L_α by a formula with constants in $A_n \cap L_\alpha$, put the subset into A_{n+1}. (There are no more such subsets than elements of A_n.)

Now show that $\cup_{n \in \omega} A_n$ is both extensional and B-closed, and has no higher cardinality than A.

Exercise 2.5. Now prove Theorem GI.

Exercise 2.6. How does Theorem GI yield Theorem G? Hint: Suppose that m is a constructible subset of L_c (c infinite). Extend $L_c \cup \{m\}$ to an extensional set K of cardinality c that is Mostowski-Shepherdson mappable to some L_α, (which can be done by Theorem GI) and show that the Mostowski-Shepherdson map carries m to itself, so that $m \in L_\alpha$, and show that $\alpha < c^*$, and hence $m \in L_{c^*}$.

§3 Some consequences of Theorem G

Before we turn to the main metamathematical consequence of Theorem G (the relative consistency of the generalized continuum hypothesis, dealt with in §4) we pause to observe some other consequences of Theorem G. To begin with, we have the following.

Corollary 3.1. *Let b be a constructible set of subsets of L_c. Then $b \in L_{c^{**}}$.*

Proof Suppose b is a constructible set of subsets of L_c. Since b is constructible, then every element of b is constructible, hence every element of b is a *constructible* subset of L_c, so by Theorem G, every element of b is an element of L_{c^*}, and so $b \subseteq L_{c^*}$. But b is constructible (by hypothesis) and so b is a *constructible* subset of L_{c^*}, and so by Theorem G again, $b \in L_{c^{**}}$. ∎

First-order Zermelo universes Let us call a set (or even a class) a first-order *Zermelo* universe if all the axioms of Zermelo set theory (the axioms of ZF minus the axioms of substitution) are true over it. An interesting consequence of Corollary 3.1 is the result (also due to Gödel) that $L_{\aleph_\omega}$ is a first-order Zermelo universe.

More generally, we can say this: A cardinal c is called a *limit cardinal* if there is no cardinal c_1 such that $c_1^* = c$. The cardinal $\aleph_0 = \omega$ is a limit cardinal and the next one after that is $\aleph_\omega$. If c is a limit cardinal then for any cardinal $c_1 < c$, there are infinitely many cardinals between c_1 and c.

Theorem 3.2. *For any limit cardinal $c > \omega$, L_c is a first-order Zermelo universe. In particular, $L_{\aleph_\omega}$ is a first-order Zermelo universe.*

Proof We need verify only the power set axiom (the others are trivial). To say that the power set axiom holds for L_c is to say that for any a in L_c, the set of all elements of L_c that are subsets of a is an element of L_c—i.e., that $\mathcal{P}(a) \cap L_c$ is an element of L_c. So let $b = \mathcal{P}(a) \cap L_c$. Clearly $b \subseteq L_c$. Then b is definable over L_c by the formula $x \subseteq a$. Hence $b \in L_{c+1}$, and so b is a constructible set. We are to show that $b \in L_c$.

Now, $a \in L_c$, and since c is a limit cardinal it is certainly a limit ordinal, hence $a \in L_\alpha$ for some $\alpha < c$. Let c_1 be the smallest cardinal greater than the cardinality of α. Then $c_1 < c$ (because c is a *limit cardinal*). Of course $\alpha < c_1$, hence $a \in L_{c_1}$, hence

$a \subseteq L_{c_1}$, so all subsets of a are subsets of L_{c_1}, hence b is a set of subsets of L_{c_1}, and since b is constructible (as we have shown), b is a constructible set of subsets of L_{c_1}. Then by Corollary 3.1, $b \in L_{c_1^{**}}$. Since $c_1 < c$ and c is a limit cardinal, then $c_1^{**} < c$, and so $b \in L_c$. ∎

§4 Metamathematical consequences of Theorem G

We have proved that if all sets are constructible, then the generalized continuum hypothesis is true—and in fact the sentence $(\forall x)\mathsf{Const}(x) \supset \mathsf{GCH}$ is a formal theorem of ZF. Hence this sentence is true over all first-order universes (since the axioms of ZF hold in all first-order universes) and so in particular, the sentence $(\forall x)\mathsf{Const}(x) \supset \mathsf{GCH}$ is true over L (since we have already proved that L is a first-order universe). But also $(\forall x)\mathsf{Const}(x)$ is true over L (as we proved in the last chapter) and so we have:

Theorem 4.1. *GCH is true over the first-order universe L.*

Therefore (assuming that there exists a Zermelo-Fraenkel universe in the first place) there exists a first-order universe L in which GCH is true. Now, if the sentence ¬GCH (the negation of the continuum hypothesis) were provable in ZF, it would be true in all first-order universes, contrary to the fact that it is *not* true in L! Thus the negation of the continuum hypothesis is not provable in ZF (assuming the existence of a Zermelo-Fraenkel universe).

The model-theoretic (or semantic) argument above can be given in a purely syntactical form, in which the notion of "truth"—or even "sets"—is not used at all, but only the notion of formal *provability* in ZF. As Gödel pointed out, given any proof of ¬GCH in ZF, one could *effectively* find a contradiction in ZF itself. And the same goes for any proof of $\neg(\forall x)\mathsf{Const}(x)$. Here is the general idea of how this can be done.

Let us write $(\exists x \in L)(\cdots)$ to abbreviate $(\exists x)(\mathsf{Const}(x) \wedge (\cdots))$. Now, given any formula φ, we define the formula φ^L (called the *relativization of φ to L*—or more precisely, the relativization of φ to the formula $\mathsf{Const}(x)$) to be the formula obtained from φ by replacing every occurrence of "$(\exists x)$" by "$(\exists x \in L)$." Thus:

(0) $(x \in y)^L = x \in y$

(1) $(\neg \varphi)^L = \neg(\varphi^L)$

(2) $(\varphi \wedge \psi)^L = (\varphi^L \wedge \psi^L)$

(3) $((\exists x)\varphi)^L = (\exists x \in L)\varphi^L$ (i.e., $(\exists x)(\mathsf{Const}(x) \wedge \varphi^L)$).

To say that a sentence X is true over L is equivalent to saying that X^L is true over V. We proved in the last chapter that $(\forall x)\mathsf{Const}(x)$ is true over L, and so $((\forall x)\mathsf{Const}(x))^L$ is true over V. This proof can be formalized in ZF–i.e., the sentence $((\forall x)\mathsf{Const}(x))^L$ is provable in ZF.

We also showed that L is a first-order universe, and therefore that every axiom X of ZF is true over L. The formal counterpart of this is that for every axiom X of ZF, the sentence X^L is *provable* in ZF. From this it follows that for every sentence X provable in ZF, the sentence X^L is also provable in ZF (because each sentence X that is provable in ZF is logically derivable from some finite set $\{X_1, \ldots, X_n\}$ of axioms of ZF, hence X^L is similarly derivable from the relativizations $X_1^L, \ldots, X_n^L$ of these axioms, but since $X_1^L, \ldots, X_n^L$ are themselves provable in ZF, so is X^L.)

To summarize:

(1) $((\forall x)\mathsf{Const}(x))^L$ is provable in ZF.

(2) For every theorem X of ZF, X^L is also provable in ZF.

(3) The sentence $(\forall x)\mathsf{Const}(x) \supset \mathsf{GCH}$ is provable in ZF.

It follows from (1) and (2) that if ZF is consistent, then the axiom of constructibility cannot be *disproved* in ZF. For suppose $\neg(\forall x)\mathsf{Const}(x)$ were provable in ZF. Then by (2), $(\neg(\forall x)\mathsf{Const}(x))^L$ would be provable, hence $\neg((\forall x)\mathsf{Const}(x))^L$ (which is the same sentence) would be provable. But also $((\forall x)\mathsf{Const}(x))^L$ is provable (by (1)), which would mean that ZF is inconsistent! And so if ZF is consistent, then the existence of a non-constructible set is not provable in ZF.

This proves the relative consistency of the axiom of constructibility with ZF. (A sentence X is called *relatively consistent* with the axioms of a system if it is consistent with the axioms, provided the axioms themselves are consistent.)

Then, by (3), we get the relative consistency of GCH, for if $\neg$GCH were provable in ZF it would follow from (3) that $\neg(\forall x)\mathsf{Const}(x)$ would be provable, which, as we have just seen, would mean that ZF would be inconsistent.

Looked at another way (and this reveals an interesting additional feature) it follows from (3) and (2) that the sentence $((\forall x)\mathsf{Const}(x) \supset \mathsf{GCH})^L$ is provable in ZF, but this sentence is $((\forall x)\mathsf{Const}(x))^L \supset \mathsf{GCH}^L$. And since $((\forall x)\mathsf{Const}(x))^L$ is provable (by (1)), so is GCH^L. Therefore (and this is the added feature), GCH is not only true over L, but the sentence GCH^L is provable in ZF! Then, if $\neg$GCH were provable in ZF, the sentence $(\neg\mathsf{GCH})^L$ would be provable, but this is the sentence $\neg\mathsf{GCH}^L$, and so ZF would be inconsistent.

We thus have the following.

Theorem 4.2. GCH *is consistent relative to* ZF.

§5 Relative consistency of the axiom of choice

By Sierpiński's theorem (§9 of Chapter 9) the generalized continuum hypothesis implies the axiom of choice. The proof of this is formalizable in ZF—i.e., GCH $\supset$ AC is a theorem of ZF. Hence if AC were disprovable in ZF, then so would GCH be disprovable in ZF, which is not the case, unless ZF is inconsistent. And so if ZF is consistent, AC is not disprovable in ZF.

Actually, the consistency of AC relative to ZF can be proved by much more elementary means. Since L can be well ordered, so can every constructible set (which is a subset of L). It can be shown that in fact every constructible set has a *constructible* well ordering, and also that the notion of being a well ordering is absolute. From this it is not hard to show that AC is true over L. Either way, we have the following.

Theorem 5.1. AC *is consistent relative to* ZF.

§6 Relative consistency of GCH and AC in class-set theory

We have shown that GCH is consistent relative to ZF. We have actually been working, not in Zermelo-Fraenkel set theory, which deals only with sets, but in *class-set* theory, using principles P_1 (extensionality) and P_2 (separation) of Chapter 2, together with axioms A_1–A_8. This system, when completely formalized, is the system *NBG* (Von Neumann, Bernays, Gödel). There are two ways such a formalization can be carried out. One way (used in Gödel's 1939 monograph) is to have one list $x_1, x_2, \ldots, x_n, \ldots$ of variables for *sets* and another, $X_1, X_2, \ldots, X_n, \ldots$, of variables for *classes*. Another way of formalizing *NBG* is by having only class variables $X_1, X_2, \ldots, X_n, \ldots$, and introducing expressions involving *set* variables $x_1, x_2, \ldots, x_n, \ldots$ only as *abbreviations* in the following manner. One first defines $\mathcal{M}(X)$ (read "X is a *set*") as $(\exists Y)(X \in Y)$ (where X, Y are any distinct class variables). Then for any formula $\varphi(X)$, one takes $(\exists x)\varphi(x)$ merely as an *abbreviation* of $(\exists X)(\mathcal{M}(X) \wedge \varphi(X))$. The second way seems to be prefered these days, but for our present exposition, the first will be a little more handy.

Now one interesting feature of *NBG* is that P_2, which consists of an infinite set of axioms—one for each formula $\varphi(X, X_1, \ldots, X_n)$—can be replaced by only finitely many axioms. This can be done in many different ways—one of which consists of asserting, for any classes A and B, the existence of the following seven classes:

B_1 The class of all ordered pairs $\langle x, y \rangle$ such that $x \in y$.

B_2 The intersection $A \cap B$.

B_3 The complement $\overline{A}$ (class of all sets not in A).

B_4 The class of all x such that $\langle x, y \rangle \in A$ for some y.

B_5 The class $A \times V$.

B_6 The class of all triples $\langle x, y, z \rangle$ such that $\langle y, z, x \rangle \in A$.

B_7 The class of all triples $\langle x, y, z \rangle$ such that $\langle x, z, y \rangle \in A$.

The fact that all instances of P_2 can be derived from $B_1 - B_7$ is known as the *class existence theorem*. We shall not prove it here; a proof can be found, for example, in (Mendelson 1987) [Chapter 4], in which a more formal treatment of *NBG* is given.

In passing, we might remark that although *NBG* is finitely axiomatizable (i.e., the infinite set P_2 of axioms can be replaced by a finite set), the system *ZF* is not. By Gödel's second incompleteness theorem (1931) *ZF* cannot prove its own consistency. However, given any *finite* set of the axioms of *ZF*–in fact, given any finite set S of sentences provable in *ZF*, the consistency of S *is* provable in *ZF* (using reflection principles). Therefore, if the set of axioms of *ZF* could be replaced by a finite set of theorems of *ZF*, then *ZF* could prove its own consistency, contrary to Gödel's second theorem.

Coming back to *NBG*, we now point out that for any first-order universe K, all the axioms of *NBG* are true if we interpret "set" to mean element of K and "class" to mean, not subclass of K, but *first-order definable* subclass of K!! Then all theorems of *NBG* are true over K, under that interpretation. And so if ¬GCH were provable in *NBG*, it would be true over L, which it isn't. Thus, assuming the existence of a first-order universe K, there is the first-order universe L in which GCH is true, hence GCH is then not refutable in *NBG* either.

Better still, there is the well known result that for any sentence X that is only about *sets*, and not about proper classes, if X is provable in *NBG*, then X is already provable in *ZF*. Thus, although one can prove things about classes that cannot even be stated in *ZF*, one cannot prove in *NBG* anything more about *sets* than can be proved in *ZF*. (This result is technically expressed by saying that *NBG* is a *conservative extension* of *ZF*.) This can be proved either model-theoretically or syntactically. By a *model* of *ZF* is meant a relational system (A, E) such that all axioms of *ZF* are true over (A, E) — i.e., true when "set" is interpreted to mean element of A, and "∈" is interpreted to mean E. In a model (A, E) of *ZF*, it is not necessary that E be the actual ∈-relation, nor that A be transitive; if these two conditions do hold, then the model is called a *standard* model of *ZF*. To say that (A, E) is a standard model is tantamount to saying that A is a first-order universe! Now, by a result of Gödel known as the *completeness theorem*, every consistent set of sentences has a model (in fact, a denumerable one), and so if *ZF* is consistent it has a denumerable model (A, E). However, it is known that the consistency of *ZF* is not enough to guarantee the existence of a *standard* model! Indeed, the consistency of *ZF* is not enough to guarantee even a well founded model, because a well founded model (A, E) can be Mostowski-Shepherdson mapped to a standard model! Thus the existence of a first-order universe, or equivalently, of a standard model, or even a well founded model, is a stronger assumption than the consistency of *ZF*.

By a *model of* NBG is meant a triple (S, C, E), such that $S \subseteq C$, E is a relation on C and all axioms of *NBG* are true, if we interpret "set" to mean element of S, "class" to mean element of C, and "∈" to be the relation E. Now, suppose $\mathcal{M}$ is a model (A, E) of *ZF*. Let $\mathcal{M}^*$ be the triple (A, A^*, E^*), where $A^* = A \cup \mathcal{F}(A)$[1] and E^* is the relation on A^* defined by the following conditions (for all x, y in A^*).

(1) If $x \notin A$, then xE^*y doesn't hold.

(2) If x and y are both in A, then xE^*y iff xEy.

(3) If $x \in A$ and $y \notin A$ (which means $y \in \mathcal{F}(A) - A$) then xE^*y iff $x \in y$.

Now, it is not difficult to verify that if $\mathcal{M}$ is a model of *ZF*, then $\mathcal{M}^*$ is a model of *NBG*. (For the verification of P$_2$, it is easiest to verify B$_1$–B$_7$ and use the class existence theorem.) Also, since for any a and b in A, the sentence $a \in b$ is true in $\mathcal{M}$ iff it is true in $\mathcal{M}^*$ (by (2) above), then it follows by induction on the structure of formulas that, for any sentence X containing only *set* variables, X is true in $\mathcal{M}$ iff X is true in $\mathcal{M}^*$, or, as is said, $\mathcal{M}^*$ is an *extension* of $\mathcal{M}$. Thus any model $\mathcal{M}$ of *ZF* can be extended to a model $\mathcal{M}^*$ of *NBG*. From this it follows that if *ZF* is consistent, so is *NBG*, because if *ZF* is consistent, it has a model $\mathcal{M}$, hence *NBG* has the model

---

[1] If A is not transitive, then A may fail to be a subset of $\mathcal{F}(A)$.

§6. RELATIVE CONSISTENCY OF GCH AND AC IN CLASS-SET THEORY 203

M^*, hence *NBG* is consistent. Thus *NBG* is consistent relative to *ZF*. More important for our present purposes, suppose X is any sentence containing only set variables and which is provable in *NBG*. Then X is true in all models of *NBG*. Hence for any model M of *ZF*, X is true in the model M^* of *NBG*. But since X contains only set variables, then X is true in M^* iff X is true in M, hence X is true in M. Thus X is true in all models of *ZF*, and hence is provable in *ZF* (by Gödel's completeness theorem).[2]

We thus have the following.

Theorem 6.1. NBG *is a conservative extension of* ZF—*i.e., any sentence X in the language of* ZF *that is provable in* NBG *is provable in* ZF.

The proof that we sketched above is due to Wang (1949). It is model theoretic and non-constructive. There is also a wholly constructive proof due to Schoenfield (1954), in which a purely mechanical recipe is given for converting any proof of a *ZF* sentence in *NBG* to a proof in *ZF*.

From Theorems 4.2, 5.1 and 6.1 we have the following.

Theorem 6.2. GCH *and* AC *are consistent relative to* NBG.

---

[2] Because if X were not provable, its negation $\neg X$ would be consistent with *ZF*, hence *ZF* would have a model in which $\neg X$ was true, contrary to the fact that X is true in all models of *ZF*.

PART III

FORCING AND INDEPENDENCE RESULTS

CHAPTER 16

FORCING, THE VERY IDEA

§1 What is forcing?

Gödel showed the continuum hypothesis is relatively consistent, and so cannot be disproved in *ZF*, using the method of *inner models*. Specifically he produced a formula, call it $L(x)$ (earlier we wrote Const(x)), that defines a transitive subclass L of V having the property that it is a first-order universe in which the continuum hypothesis, the axiom of choice, and the axiom of constructibility all are true. Unfortunately this method won't work in proving the *independence* of the continuum hypothesis. The reason is quite simple.

Suppose we had a formula $M(x)$ that defined a transitive proper class M so that M is a first-order universe in which the negation of the continuum hypothesis is true. (We assume all this can be carried out within *ZF* itself, so each axiom of *ZF*, relativized to $M(x)$, is a theorem of *ZF*, and similarly for the negation of the continuum hypothesis.) Since L is the smallest proper class that is a first-order universe, $L \subseteq M$. But the continuum hypothesis, relativized to L, is provable in *ZF*, while its negation, relativized to M, would also be provable in *ZF*. It follows that in *ZF* we would be able to prove $L \neq M$ and hence that not every set is constructible. But this contradicts Gödel's result that the axiom of constructibility is consistent with *ZF*.

Cohen (1963) invented a technique called *forcing* which can be used to prove the independence of the continuum hypothesis—see also (Cohen 1964, Cohen 1966). It is much less transparent than Gödel's use of constructible sets. In fact, it is even a little difficult to say just what it is. There have been several approaches to forcing over the years, and what their common elements may be is something not always apparent, though on close study all approaches are similar to each other as far as "deep structure" is concerned. We will present forcing from a somewhat unusual perspective, then we briefly sketch the now-standard approach. We will also give references to other treatments.

How can we get around the problem mentioned above—that inner models cannot establish independence? There are two essentially different ways, both deriving directly from the work of Cohen: change the notion of model, or change the notion of inner. Let us elaborate on each of these possibilities.

Change the meaning of model

There are many logics besides classical logic—they are lumped together rather negatively under the name "non-classical logics." For many years such logics were mostly peripheral to the concerns of logicians, and certainly to those of mathematicians. Recently, though, interest in non-classical logics has been growing, in part since they have found many applications in computer science and in artificial intelligence, in part because they have provided fresh insight into the ideas underlying Gödel's famous incompleteness theorems (Boolos 1979, Boolos 1993). Notions of semantics have been developed for these non-classical logics, utilizing quite a variety of devices that are interesting both intuitively and mathematically. One thing they all have in common is that they involve considerably more machinery than the semantics for classical logic does. This additional machinery provides a degree of control that is lacking classically. The result about inner models cited above does not rule out the possibility of producing an inner *non-classical* model in which the continuum hypothesis fails—and the extra machinery of non-classical semantics gives us something to use in trying to achieve this end.

Suppose we succeed—what have we accomplished? We will have established the independence of the continuum hypothesis but *using a non-classical logic*. So what? Well, as it happens, there are several non-classical logics into which classical logic can be embedded in non-trivial ways. That is, for a particular non-classical logic N there may be a way of translating each classical logic formula φ into a formula φ^* in the language of N in such a way that φ is a classical theorem if and only if φ^* is a theorem of N. Then if we could show that CH^* does not follow from ZF^* in the non-classical logic N, this would be enough to establish that CH does not follow from ZF classically.

This was more-or-less Cohen's original approach, though it was not described quite as we did. In effect a non-classical logic was invented for the occasion—its negation behaved differently than in classical logic. A semantics was created, a non-classical inner model was produced, and used to establish the classical independence of the continuum hypothesis. Later other more familiar non-classical logics were used for the purpose. Intuitionistic logic was the basis for (Fitting 1969). Boolean-valued logic—a kind of many-valued logic—was used by Scott & Solovay (1971)—see also (Rosser 1969, Bell 1977). We are going to use a particular *modal* logic called $S4$—we will explain what modal logics are generally, and what $S4$ is particularly, in the next few sections. Meanwhile, what about the other approach?

Change the meaning of inner

Gödel's introduction of constructible sets not only established the consistency of the continuum hypothesis, but also produced an interesting classical model, L, worth studying for its own sake. One problem with the non-classical approach just described is that it produces no classical models at all. One is left with the feeling that something more concrete ought to be possible. Well, it is, but with some curious limitations.

Cohen understood this desire for classical models and so, although the non-classical approach sketched above was the original version of forcing, the published version was along different lines. If an inner model approach cannot work, what about constructing a new model, M', "next to" a given first-order model M, but in such a way that the members of M' have "names" in M and so the truth or falsity of assertions about M' can be considered from inside M even though M' itself is not a submodel of M. It turns out this is possible, and one has considerable control over the properties of M', using essentially the same machinery described above as the non-classical approach. That is, a non-classical model created within M tells us things about the classical model M' that is outside M.

But there is a price to be paid for this. The creation of a classical model M' in the way just described requires the additional assumption that M is not just a transitive model for *ZF*, but is *countable*. We will see later on just how strong an assumption this is. Still, it is an assumption that is not unacceptable, and these models M' have been extensively studied. We will discuss the basic ideas only, leaving the rest to the specialists.

§2 What is modal logic?

In classical logic formulas are simply true or false in models. In modal logic truth can be qualified: necessarily true, known to be true, proved to be true, etc. These qualifications have some features in common, others not, so there is a multiplicity of modal logics. The history of modal logics goes back at least to Aristotle, but formal proof procedures were not introduced until early in this century, and semantics later still. We do not present a general treatment of modal logics—Hughes & Cresswell (1968) and Chellas (1980) are standard references while Bull & Segerberg (1984) and Fitting (1993) are summary versions. We only present as much as we need for our application to set theory—plus a little more for lagniappe.[1]

Syntactically, the language of classical logic is augmented with the symbol $\square$, and the additional formation rule: if X is a formula, so is $\square X$. The formula $\square X$ is read "necessarily X." A dual operator $\lozenge$ is generally introduced by definition: $\lozenge X$ abbreviates $\neg\square\neg X$. (It will follow that $\square X$ is equivalent to $\neg\lozenge\neg X$.) $\lozenge X$ is read "possibly X."

The most common semantics for modal logic is a *possible worlds* semantics, often called a *Kripke semantics* (there are other kinds of semantics as well—we will not need them). The idea, which goes back to Leibnitz, is that necessary truth means truth in all possible worlds. This is modified by the observation: which worlds are possible depends on which world you are in.

Definition 2.1. A *frame* is a pair $\langle \mathcal{G}, \mathcal{R} \rangle$ where $\mathcal{G}$ is a non-empty set (of possible worlds) and $\mathcal{R}$ is a binary relation on $\mathcal{G}$. For $p, q \in \mathcal{G}$, if $p\mathcal{R}q$ we say "q is accessible from p."

The idea is, $p\mathcal{R}q$ informally says world q is possible relative to world p. To turn a frame into a model for a first-order modal language, domains of quantification must be introduced. It is common for these domains to be world dependent, and in many applications this is most natural. We will not need such complexity for our purposes, so we can associate the same domain with each possible world of a frame.

Definition 2.2. An *augmented frame* is a triple $\langle \mathcal{G}, \mathcal{R}, \mathcal{D} \rangle$ where $\langle \mathcal{G}, \mathcal{R} \rangle$ is a frame and $\mathcal{D}$ is a non-empty set (or class), called the *domain* of the frame.

As with classical formulas in Chapter 11, we will make use of *formulas with constants*, where the constants are members of some class A—in our case this class will be $\mathcal{D}$, the domain of a frame. All the syntactical machinery from that chapter is carried

---
[1] Look it up!

over to the present setting with obvious modifications. Thus the notion of substitution has an additional clause: $(\Box\varphi)_a^x = \Box(\varphi)_a^x$. Likewise the definition of degree is changed to count occurrences of $\Box$ as well as of $\neg$, $\wedge$ and $\exists$. These simple extensions are straightforward.

Definition 2.3. A *modal model* is a structure $\langle \mathcal{G}, \mathcal{R}, \mathcal{D}, \mathcal{V} \rangle$ where $\langle \mathcal{G}, \mathcal{R}, \mathcal{D} \rangle$ is an augmented frame and $\mathcal{V}$ is a mapping from worlds (members of $\mathcal{G}$) to sets of closed atomic sentences (with constants from $\mathcal{D}$).

The idea is that if $A \in \mathcal{V}(p)$, the atom A is considered to be true at the world p. Now we extend the notion of truth at a world to all closed formulas. We symbolize truth of formula X at world p of model $\mathcal{M}$ by $p \Vdash_\mathcal{M} X$ (sometimes read "p forces X"). We write $p \nVdash_\mathcal{M} X$ to indicate that the relation $p \Vdash_\mathcal{M} X$ does not hold.

Definition 2.4. Let $\mathcal{M} = \langle \mathcal{G}, \mathcal{R}, \mathcal{D}, \mathcal{V} \rangle$ be a modal model, and let $p \in \mathcal{G}$.

(1) For a closed atomic formula A, $p \Vdash_\mathcal{M} A$ iff $A \in \mathcal{V}(p)$.

(2) For closed formulas X and Y,

  (a) $p \Vdash_\mathcal{M} \neg X$ iff $p \nVdash_\mathcal{M} X$;

  (b) $p \Vdash_\mathcal{M} (X \wedge Y)$ iff $p \Vdash_\mathcal{M} X$ and $p \Vdash_\mathcal{M} Y$.

(3) For a formula $\varphi(x)$ with at most x free, $p \Vdash_\mathcal{M} (\exists x)\varphi(x)$ iff $p \Vdash_\mathcal{M} \varphi(d)$ for some $d \in \mathcal{D}$.

(4) For a closed formula X, $p \Vdash_\mathcal{M} \Box X$ iff for every $q \in \mathcal{G}$ with $p\mathcal{R}q$, $q \Vdash_\mathcal{M} X$.

Remark It is an easy consequence of the definition above that the defined connectives, quantifiers, and modal operators have the following behavior in modal models.

(1) $p \Vdash_\mathcal{M} (X \vee Y)$ iff $p \Vdash_\mathcal{M} X$ or $p \Vdash_\mathcal{M} Y$;

(2) $p \Vdash_\mathcal{M} (X \supset Y)$ iff $p \nVdash_\mathcal{M} X$ or $p \Vdash_\mathcal{M} Y$;

(3) $p \Vdash_\mathcal{M} (\forall x)\varphi(x)$ iff $p \Vdash_\mathcal{M} \varphi(d)$ for each $d \in \mathcal{D}$;

(4) $p \Vdash_\mathcal{M} \Diamond X$ iff for some $q \in \mathcal{G}$ with $p\mathcal{R}q$, $q \Vdash_\mathcal{M} X$.

Most of the conditions above are quite straightforward, and essentially say that the classical connectives and quantifiers behave classically at each world. In fact it is easy to see that if X is classically valid, $p \Vdash_\mathcal{M} X$ for every p. The condition for $\Box$ is the key new thing. It says necessary truth amounts to truth at all possible worlds that are accessible—Leibniz's idea in formal disguise. The condition for $\Diamond$ is dual to this.

Notation conventions We will generally drop the subscript from $\Vdash_\mathcal{M}$ when it is clear which model we have in mind. Also we will sometimes say X is *true!at* p if $p \Vdash X$. And finally, rather than specifying truth at the atomic level by writing "$A \in \mathcal{V}(p)$," we will simply write "$p \Vdash A$," since we find this easier to read.

§2. WHAT IS MODAL LOGIC?

Definition 2.5. A closed formula is *valid* in a modal model if it is true at each possible world of the model. A formula with free variables is *valid* in a model with domain $\mathcal{D}$ if each closed instance of it, using constants from $\mathcal{D}$, is valid in that model. A formula is *valid* in a collection of models if it is valid in each one of them, and is simply valid if it is valid in all modal models whatsoever.

Incidentally, we have a simple treatment of validity for formulas allowing free variables because we are using models in which the domain does not vary from possible world to possible world. The general situation requires considerably more care. Now here are some simple facts we will use many times.

Proposition 2.6.

(1) If X is valid in a modal model so is $\Box X$.

(2) If $X \supset Y$ is valid in a modal model so is $\Box X \supset \Box Y$.

(3) If $X \supset Y$ is valid in a modal model so is $\Diamond X \supset \Diamond Y$.

Proof Without loss of generality we can assume X has no free variables, since the case in which it does is an easy consequence of the one in which it does not. Now, if X is valid in modal model $\mathcal{M}$, it is true at each world of $\mathcal{M}$. But then, for any particular world p of that model, X is true at every world q that is accessible from p, so $\Box X$ is true at p. Since p is an arbitrary world, $\Box X$ is valid in $\mathcal{M}$. This establishes part (1).

If $X \supset Y$ is valid in $\mathcal{M}$, so is $\Box(X \supset Y)$, by the first part. Then validity of $\Box X \supset \Box Y$ follows immediately (see Exercise 2.2).

Finally, if $X \supset Y$ is valid in $\mathcal{M}$, so is $\neg Y \supset \neg X$. Then by the previous part, the validity of $\Box \neg Y \supset \Box \neg X$ follows, and hence that of $\neg \Box \neg X \supset \neg \Box \neg Y$, that is, of $\Diamond X \supset \Diamond Y$. ∎

Our use of modal logic will be entirely semantic. But for the record, there is a well-developed proof theory too—older than the semantics in fact. The set of modal formulas that are valid in all modal models can be characterized by adding to a standard axiomatization of classical first-order logic (with modus ponens and universal generalization as rules—it is convenient to take ∀ as basic rather than ∃ just now) two axioms and a rule of inference. These additions to the classical axiomatization are the following.

K axiom $\Box(X \supset Y) \supset (\Box X \supset \Box Y)$.

Barcan formula $(\forall x)\Box\varphi(x) \supset \Box(\forall x)\varphi(x)$.

Necessitation rule Conclude $\Box X$ from X.

We have the same quantifier domain for every possible world. Axiomatically, this corresponds directly to the Barcan formula. Presentations of modal logic in the literature often do not take the Barcan formula as an axiom, and allow quantifier domains to vary from world to world.

The replacement theorem of classical logic carries over in a simple way. It will be useful to us.

Theorem 2.7 (Replacement). *Suppose $A \equiv B$ is valid in a modal model, X and Y are formulas, and Y is like X except that an occurence of A as a subformula has been replaced by B. Then $X \equiv Y$ is also valid in that model.*

Proof The proof of the corresponding theorem for classical logic proceeds by induction on the complexity of X, with one case for each connective and quantifier. Any standard text on classical logic can be consulted for this. This proof carries over to modal logic with the addition of one more case, for the modal operator. For this case, the hypothesis is as follows. Assume the validity in modal model $\mathcal{M}$ of:

$$\varphi(x_1, \ldots, x_n) \equiv \psi(x_1, \ldots, x_n).$$

Then we have the validity of each of

$$\varphi(x_1, \ldots, x_n) \supset \psi(x_1, \ldots, x_n) \text{ and } \psi(x_1, \ldots, x_n) \supset \varphi(x_1, \ldots, x_n).$$

Using Proposition 2.6, we have the validity in $\mathcal{M}$ of each of:

$$\Box\varphi(x_1, \ldots, x_n) \supset \Box\psi(x_1, \ldots, x_n) \text{ and } \Box\psi(x_1, \ldots, x_n) \supset \Box\varphi(x_1, \ldots, x_n),$$

and consequently the validity in $\mathcal{M}$ of

$$\Box\varphi(x_1, \ldots, x_n) \equiv \Box\psi(x_1, \ldots, x_n).$$

∎

A note on sets and classes So far, a modal model is a structure $\langle \mathcal{G}, \mathcal{R}, \mathcal{D}, \mathcal{V} \rangle$ meeting certain conditions, but we will need to relax this somewhat. In our constructions the domain, $\mathcal{D}$, will often be a *proper class*, and similarly for $\mathcal{V}$, though not for $\mathcal{G}$ or $\mathcal{R}$. If so, a model cannot be a 4-tuple in the formal sense. In such a case we can still talk about the mathematical structure consising of a frame, a domain, and a truth function, and we will denote it by $(\mathcal{G}, \mathcal{R}, \mathcal{D}, \mathcal{V})$.

Something of this sort came up before when we introduced the notion of *relational structure* in Chapter 10. It would be useful for you to review the beginning of that chapter. We follow exactly the same conventions—using round parentheses to indicate an informal grouping, instead of the angled parentheses that indicate the pairing function of formal set theory. Thus from now on, a modal model is a collection of four related things, $(\mathcal{G}, \mathcal{R}, \mathcal{D}, \mathcal{V})$, some of which may be proper classes. Nothing we have said so far (or will say) is affected by this broadening of the notion of modal model.

EXERCISES

Exercise 2.1. Show the conditions for $\lor, \supset, \forall$ and $\Diamond$, listed following Definition 2.4, are correct.

Exercise 2.2. Show that Axiom **K** is valid in every modal model. Likewise for the Barcan formula.

Exercise 2.3. Show the converse of the Barcan formula is also valid, and hence we have the validity of: $(\forall x)\Box\varphi(x) \equiv \Box(\forall x)\varphi(x)$.

Exercise 2.4. As defined, closed formulas have truth values at worlds of modal models, but we could extend the definition and say that $p \Vdash \varphi(x_1,\ldots,x_n)$ means $p \Vdash (\forall x_1)\ldots(\forall x_n)\varphi(x_1,\ldots,x_n)$. Show that $p \Vdash \Box\varphi(x_1,\ldots,x_n)$ iff for all $q \in \mathcal{G}$ with $p\mathcal{R}q$, $q \Vdash \varphi(x_1,\ldots,x_n)$.

Exercise 2.5.

(a) Show that $\Box(X \wedge Y) \supset (\Box X \wedge \Box Y)$ is valid.

(b) Show that $(\Box X \wedge \Box Y) \supset \Box(X \wedge Y)$ is valid. (Note that we now have $\Box(X \wedge Y) \equiv (\Box X \wedge \Box Y)$ is valid.)

(c) Show that $(\Diamond X \vee \Diamond Y) \equiv \Diamond(X \vee Y)$ is valid.

(d) Show that $(\Diamond X \wedge \Diamond Y) \supset \Diamond(X \wedge Y)$ is not valid, and $\Box(X \vee Y) \supset (\Box X \vee \Box Y)$ is not valid.

Exercise 2.6. Show the following is valid in every modal model: $\Box(X \equiv Y) \supset (\Box X \equiv \Box Y)$.

Exercise 2.7. Using the axiom system outlined in this section, prove the following.

(a) $\Box(X \wedge Y) \supset \Box X$.

(b) $\Box(X \wedge Y) \supset (\Box X \wedge \Box Y)$.

(c) $(\Box X \wedge \Box Y) \supset \Box(X \wedge Y)$.

§3 What is $S4$ and why do we care?

Several interesting modal logics can be characterized by putting simple conditions on accessibility relations. For instance, the well-known modal logic $S5$ can be described as the set of formulas valid in all modal models whose accessibility relation is an *equivalence relation*. Symmetry alone, transitivity alone, and various combinations, all give rise to well-investigated modal logics. Now, the only modal logic we will need for applications to set theory is usually called $S4$ and is characterized semantically by models whose accessibility relations are *preorderings*: reflexive and transitive. (In fact, the $S4$ models we will construct involve not just preorderings, but *partial orderings*: reflexive, transitive and anti-symmetric.)

Definition 3.1. An $S4$ *model* is a modal model whose accessibility relation is reflexive and transitive—a preordering. A formula is $S4$ *valid* if it is valid in all $S4$ models.

The logic $S4$ is widely used, with a long history—it is a logic that is discussed in every text on modal logic. But why does it come up here? Let us speak very informally for a bit. Suppose we are trying to create a model of set theory having some particular property — for instance, we want the model to contain a non-constructible set a. Then we have a variety of "conditions" that must be met in order to guarantee a is different from each constructible set: for each constructible set c we want a and c to differ on some x, and we do not really care what the set x is. In the entire space of conditions (whatever they are) many different ones could ensure that a and c differ, since the choice of x is open, but of course, some conditions might conflict with others. We want to find a family of conditions that are compatible, that collectively describe a single model, and that guarantee a and c differ for each constructible set c. We have the beginnings of a modal model: take the collection of possible worlds $\mathcal{G}$ to be the set of all conditions.

There is a natural notion of one condition being stronger than another. For instance, a condition that ensures a has both x and y as members is stronger than one only ensuring that a contains x. A little thought suggests this notion of relative strength is really a preordering, and this gives us the next component of a modal model: consider condition q to be accessible from condition p—$p\mathcal{R}q$—if q is stronger than p. We thus have a frame $\langle \mathcal{G}, \mathcal{R} \rangle$ and since accessibility is a preordering, it is a frame for $S4$. The modal language associated with any $S4$ model based on this frame gives us a simple means for making assertions about the space of conditions using the machinery of formal logic—we want to be able to talk about conditions and stronger versions of them, exactly what $\square$ and $\Diamond$ will allow us to do. (In standard treatments of forcing, modal operators are not used and assertions about conditions and strengthenings of them are made from the outside—in the metalanguage.)

Before we bring set theory more explicitly into the picture, we pause to discuss $S4$ for its own sake a bit, to help you familiarize yourselves with it. In the previous section we gave an axiomatic characterization of the family of closed formulas valid in *all* modal models. To get from that to $S4$ it is sufficient to add the following two axiom schemas.

Reflexivity axiom $\square X \supset X$.

Transitivity axiom $\square X \supset \square \square X$.

The reflexivity schema is valid in all modal models whose accessibility relations are reflexive; the transitivity schema is valid in all modal models whose accessibility relations are transitive. We leave the verification of this—soundness of the axiom system—to you as an exercise. It is quite straightforward. The converse is much less obvious—every closed formula that is $S4$ valid is provable using the axiom system of the previous section together with the axioms above. A proof of this completeness theorem can be found in any standard book on modal logic. We do not need it here, and do not prove it.

Finally a simple observation about $S4$ that will be of much use. In an $S4$ model, if $p \Vdash \square X$ and $p\mathcal{R}p'$, then $p' \Vdash \square X$ as well. This follows immediately from the transitivity of the accessibility relation. Suppose $p \Vdash \square X$ and $p\mathcal{R}p'$. If q is any world accessible from p', by transitivity q is also accessible from p, hence $q \Vdash X$. Since q is arbitrary, $p' \Vdash \square X$.

EXERCISES

Exercise 3.1. Show that $\Box(\Box A \supset \Box A)$ is $S4$ valid, but $\Box(\Box A \vee \Box \neg \Box A)$ is not.

Exercise 3.2. Show $\Box X \supset X$ and $\Box X \supset \Box \Box X$ are $S4$ valid.

Exercise 3.3. Show $\Diamond X \supset \Box \Diamond X$ is not $S4$ valid.

Exercise 3.4. Show the $S4$ validity of $\Box \Box X \equiv \Box X$ and $\Diamond \Diamond X \equiv \Diamond X$.

§4 A classical embedding

So far, we have sketched why $S4$ comes up—we are interested in meeting various conditions; conditions are naturally preordered; and $S4$ is the modal logic that a preordering gives us. Now let us take things a little further. Under what circumstances can we say that meeting a condition is enough to "ensure the truth" of some formula X? The problem is, if X is true given condition p—or at possible world p—perhaps more information will convince us X is false after all. We want a certain stability here; additional information should preserve truth. Let us say "p forces X" if X is true at p and also at every condition that strengthens p. Then, using modal notation, saying X is forced by p simply amounts to saying $\Box X$ is true at p since the condition for forcing corresponds exactly to the behavior of $\Box$ in $S4$ models. But it is not enough to have stability just for X, we should also have it for the constituents of X as well. This suggests we should consider translating a formula from the language of classical logic into a formula in a modal language by simply putting $\Box$ in front of it and every subformula of it. For instance, if A is atomic, $A \wedge \neg A$ translates to $\Box(\Box A \wedge \Box \neg \Box A)$.

The mapping just described is an interesting translation—one that has been much studied. Unfortunately it is not quite what we want. For instance, it makes a difference whether we treat $\vee$ as primitive or defined, and similarly for the other connectives and quantifier. If we treat them as primitive, this simply does not embed classical logic into $S4$. $A \supset A$ is classically valid, and its translate, $\Box(\Box A \supset \Box A)$, is valid in all $S4$ models, but $A \vee \neg A$ is also classicaly valid, and its translate, $\Box(\Box A \vee \Box \neg \Box A)$, is *not* valid in all $S4$ models (Exercise 3.1). In fact, this mapping corresponds to *intuitionistic*, rather than classical logic (with a minor deviation because we have made domains of quantification independent of worlds, a point of no importance here).

Suppose we aim our sights a little lower. Instead of asking whether a condition guarantees X no matter what additional information we get, let us allow for the possibility that further information could temporarily convince us X is not the case, provided X is "recoverable" in the sense that there will always be additional information that suffices to restore X to us. More precisely, we might say a condition p *weakly forces* X provided, no matter what stronger set of conditions q we consider, there is a still stronger set r that establishes the truth of X. That is, p weakly forces X if, for every q accessible from p, there is an r accessible from q at which X is true. But this too can be said easily using modal operators: it amounts to asserting that $\Box \Diamond X$ is true at p. And once again, we want to consider not just X itself, but its subformulas as well. Now we have the essential idea of the embedding we really will use.

Definition 4.1. We specify a translation from non-modal to modal formulas as follows.

(1) For A atomic, $\|A\| = \square\lozenge A$.

(2) $\|\neg X\| = \square\lozenge\neg\|X\|$.

(3) $\|X \wedge Y\| = \square\lozenge(\|X\| \wedge \|Y\|)$.

(4) $\|(\exists x)\varphi\| = \square\lozenge(\exists x)\|\varphi\|$.

Thus the translation from X to $\|X\|$ amounts to putting $\square\lozenge$ in front of every subformula of X. This assumes we take $\neg$, $\wedge$, and $\exists$ as basic, though in fact this does not really matter—something we show after a few preliminary results. In fact, once we establish the basic properties of the $\|\ \|$ translation, we will rarely find ourselves working with $\square$ and $\lozenge$ again. This is the only section in which direct manipulation of the modal operators plays a significant role.

Rather than working with models directly, it is generally easier to make use of previously established validities and validity preserving transformations. This amounts to a "pseudo-axiomatic" approach. First, a lemma that will help to prove the more interesting items that follow.

Lemma 4.2. *The following formulas are all $S4$ valid.*

(1) $\square\lozenge\square\lozenge X \equiv \square\lozenge X$.

(2) $\square\lozenge(\square X \supset \square\lozenge Y) \supset (\square X \supset \square\lozenge Y)$.

(3) $\square\lozenge(\square\lozenge X \wedge \square\lozenge Y) \supset (\square\lozenge X \wedge \square\lozenge Y)$.

(4) $(\square\lozenge X \wedge \square\lozenge Y) \supset \square\lozenge(\square\lozenge X \wedge \square\lozenge Y)$.

(5) $(\square\lozenge X \wedge \square\lozenge Y) \equiv \square\lozenge(\square\lozenge X \wedge \square\lozenge Y)$.

Proof

(1) First, $\square\lozenge X \supset \lozenge X$ is $S4$ valid (since $\square Z \supset Z$ is for any Z), hence by Proposition 2.6 we have the $S4$ validity of $\lozenge\square\lozenge X \supset \lozenge\lozenge X$ and $\square\lozenge\square\lozenge X \supset \square\lozenge\lozenge X$. Finally since $\lozenge\lozenge X \equiv \lozenge X$ is $S4$ valid, the $S4$ validity of $\square\lozenge\square\lozenge X \supset \square\lozenge X$ follows. We leave the converse as Exercise 4.2 below.

(2) The following is a list of modal formulas. The implication between each formula and the next is $S4$ valid—something we leave to you to verify.

$$\square\lozenge(\square X \supset \square\lozenge Y)$$
$$\square\lozenge(\neg\square X \vee \square\lozenge Y)$$
$$\square(\lozenge\neg\square X \vee \lozenge\square\lozenge Y)$$
$$\square(\neg\square\square X \vee \lozenge\square\lozenge Y)$$
$$\square(\square\square X \supset \lozenge\square\lozenge Y)$$
$$\square\square\square X \supset \square\lozenge\square\lozenge Y$$
$$\square X \supset \square\lozenge Y.$$

Item (5) is an immediate consequence of items (3) and (4). We leave these as Exercise 4.3. ∎

§4. A CLASSICAL EMBEDDING

Now some fundamental properties of our embedding.

Proposition 4.3. *The following are all S4 valid.*

(1) $[\![X]\!] \equiv \Box[\![X]\!]$.

(2) $[\![X]\!] \equiv \Box\Diamond[\![X]\!]$.

(3) $[\![X \wedge Y]\!] \equiv ([\![X]\!] \wedge [\![Y]\!])$.

(4) $[\![\neg X]\!] \supset \neg[\![X]\!]$.

(5) $(\exists x)[\![\varphi(x)]\!] \supset [\![(\exists x)\varphi(x)]\!]$.

Proof

(1) is trivial since $\Box Z \equiv \Box\Box Z$ in $S4$, and $[\![X]\!]$ begins with an occurrence of $\Box$.

(2) is an easy consequence of the definition of $[\![X]\!]$ and part (1) of the lemma above.

(3) Follows using part (5) of the previous lemma.

We leave the remaining parts as Exercise 4.4. ∎

Remark We know that if $p \Vdash \Box X$ and $p \mathcal{R} p'$ then $p' \Vdash \Box X$. Then by part (1) of the proposition above, if $p \Vdash [\![X]\!]$ and $p\mathcal{R}p'$, it follows that $p' \Vdash [\![X]\!]$. This is a very useful fact, and is one that we will use over and over, generally without comment.

Now we show that, with respect to this translation, the behavior of the connectives and quantifiers we have taken as defined is the same as if we had taken than as primitive.

Proposition 4.4. *The following are S4-valid:*

(1) $[\![X \vee Y]\!] \equiv \Box\Diamond([\![X]\!] \vee [\![Y]\!])$.

(2) $[\![X \supset Y]\!] \equiv \Box\Diamond([\![X]\!] \supset [\![Y]\!])$.

(3) $[\![(\forall x)\varphi]\!] \equiv \Box\Diamond(\forall x)[\![\varphi]\!]$.

Proof

(1) By definition of the connective $\vee$, $[\![X \vee Y]\!]$ is $[\![\neg(\neg X \wedge \neg Y)]\!]$, and by the definition of the translation, this is $\Box\Diamond\neg[\![\neg X \wedge \neg Y]\!]$. By part (3) of Proposition 4.3 (and the replacement theorem), this is equivalent to $\Box\Diamond\neg([\![\neg X]\!] \wedge [\![\neg Y]\!])$. Again using the definition of the translation, this is $\Box\Diamond\neg(\Box\Diamond\neg[\![X]\!] \wedge \Box\Diamond\neg[\![Y]\!])$. Now $\Box$ distributes over a conjunction (Exercise 2.5) so this in turn is equivalent to the formula $\Box\Diamond\neg\Box(\Diamond\neg[\![X]\!] \wedge \Diamond\neg[\![Y]\!])$. Next, $\Diamond\neg Z \equiv \neg\Box Z$, so this is equivalent to $\Box\Diamond\Diamond\neg(\neg\Box[\![X]\!] \wedge \neg\Box[\![Y]\!])$. Also $\Diamond\Diamond Z \equiv \Diamond Z$, and we have part (1) of Proposition 4.3, so finally, this formula is equivalent to $\Box\Diamond\neg(\neg[\![X]\!] \wedge \neg[\![Y]\!])$, or $\Box\Diamond([\![X]\!] \vee [\![Y]\!])$.

(2) We leave this as Exercise 4.5.

(3) By definition of the universal quantifier, $[\![(\forall x)\varphi]\!]$ is equivalent to $[\![\neg(\exists x)\neg\varphi]\!]$ which expands to $\Box\Diamond\neg\Box\Diamond(\exists x)\Box\Diamond\neg[\![\varphi]\!]$. Now, using the fact that $\Box$ and $\Diamond$ are dual, as are $\forall$ and $\exists$, this converts to $\Box\Diamond\Diamond\Box(\forall x)\Diamond\Box[\![\varphi]\!]$. But $\Box$ and $\forall$ commute (Exercise 2.3), so this is equivalent to $\Box\Diamond\Diamond(\forall x)\Box\Diamond\Box[\![\varphi]\!]$. Finally, since $\Diamond\Diamond Z \equiv \Diamond Z$, $\Box[\![\varphi]\!] \equiv [\![\varphi]\!]$, and $\Box\Diamond[\![\varphi]\!] \equiv [\![\varphi]\!]$, this is equivalent to $\Box\Diamond(\forall x)[\![\varphi]\!]$. ∎

Now that we know all the connectives and quantifiers are "safe," it is useful to have some properties of the defined ones available.

Proposition 4.5. *The following formulas are S4 valid.*

(1) $[\![X \supset Y]\!] \supset ([\![X]\!] \supset [\![Y]\!])$.

(2) $([\![X]\!] \vee [\![Y]\!]) \supset [\![X \vee Y]\!]$.

(3) $[\![(\forall x)\varphi(x)]\!] \equiv (\forall x)[\![\varphi(x)]\!]$.

Proof

(1) Using part (2) of Lemma 4.2,

$$\begin{aligned}
[\![X \supset Y]\!] &= \Box\Diamond([\![X]\!] \supset [\![Y]\!]) \\
&\equiv \Box\Diamond(\Box\Diamond[\![X]\!] \supset \Box\Diamond[\![Y]\!]) \\
&\supset \Box\Diamond[\![X]\!] \supset \Box\Diamond[\![Y]\!] \\
&\equiv [\![X]\!] \supset [\![Y]\!].
\end{aligned}$$

Part (2) is similar.

(3) In one direction,

$$(\forall x)[\![\varphi(x)]\!] \equiv (\forall x)\Box\Diamond[\![\varphi(x)]\!] \supset \Box\Diamond[\![\varphi(x)]\!]$$

from which follows (by Proposition 2.6)

$$\Box\Diamond(\forall x)[\![\varphi(x)]\!] \supset \Box\Diamond\Box\Diamond[\![\varphi(x)]\!] \equiv \Box\Diamond[\![\varphi(x)]\!]$$

and then by universal generalization,

$$\Box\Diamond(\forall x)[\![\varphi(x)]\!] \supset (\forall x)\Box\Diamond[\![\varphi(x)]\!].$$

Hence
$$[\![(\forall x)\varphi(x)]\!] \supset (\forall x)[\![\varphi(x)]\!].$$

In the other direction,

$$(\forall x)\Box\Diamond[\![\varphi(x)]\!] \supset \Diamond(\forall x)\Box\Diamond[\![\varphi(x)]\!]$$

and so
$$\Box(\forall x)\Box\Diamond[\![\varphi(x)]\!] \supset \Box\Diamond(\forall x)\Box\Diamond[\![\varphi(x)]\!]$$
but using the Barcan formula,
$$\begin{aligned} \Box(\forall x)\Box\Diamond[\![\varphi(x)]\!] &\equiv (\forall x)\Box\Box\Diamond[\![\varphi(x)]\!] \\ &\equiv (\forall x)\Box\Diamond[\![\varphi(x)]\!] \\ &\equiv (\forall x)[\![\varphi(x)]\!] \end{aligned}$$

and
$$\begin{aligned} \Box\Diamond(\forall x)\Box\Diamond[\![\varphi(x)]\!] &\equiv \Box\Diamond(\forall x)[\![\varphi(x)]\!] \\ &\equiv [\![(\forall x)\varphi(x)]\!]. \end{aligned}$$

So
$$(\forall x)[\![\varphi(x)]\!] \supset [\![(\forall x)\varphi(x)]\!].$$

∎

As we noted earlier, specific manipulation of $\Box$ and $\Diamond$ will rarely appear after this section; rather, properties of these modal connectives have been used to establish the behavior of the translation $[\![\ ]\!]$, and it is this behavior that will be used directly. Note that some of the items in the propositions above were implications, not equivalences. The following deals with this, along with some additional material.

Proposition 4.6. *In an $S4$ model, for any possible world p:*

(1) *if $p \Vdash \neg[\![X \supset Y]\!]$ then for some p' with $p\mathcal{R}p'$, $p' \Vdash [\![X]\!]$ and $p' \Vdash [\![\neg Y]\!]$;*

(2) *if $p \Vdash \neg[\![X]\!]$ then for some p' with $p\mathcal{R}p'$, $p' \Vdash [\![\neg X]\!]$;*

(3) *if $p \Vdash [\![X \vee Y]\!]$ then for some p' with $p\mathcal{R}p'$, $p' \Vdash [\![X]\!] \vee [\![Y]\!]$;*

(4) *if $p \Vdash [\![(\exists x)\varphi(x)]\!]$ then for some p' with $p\mathcal{R}p'$, $p' \Vdash (\exists x)[\![\varphi(x)]\!]$.*

Proof We start with (2). Suppose $p \Vdash \neg[\![X]\!]$. Then $p \Vdash \neg\Box\Diamond[\![X]\!]$ so $p \Vdash \Diamond\Box\neg[\![X]\!]$. Then for some p' with $p\mathcal{R}p'$, $p' \Vdash \Box\neg[\![X]\!]$. From this it follows that $p' \Vdash \Box\Diamond\neg[\![X]\!]$, that is, $p' \Vdash [\![\neg X]\!]$.

Item (1) is similar, and we leave the rest as Exercise 4.6. ∎

Now, finally, we show we really do have an embedding in the technical sense we need.

Proposition 4.7. *Let X be a closed formula in the language of classical logic; X is classically valid if and only if $[\![X]\!]$ is valid in all $S4$ models.*

Proof One way is rather trivial. Suppose X is not true in some classical model M. Construct an $S4$ "one-world" model $(\mathcal{G}, \mathcal{R}, \mathcal{D}, \mathcal{V})$ as follows. Let $\mathcal{G}$ be $\{p\}$ where p is some arbitrary object. Set $p\mathcal{R}p$. Take for $\mathcal{D}$ the domain of the classical model M. And finally, for an atomic formula A, set $p \Vdash A$ iff A is true in M. In such a one-world model, for any formula Z, $p \Vdash Z$ iff $p \Vdash \Box Z$ iff $p \Vdash \Diamond Z$. Then it follows easily, by induction on degree, that for each closed formula Z, Z is true in M iff $p \Vdash \llbracket Z \rrbracket$. Since X is not true in M, $\llbracket X \rrbracket$ is not true at p in the $S4$ model we constructed. Contrapositively, if $\llbracket X \rrbracket$ is valid in all $S4$ models, X is classically valid.

The other direction is more work, though conceptually not deep. Suppose X is classically valid. By the completeness theorem for classical logic, X is provable using any standard axiom system, so it is enough to show that the translate of each axiom is valid in all $S4$ models; and that the validity of translates of the premises of a rule of inference implies validity of the translate of the conclusion of the rule. Any axiomatization of classical logic will do. We use the following (which happens to take $\neg$, $\supset$, and $\forall$ as primitive, though as we have seen, this does not matter).

Axioms All formulas of the following forms

(1) $A \supset (B \supset A)$

(2) $(A \supset (B \supset C)) \supset ((A \supset B) \supset (A \supset C))$

(3) $(\neg A \supset \neg B) \supset (B \supset A)$

(4) $(\forall x)\varphi(x) \supset \varphi(y)$ where y is a variable such that x does not occur in the scope of $(\forall y)$ in $\varphi(x)$.

(5) $(\forall x)(\psi \supset \varphi(x)) \supset (\psi \supset (\forall x)\varphi(x))$ where x does not occur free in ψ.

Rules of inference

Modus ponens From A and $A \supset B$ infer B.

Universal generalization From $\varphi(x)$ infer $(\forall x)\varphi(x)$.

In verifying the validity of translates of the classical axioms above, which may contain free variables, we work with instances and verify validity of them. And we only consider one classical axiom, as a representative example: $A \supset (B \supset A)$. We show the $S4$-validity of $\llbracket A' \supset (B' \supset A') \rrbracket$ (where this is an instance of the translation). Now, if there were an $S4$ model and a possible world p of it at which $\llbracket A' \supset (B' \supset A') \rrbracket$ were not true, by part (1) of Proposition 4.6 there would be some possible world p', with $p\mathcal{R}p'$, such that $p' \Vdash \llbracket A' \rrbracket$ and $p' \Vdash \llbracket \neg(B' \supset A') \rrbracket$. Then by part (4) of Proposition 4.3, $p' \Vdash \neg \llbracket B' \supset A' \rrbracket$, so again, for some p'' with $p'\mathcal{R}p''$, $p'' \Vdash \llbracket B' \rrbracket$ and $p'' \Vdash \llbracket \neg A' \rrbracket$. From this, $p'' \Vdash \neg \llbracket A' \rrbracket$ and so $p'' \nVdash \llbracket A' \rrbracket$. But we also have that $p' \Vdash \llbracket A' \rrbracket$ and since $p'\mathcal{R}p''$, then $p'' \Vdash \llbracket A' \rrbracket$, and this is a contradiction.

That validity of closures of translates is preserved under applications of modus ponens is an easy consequence of part (1) of Proposition 4.5. The universal generalization rule is handled using part (3) of the same proposition. ∎

Corollary 4.8. *If $X \equiv Y$ is classically valid then $[\![X]\!] \equiv [\![Y]\!]$ is $S4$ valid.*

Proof If $X \equiv Y$ is classically valid then $[\![X \equiv Y]\!]$ is $S4$ valid. Then, using earlier results,
$$\begin{aligned}[\![X \equiv Y]\!] &= [\![(X \supset Y) \wedge (Y \supset X)]\!] \\ &\equiv ([\![X \supset Y]\!] \wedge [\![Y \supset X]\!]) \\ &\sqsupset (([\![X]\!] \supset [\![Y]\!]) \wedge ([\![Y]\!] \supset [\![X]\!])) \\ &= ([\![X]\!] \equiv [\![Y]\!]).\end{aligned}$$

∎

Remark By virtue of this result we will freely leave some parentheses out, writing $[\![X \vee Y \vee Z]\!]$ for instance, since classically any parenthesization is equivalent to any other.

EXERCISES

Exercise 4.1. Show the validity of the translates of the other axioms, for the classical axiomatization given above.

Exercise 4.2. Show the $S4$ validity of $\Box\Diamond X \supset \Box\Diamond\Box\Diamond X$.

Exercise 4.3. Show the $S4$ validity of the following.

(a) $\Box\Diamond(\Box\Diamond X \wedge \Box\Diamond Y) \supset (\Box\Diamond X \wedge \Box\Diamond Y)$.

(b) $(\Box\Diamond X \wedge \Box\Diamond Y) \supset \Box\Diamond(\Box\Diamond X \wedge \Box\Diamond Y)$.

Exercise 4.4. Show parts (4) and (5) of Proposition 4.3.

Exercise 4.5. Show part (2) of Proposition 4.4

Exercise 4.6. Show parts (3) and (4) of Proposition 4.6.

Exercise 4.7. Which of the following are $S4$ valid, which are not:

(a) $[\![X \vee Y]\!] \supset ([\![X]\!] \vee [\![Y]\!])$;

(b) $([\![X]\!] \vee [\![Y]\!]) \supset [\![X \vee Y]\!]$.

Exercise 4.8. Extend Proposition 4.7 and its proof from validity to logical consequence.

§5 The basic idea

We can now set up the goal for much of what follows.

Definition 5.1. We call an $S4$ model $\mathcal{M}$ an *$S4$-ZF model* if, for each axiom A of ZF, and for A being the axiom of well foundedness, $[\![A]\!]$ is valid in $\mathcal{M}$.

(In the interests of simple terminology, we have thrown well foundedness in with the usual ZF axioms. Of course proving consistency and independence relative to ZF plus well foundedness is a stronger result than proving it with respect to *ZF* alone.) Now, suppose we can construct an *S*4-*ZF* model $\mathcal{M}$ in which $[\![\text{CH}]\!]$ is *not* valid, where CH is the continuum hypothesis. From this the *classical* independence is immediate, by the following argument. If CH were provable from the axioms of *ZF* together with the axiom of well foundedness, only a finite number of axioms, say $A_1, \ldots, A_n$, would be needed. Then we would have the classical validity of $(A_1 \wedge \cdots \wedge A_n) \supset \text{CH}$. Now the *S*4 validity of the translate is a consequence: $[\![(A_1 \wedge \cdots \wedge A_n) \supset \text{CH}]\!]$. But using Propositions 4.3 and 4.5, we would then have the *S*4 validity of the following:

$$([\![A_1]\!] \wedge \cdots \wedge [\![A_n]\!]) \supset [\![\text{CH}]\!]$$

and this is impossible because $\mathcal{M}$ is a countermodel. Thus the construction of suitable *S*4-*ZF* models is the goal.

CHAPTER 17

THE CONSTRUCTION OF $S4$ MODELS FOR ZF

§1 What are the models?

We have the necessary background out of the way, and it is time to begin our real work. We say what $S4$ model, or rather, what family of $S4$ models we will be investigating, and we start the investigation off. We have a family of models, rather than a single one, because one model is needed to show the independence of the continuum hypothesis, another to show the independence of the axiom of choice, and so on. But all these models have much in common, and it is to this common part that the present chapter is devoted.

Important note Let M be a transitive subclass of V that is a well founded first-order universe. M could be R_Ω, or L, for instance. (Since it is transitive, the things M thinks are ordinals are simply the ordinals of V that happen to be in M, and similarly for other absolute notions.) Almost all the work in this and subsequent chapters will take place within M. Since, as it happens, forcing is not absolute, the choice of M can be significant. We gain fine control over what happens by making explicit what assumptions about M we need for particular results. Since so much takes place in M we will not reflect it in our notation, but merely remark in our explanations when we are and when we are not working in M. We will create, within M, a family of $S4$ models and show translates of the ZF axioms are valid in them. This includes the axiom of well foundedness, which we are assuming is true in M, and the axiom of choice, provided we assume it is true in M.

Let $\mathcal{G}$ be some collection of items (called possible worlds or forcing conditions, as appropriate), and let $\mathcal{R}$ be a relation on $\mathcal{G}$ that is reflexive and transitive. Then $\langle \mathcal{G}, \mathcal{R} \rangle$ is an $S4$ frame. For the entire of Part III, the frame $\langle \mathcal{G}, \mathcal{R} \rangle$ is assumed to be a member of M, or as we will sometimes say, an M-set.

To turn a frame into an augmented frame we need a domain. For this we follow, and extend, the definition of well founded set in almost a direct way—simply taking into account the multiplicity of worlds. As a first approximation, we could say that a member f of the domain is to be a function that assigns to each world p a set $f(p)$, and for each p, $f(p)$ is a set of "previously constructed" items. The idea is, given the "information" p, we know f has the things in $f(p)$ as members. There is a technical problem with this, however. In Chapter 7 we introduced a notion of ordinal hierarchy, and it would be nice if we could make use of it now, instead of proving new items that are almost the same as what we proved earlier. To get around this difficulty we use a simple device that avoids the problem—we work with relations instead of with functions.

Definition 1.1. For each ordinal α (in M) we define a set $R_\alpha^{\mathcal{G}}$ as follows.

(1) $R_0^{\mathcal{G}} = \emptyset$.

(2) $R_{\alpha+1}^{\mathcal{G}}$ is the set of all subsets (in M) of $\mathcal{G} \times R_\alpha^{\mathcal{G}}$.

(3) For a limit ordinal λ,
$$R_\lambda^{\mathcal{G}} = \cup_{\alpha < \lambda} R_\alpha^{\mathcal{G}}.$$

Now let
$$\mathcal{D}^{\mathcal{G}} = \cup_\alpha R_\alpha^{\mathcal{G}}$$
where the union is taken over the ordinals in M. $(\mathcal{G}, \mathcal{R}, \mathcal{D}^{\mathcal{G}})$ is an augmented frame.

If $f \subseteq \mathcal{D}^{\mathcal{G}}$ and $p \in \mathcal{G}$, $f''(\{p\})$ is the proper version of the function treatment suggested earlier.

Remarks There are several important items concerning this definition.

(1) Each $R_\alpha^{\mathcal{G}}$ is a member of M, and $\mathcal{D}^{\mathcal{G}}$ is a subclass of M.

(2) The sequence $R_0^{\mathcal{G}}, R_1^{\mathcal{G}}, \ldots, R_\alpha^{\mathcal{G}}, \ldots$ is an *ordinal hierarchy* as defined in Chapter 7, §2 (you are asked to show this in Exercise 1.2) and hence all the results about such hierarchies apply.

(3) The ordinal hierarchy is *strict*, that is, $R_{\alpha+1}^{\mathcal{G}} \neq R_\alpha^{\mathcal{G}}$ (we will show this later). It follows that $\mathcal{D}^{\mathcal{G}}$ is a proper class in M.

(4) If $f \subseteq \mathcal{G} \times \mathcal{D}^{\mathcal{G}}$ and f is an M-set then $f \in \mathcal{D}^{\mathcal{G}}$. This is an easy consequence of the axiom of substitution.

(5) If $\mathcal{G}$ has only a single world, the construction above is isomorphic to that of the R_α sequence.

Since we have an ordinal hierarchy, a notion of rank can be introduced.

Definition 1.2. We say $f \in \mathcal{D}^{\mathcal{G}}$ has $\mathcal{G}$-*rank* α if $f \in R_{\alpha+1}^{\mathcal{G}}$ but $f \notin R_\alpha^{\mathcal{G}}$.

Note that if $f \in \mathcal{D}^{\mathcal{G}}$ and $\langle p, g \rangle \in f$ for some $p \in \mathcal{G}$, then $g \in \mathcal{D}^{\mathcal{G}}$ and g must have lower $\mathcal{G}$-rank than f. This observation is the heart of transfinite induction proofs about $\mathcal{D}^{\mathcal{G}}$.

We still must define a truth assignment, $\mathcal{V}^{\mathcal{G}}$, and this is the trickiest part. We define truth at the atomic level in three stages. For starters we introduce a first approximation to the "real" membership relation. We write $x \, \varepsilon \, y$, and read it as x is *immediately in* y. Its truth definition is straightforward.

Definition 1.3. For $p \in \mathcal{G}$ and $f, g \in \mathcal{D}^{\mathcal{G}}$, set $f \, \varepsilon \, g$ to be true at p if $\langle p, f \rangle \in g$, that is if $f \in g''(\{p\})$.

§1. WHAT ARE THE MODELS?

The problem with using ε as our membership notion is that it is not extensional. Consider the following example. $\mathcal{G} = \{p, q\}$ has two worlds; the accessibility relation does not matter for the example. Suppose a, f, g, and h are members of $\mathcal{D}^{\mathcal{G}}$, and we have the following simple set-up:

$$\begin{aligned} f &= \{\langle q, a \rangle\} \\ g &= \{\langle p, a \rangle, \langle q, a \rangle\} \\ h &= \{\langle q, f \rangle\}. \end{aligned}$$

Now, $(f \varepsilon h)$ is true at q. Also, f and g have the same members as far as q is concerned. (They are different because they differ at p.) Yet $(g \varepsilon h)$ is not true at q.

If we want extensionality we have some work to do, but the basic idea is simple. Take as members of h at world q those things that are immediately in h at q, and also those things that q "thinks" are equal to things immediately in h. This means we should begin by defining a notion of equality, then the membership notion we really want. The center of the development lies in characterizing equality, and we approach it "by approximation."

Notation convention Let $\mathcal{L}_C$ be the classical first-order language of set theory, as defined in Chapter 11. This has $\in$ as its only relation symbol—it is intended to represent the "real" membership relation. We now introduce a richer second language, $\mathcal{L}_M$, for use in our modal models. First of all, of course, it allows occurrences of $\square$. Secondly, there are infinitely many relation symbols in addition to $\in$—specifically we have ε, which was used above, $\approx$, intended to represent set equality (in the next section we see to what extent it does so), and also for each ordinal α in M there is $\approx_\alpha$, the αth approximation to $\approx$. We use $\mathcal{L}_M$ to give our basic definitions. We also use the notation $\mathcal{L}_C^*$ for the sublanguage of $\mathcal{L}_M$ having only $\in$ as a relation symbol (but it allows modal operators). Though it is $\mathcal{L}_C^*$ we are basically interested in, we need the entire of $\mathcal{L}_M$ in order to give the fundamental truth definitions.

We now give the definition of $\Vdash$ for atomic formulas of $\mathcal{L}_M$. We have already done so for those of the form $f \varepsilon g$, that is, $p \Vdash (f \varepsilon g)$ iff $\langle p, f \rangle \in g$. We turn to $\approx_\alpha$ next, where α is an ordinal in M, and the definition is an inductive one.

Definition 1.4. For $p \in \mathcal{G}$ and $f, g \in \mathcal{D}^{\mathcal{G}}$:

(1) $p \nVdash (f \approx_0 g)$.

(2) $p \Vdash (f \approx_{\alpha+1} g)$ if

$$p \Vdash [(\forall x)[(x \varepsilon f) \supset (\exists y)(y \varepsilon g \land y \approx_\alpha x)] \land \\ (\forall x)[(x \varepsilon g) \supset (\exists y)(y \varepsilon f \land y \approx_\alpha x)]].$$

(3) For a limit ordinal λ, $p \Vdash (f \approx_\lambda g)$ if $p \Vdash (f \approx_\alpha g)$ for some $\alpha < \lambda$.

Remark Whether or not $p \Vdash (f \approx_{\alpha+1} g)$ depends on whether or not a formula of the form $[\![Z]\!]$ is true at p. Since $[\![Z]\!] \equiv \square\Diamond [\![Z]\!]$ is $S4$-valid, it follows that $(f \approx_{\alpha+1} g) \equiv$

$\Box\Diamond(f \approx_{\alpha+1} g) \equiv [\![f \approx_{\alpha+1} g]\!]$ is also $S4$-valid. This is trivially also the case for $(f \approx_0 g)$. And it is easy to see that for a limit ordinal λ we have that $(f \approx_\lambda g) \supset [\![f \approx_\lambda g]\!]$ is valid. For uniformity's sake, we generally write $[\![f \approx_\alpha g]\!]$ from now on, instead of $f \approx_\alpha g$.

The key item we need in this section is that the sequence of equality approximations is cumulative, after which a treatment of $\approx$ is easy. We will show this shortly, and in the next section additional properties will be proved. The primary tools used in the proof are the following fundamental features of the $[\![\ ]\!]$ mapping. These are either given directly in Propositions 4.3, 4.4, 4.5, and 4.6 of Chapter 16, or are easy consequences of these propositions.

Embedding properties For any possible world $p \in \mathcal{G}$:

P$_1$ if $[\![X]\!]$ is true at p, it is true at any world accessible from p;

P$_2$ if $[\![X]\!]$ is not true at p, then for some world p' accessible from p, $[\![\neg X]\!]$ is true at p';

P$_3$ if $[\![\neg X]\!]$ is true at p, $[\![X]\!]$ is not;

P$_4$ if $[\![\neg X]\!]$ is not true at p, then for some world p' accessible from p, $[\![X]\!]$ is true at p';

P$_5$ $[\![X \wedge Y]\!]$ is true at p iff both $[\![X]\!]$ and $[\![Y]\!]$ are true at p;

P$_6$ if either $[\![X]\!]$ or $[\![Y]\!]$ is true at p, so is $[\![X \vee Y]\!]$;

P$_7$ if $[\![X \vee Y]\!]$ is true at p, then for some world p' accessible from p, at least one of $[\![X]\!]$ or $[\![Y]\!]$ is true at p';

P$_8$ if $[\![X \supset Y]\!]$ and $[\![X]\!]$ are true at p, so is $[\![Y]\!]$;

P$_9$ if $[\![X \supset Y]\!]$ is not true at p, then for some world p' accessible from p, $[\![X]\!]$ and $[\![\neg Y]\!]$ are true at p';

P$_{10}$ if some instance of $[\![\varphi(x)]\!]$ is true at p, so is $[\![(\exists x)\varphi(x)]\!]$;

P$_{11}$ if $[\![(\exists x)\varphi(x)]\!]$ is true at p, then for some world p' accessible from p, some instance of $[\![\varphi(x)]\!]$ is true at p';

P$_{12}$ $[\![(\forall x)\varphi(x)]\!]$ is true at p iff each instance of $[\![\varphi(x)]\!]$ is true at p.

§1. WHAT ARE THE MODELS?   227

Example Generally we will use the items above without explicitly mentioning them, since they turn up so often. So to make sure you are familiar with their uses, we present an example in detail. It is used in the proof of the proposition immediately following.

Suppose $p \not\Vdash [\![(\forall x)[(x \,\varepsilon\, f) \supset (\exists y)(y \,\varepsilon\, g \wedge y \approx_\beta x)]]\!]$. Then by item $\mathbf{P}_{12}$, for some a, $p \not\Vdash [\![(a \,\varepsilon\, f) \supset (\exists y)(y \,\varepsilon\, g \wedge y \approx_\beta a)]]\!]$. Next, by $\mathbf{P}_9$, for some p' with $p\mathcal{R}p'$, $p' \Vdash [\![a \,\varepsilon\, f]\!]$ and $p' \Vdash [\![\neg(\exists y)(y \,\varepsilon\, g \wedge y \approx_\beta a)]\!]$, which is equivalent to $p' \Vdash [\![(\forall y)((y \,\varepsilon\, g) \supset \neg(y \approx_\beta a))]\!]$ (Corollary 4.8 of Chapter 16 comes in here). Finally, by $\mathbf{P}_{12}$ again, this in turn is equivalent to $p' \Vdash (\forall y)[\![(y \,\varepsilon\, g) \supset \neg(y \approx_\beta a)]\!]$.

In the future we will simply say that if $p \not\Vdash [\![(\forall x)[(x \,\varepsilon\, f) \supset (\exists y)(y \,\varepsilon\, g \wedge y \approx_\beta x)]]\!]$, then for some world p' accessible from p, and for some a, $[\![a \,\varepsilon\, f]\!]$ and $(\forall y)[\![(y \,\varepsilon\, g) \supset \neg(y \approx_\beta a)]\!]$ are both true at p'. We will leave it to you to formally apply the items above if you feel transitions like this need justification.

Now the cumulativity result we need to complete the definition of our models.

Proposition 1.5. *If* $p \Vdash [\![f \approx_\alpha g]\!]$ *and* $\alpha < \beta$ *then* $p \Vdash [\![f \approx_\beta g]\!]$.

Proof Call α *good* if $p \Vdash [\![f \approx_\alpha g]\!]$ implies $p \Vdash [\![f \approx_\beta g]\!]$ for every $\beta > \alpha$ (and every $p \in \mathcal{G}$ and every $f, g \in \mathcal{D}^\mathcal{G}$). We show by transfinite induction that every ordinal is good, which will establish the proposition.

Trivially 0 is good, since $p \Vdash [\![f \approx_0 g]\!]$ never holds.

The limit ordinal case is almost as easy. Assume λ is a limit ordinal and every $\alpha < \lambda$ is good. Now suppose $p \Vdash [\![f \approx_\lambda g]\!]$ and $\lambda < \beta$. By definition of $\approx_\lambda$, for some $\alpha < \lambda$, $p \Vdash [\![f \approx_\alpha g]\!]$. Since $\alpha < \lambda$, α is good, and since $\alpha < \lambda < \beta$, it follows that $p \Vdash [\![f \approx_\beta g]\!]$. It follows that λ is good.

The successor ordinal case is the heart of the matter. Assume α is good; we show $\alpha + 1$ is good.

Suppose $p \Vdash [\![f \approx_{\alpha+1} g]\!]$ and $\alpha + 1 < \gamma$; we show $p \Vdash [\![f \approx_\gamma g]\!]$. If γ is a limit ordinal, for $p \Vdash [\![f \approx_\gamma g]\!]$ to hold a similar condition must hold for a smaller ordinal. Consequently it is enough to treat the case where γ is a successor ordinal. So, suppose $\alpha + 1 < \beta + 1$ and $p \Vdash [\![f \approx_{\alpha+1} g]\!]$, but $p \not\Vdash [\![f \approx_{\beta+1} g]\!]$; we derive a contradiction.

Since $p \not\Vdash [\![f \approx_{\beta+1} g]\!]$, either $p \not\Vdash [\![(\forall x)[(x \,\varepsilon\, f) \supset (\exists y)(y \,\varepsilon\, g \wedge y \approx_\beta x)]]\!]$ or $p \not\Vdash [\![(\forall x)[(x \,\varepsilon\, g) \supset (\exists y)(y \,\varepsilon\, f \wedge y \approx_\beta x)]]\!]$; say the first. It follows that for some p' with $p\mathcal{R}p'$, and for some a, $p' \Vdash [\![a \,\varepsilon\, f]\!]$ and $p' \Vdash (\forall y)[\![(y \,\varepsilon\, g) \supset \neg(y \approx_\beta a)]\!]$.

Since $p \Vdash [\![f \approx_{\alpha+1} g]\!]$, then $p' \Vdash [\![f \approx_{\alpha+1} g]\!]$, and from the definition of $\approx_{\alpha+1}$, $p' \Vdash [\![(\forall x)[(x \,\varepsilon\, f) \supset (\exists y)(y \,\varepsilon\, g \wedge y \approx_\alpha x)]]\!]$. Since $p' \Vdash [\![a \,\varepsilon\, f]\!]$ then $p' \Vdash [\![(\exists y)(y \,\varepsilon\, g \wedge y \approx_\alpha a)]\!]$. It follows that for some p'' with $p'\mathcal{R}p''$, and for some $b \in \mathcal{D}^\mathcal{G}$, $p'' \Vdash [\![b \,\varepsilon\, g]\!]$ and $p'' \Vdash [\![b \approx_\alpha a]\!]$. We are assuming $\alpha + 1 < \beta + 1$, consequently $\alpha < \beta$. Then since α is good, $p'' \Vdash [\![b \approx_\beta a]\!]$.

Finally, $p' \Vdash (\forall y)[\![(y \varepsilon g) \supset \neg(y \approx_\beta a)]\!]$ so $[\![(b \varepsilon g) \supset \neg(b \approx_\beta a)]\!]$ is true at p', and hence also at p''. Since $p'' \Vdash [\![b \varepsilon g]\!]$, then $p'' \Vdash [\![\neg(b \approx_\beta a)]\!]$, so $[\![b \approx_\beta a]\!]$ is not true at p'', and this is a contradiction. ∎

With cumulativity established, we can give the key definition for equality.

Definition 1.6. For $f, g \in \mathcal{D}^{\mathcal{G}}$, $f \approx g$ is true at p if $f \approx_\alpha g$ is true at p for some ordinal α in M.

And with equality out of the way, we can finally introduce the formal notion of membership we really want, denoted by $x \in y$.

Definition 1.7. For $p \in \mathcal{G}$ and $f, g \in \mathcal{D}^{\mathcal{G}}$, set $f \in g$ to be true at p if for some $h \in \mathcal{D}^{\mathcal{G}}$, $p \Vdash [\![h \approx f]\!]$ and $p \Vdash [\![h \varepsilon g]\!]$.

Remark The definition above could have appeared in other equivalent forms:

$$(f \in g) \equiv (\exists x)[[\![x \approx f]\!] \wedge [\![x \varepsilon g]\!]] \equiv (\exists x)[\![x \approx f \wedge x \varepsilon g]\!].$$

Note also that

$$[\![f \in g]\!] \equiv [\![(\exists x)[x \approx f \wedge x \varepsilon g]]\!],$$

since the left hand side is equivalent to $\Box\Diamond[\![f \in g]\!]$ and the right hand side is equivalent to $\Box\Diamond(\exists x)[\![x \approx f \wedge x \varepsilon g]\!]$.

Since we have now given a definition of truth at possible worlds for the atomic formulas of $\mathcal{L}_M$, we have completed the definition of the family of $S4$ models we will be investigating. Note, incidentally, that since the truth of $f \in g$ at a world depends on the truth of two formulas $[\![h \approx f]\!]$ and $[\![h \varepsilon g]\!]$, both of which are of the form $[\![Z]\!]$, it follows that $(f \in g) \equiv [\![f \in g]\!]$. This is something we will make considerable use of.

Definition 1.8. An $S4$-*regular* model over M is an $S4$ model $(\mathcal{G}, \mathcal{R}, \mathcal{D}^{\mathcal{G}}, \mathcal{V}^{\mathcal{G}})$ where $\langle \mathcal{G}, \mathcal{R} \rangle$ is a member of M, $\mathcal{D}^{\mathcal{G}}$ is as in Definition 1.1, and $\mathcal{V}^{\mathcal{G}}$ is as defined above. We generally write $(\mathcal{G}, \mathcal{R}, \mathcal{D}^{\mathcal{G}}, \Vdash)$, displaying $\Vdash$ instead of $\mathcal{V}^{\mathcal{G}}$.

The well founded sets are sometimes called *regular* sets. This is why we have chosen to use the terminology $S4$-regular above.

EXERCISES

Exercise 1.1. Suppose $\mathcal{G} = \{p\}$, and a mapping m is defined by

$$m(f) = \{m(g) \mid \langle p, g \rangle \in f\}.$$

Show that for each ordinal α, m is a 1-1 correspondence between $\mathcal{R}_\alpha^{\mathcal{G}}$ and R_α.

Exercise 1.2. Show the notion of $\mathcal{G}$-rank is an ordinal hierarchy, i.e., show that $\mathcal{R}_\alpha^{\mathcal{G}} \subseteq \mathcal{R}_{\alpha+1}^{\mathcal{G}}$.

Exercise 1.3. Show that for $f, g \in R_\alpha^{\mathcal{G}}$, if $p \Vdash (f \approx g)$ then $p \Vdash (f \approx_\alpha g)$.

§2 About equality

Let $\mathcal{M} = (\mathcal{G}, \mathcal{R}, \mathcal{D}^{\mathcal{G}}, \Vdash)$ be an $S4$-regular model over M, fixed for this section. All work is relative to this. In order to specify when $f \in g$ is true at a world of $\mathcal{M}$ we found it necessary to introduce a version of equality, $f \approx g$, first. Though this plays a subsidiary role, it is useful to determine to what extent it does represent equality. In this section we show a variety of results that, collectively, say two things. First, $\approx$ has the logical properties one would expect of equality: reflexivity, symmetry, transitivity, and substitutivity. Second, $\approx$ has the set-theoretic properties one would expect of equality: sets with the same members are equal (something which must be carefully stated, of course). One consequence of this will be the validity, in all $S4$-regular models, of the translate of the axiom of extensionality. This is the first step in showing that translates of *all* the axioms of ZF are valid in $S4$-regular models.

Lemma 2.1. *For each ordinal α, if $p \Vdash [\![f \approx_\alpha g]\!]$ and $p \Vdash [\![g \approx_\alpha h]\!]$ then $p \Vdash [\![f \approx_\alpha h]\!]$.*

Proof By induction on α. The cases where α is 0 or a limit ordinal are simple, so we go directly to the successor case. And incidentally, we recall once more that $(f \approx_\alpha g) \equiv [\![f \approx_\alpha g]\!]$.

Suppose the result is known for α, for all worlds and all members of $\mathcal{D}^{\mathcal{G}}$, and suppose also that $p \Vdash [\![f \approx_{\alpha+1} g]\!]$ and $p \Vdash [\![g \approx_{\alpha+1} h]\!]$, but $p \not\Vdash [\![f \approx_{\alpha+1} h]\!]$. We derive a contradiction.

Since $p \not\Vdash [\![f \approx_{\alpha+1} h]\!]$, one of the two conjuncts of the definition of $\approx_{\alpha+1}$ must fail. The argument is similar either way, so say we have $p \not\Vdash (\forall x)[\![x \varepsilon f \supset (\exists y)(y \varepsilon h \wedge y \approx_\alpha x)]\!]$. Then for some p' with $p\mathcal{R}p'$, and for some $a \in \mathcal{D}^{\mathcal{G}}$, $p' \Vdash [\![a \varepsilon f]\!]$ and $p' \Vdash [\![\neg(\exists y)(y \varepsilon h \wedge y \approx_\alpha a)]\!]$.

Next, since $p \Vdash [\![f \approx_{\alpha+1} g]\!]$, also $p' \Vdash [\![f \approx_{\alpha+1} g]\!]$, and hence $p' \Vdash (\forall x)[\![x \varepsilon f \supset (\exists y)(y \varepsilon g \wedge y \approx_\alpha x)]\!]$. Since $p' \Vdash [\![a \varepsilon f]\!]$, then $p' \Vdash [\![(\exists y)(y \varepsilon g \wedge y \approx_\alpha a)]\!]$, so for some p'' with $p'\mathcal{R}p''$, and for some $b \in \mathcal{D}^{\mathcal{G}}$, $p'' \Vdash [\![b \varepsilon g]\!]$ and $p'' \Vdash [\![b \approx_\alpha a]\!]$.

We also have that $p \Vdash [\![g \approx_{\alpha+1} h]\!]$, so $p'' \Vdash [\![g \approx_{\alpha+1} h]\!]$. Then from the definition, $p'' \Vdash [\![b \varepsilon g \supset (\exists y)(y \varepsilon h \wedge y \approx_\alpha b)]\!]$. Since $p'' \Vdash [\![b \varepsilon g]\!]$ we have $p'' \Vdash [\![(\exists y)(y \varepsilon h \wedge y \approx_\alpha b)]\!]$. Then for some p''' with $p''\mathcal{R}p'''$, and for some $c \in \mathcal{D}^{\mathcal{G}}$, $p''' \Vdash [\![c \varepsilon h]\!]$ and $p''' \Vdash [\![c \approx_\alpha b]\!]$.

Now, $p'' \Vdash [\![b \approx_\alpha a]\!]$, so also $p''' \Vdash [\![b \approx_\alpha a]\!]$. We also have $p''' \Vdash [\![c \approx_\alpha b]\!]$, so by the induction hypothesis, $p''' \Vdash [\![c \approx_\alpha a]\!]$. Further, we have $p' \Vdash [\![\neg(\exists y)(y \varepsilon h \wedge y \approx_\alpha a)]\!]$ so $p''' \Vdash [\![\neg(\exists y)(y \varepsilon h \wedge y \approx_\alpha a)]\!]$, or equivalently $p''' \Vdash (\forall y)[\![(y \varepsilon h \supset \neg(y \approx_\alpha a))]\!]$. Since we have $p''' \Vdash [\![c \varepsilon h]\!]$, we have $p''' \Vdash [\![\neg(c \approx_\alpha a)]\!]$, and this is impossible. ∎

Theorem 2.2. *The equality relation is transitive. That is, if $p \Vdash [\![f \approx g]\!]$ and $p \Vdash [\![g \approx h]\!]$ then $p \Vdash [\![f \approx h]\!]$.*

Proof Suppose $p \Vdash [\![f \approx g]\!]$. Then for some α, $p \Vdash [\![f \approx_\alpha g]\!]$. Likewise suppose we have $p \Vdash [\![g \approx h]\!]$, so for some ordinal β, $p \Vdash [\![g \approx_\beta h]\!]$. Let γ be the larger of α and β. By cumulativity, we have both $p \Vdash [\![f \approx_\gamma g]\!]$ and $p \Vdash [\![g \approx_\gamma h]\!]$, so by the lemma, $p \Vdash [\![f \approx_\gamma h]\!]$, and hence $p \Vdash [\![f \approx h]\!]$. ∎

Theorem 2.3. *The equality relation is reflexive and symmetric. That is, $p \Vdash [\![f \approx f]\!]$, and if $p \Vdash [\![f \approx g]\!]$ then $p \Vdash [\![g \approx f]\!]$.*

Proof Symmetry is obvious from the definition of $\approx_\alpha$. Reflexivity is more work, and is left as Exercise 2.1. ∎

Corollary 2.4. *If $p \Vdash [\![f \,\varepsilon\, g]\!]$ then $p \Vdash [\![f \in g]\!]$.*

Proof Suppose $p \Vdash [\![f \,\varepsilon\, g]\!]$. By reflexivity, $p \Vdash [\![f \approx f]\!]$, so by the definition, $p \Vdash [\![f \in g]\!]$. ∎

With these basic items out of the way we turn to the substitutivity of equality, starting at the atomic level.

Lemma 2.5. *If $p \Vdash [\![f \approx g]\!]$ and $p \Vdash [\![f \in h]\!]$ then $p \Vdash [\![g \in h]\!]$.*

Proof Suppose $p \Vdash [\![f \approx g]\!]$ and $p \Vdash [\![f \in h]\!]$, but $p \not\Vdash [\![g \in h]\!]$. We derive a contradiction.

Since $p \not\Vdash [\![g \in h]\!]$, $p \not\Vdash [\![(\exists x)[x \approx g \wedge x \,\varepsilon\, h]]\!]$. Then for some p' with $p\mathcal{R}p'$, $p' \Vdash (\forall x)[\![(x \approx g) \supset \neg(x \,\varepsilon\, h)]\!]$.

Now, $p \Vdash [\![f \in h]\!]$, so $p' \Vdash [\![f \in h]\!]$. Then for some a, $p' \Vdash [\![a \approx f]\!]$ and $p' \Vdash [\![a \,\varepsilon\, h]\!]$. We also have $p \Vdash [\![f \approx g]\!]$, so $p' \Vdash [\![f \approx g]\!]$, and by transitivity, $p' \Vdash [\![a \approx g]\!]$. By the previous paragraph, $p' \Vdash [\![(a \approx g) \supset \neg(a \,\varepsilon\, h)]\!]$, and since $p' \Vdash [\![a \approx g]\!]$, then $p' \Vdash [\![\neg(a \,\varepsilon\, h)]\!]$ and this is impossible since $p' \Vdash [\![a \,\varepsilon\, h]\!]$. ∎

Next a similar result, but concerning substitution on the right side of a membership formula instead of on the left.

Lemma 2.6. *If $p \Vdash [\![f \approx g]\!]$ and $p \Vdash [\![h \in f]\!]$ then $p \Vdash [\![h \in g]\!]$.*

Proof Once again, a proof by contradiction. Assume $p \Vdash [\![f \approx g]\!]$, $p \Vdash [\![h \in f]\!]$, but $p \not\Vdash [\![h \in g]\!]$—we show this is impossible.

Since $p \not\Vdash [\![h \in g]\!]$, for some p' with $p\mathcal{R}p'$, $p' \Vdash [\![\neg(h \in g)]\!]$. And of course both $[\![f \approx g]\!]$ and $[\![h \in f]\!]$ are true at p'.

Since $p' \Vdash [\![h \in f]\!]$, for some a, $p' \Vdash [\![a \approx h]\!]$ and $p' \Vdash [\![a \,\varepsilon\, f]\!]$.

Now, $[\![f \approx g]\!]$ is true at p' so for some α, $[\![f \approx_{\alpha+1} g]\!]$ is true at p'. It follows from the definition of $\approx_{\alpha+1}$ that $p' \Vdash (\forall x)[\![(x \,\varepsilon\, f) \supset (\exists y)(y \,\varepsilon\, g \wedge y \approx_\alpha x)]\!]$. Since

$p' \Vdash \llbracket a \, \varepsilon \, f \rrbracket$, it follows that $p' \Vdash \llbracket (\exists y)(y \, \varepsilon \, g \wedge y \approx_\alpha a) \rrbracket$. But then for some p'' with $p'\mathcal{R}p''$, and for some b, $p'' \Vdash \llbracket b \, \varepsilon \, g \rrbracket$ and $p'' \Vdash \llbracket b \approx_\alpha a \rrbracket$, and so $p'' \Vdash \llbracket b \approx a \rrbracket$. But $\llbracket a \approx h \rrbracket$ is true at p', hence also at p'', and by transitivity of $\approx$, $p'' \Vdash \llbracket b \approx h \rrbracket$. Since $p'' \Vdash \llbracket b \, \varepsilon \, g \rrbracket$, by definition of $\in$, $p'' \Vdash \llbracket h \in g \rrbracket$, contradicting the fact that $\llbracket \neg(h \in g) \rrbracket$ is true at p', hence also at p''. ∎

Now the main result about the substitutivity of equality in $S4$ models.

Theorem 2.7. *Suppose $f, g \in \mathcal{D}^\mathcal{G}$ and X and X' are closed formulas of the (classical) language $\mathcal{L}_\mathcal{C}$, differing in that X' has an occurrence of g in a location where X has an occurrence of f. If $p \Vdash \llbracket f \approx g \rrbracket$ then $p \Vdash \llbracket X \rrbracket \equiv \llbracket X' \rrbracket$.*

Proof The argument is by induction on the complexity of X, and the atomic case is already covered by the two lemmas above. Classically, every formula is equivalent to one with only $\neg$, $\wedge$ and $\forall$ as connectives and quantifiers, and by Corollary 4.8 of Chapter 16, if A and B are classically equivalent, then $\llbracket A \rrbracket$ and $\llbracket B \rrbracket$ are $S4$ equivalent. Consequently we only need to consider the $\neg$, $\wedge$ and $\forall$ cases in the induction. And two of these cases are simple since $\llbracket A \wedge B \rrbracket \equiv \llbracket \llbracket A \rrbracket \wedge \llbracket B \rrbracket \rrbracket$ is $S4$ valid, and similarly for the universal quantifier. Consequently it is only the negation case that we present in detail.

Suppose we know that whenever $\llbracket f \approx g \rrbracket$ is true at a possible world, so is $\llbracket X \rrbracket \equiv \llbracket X' \rrbracket$. Suppose further that $p \Vdash \llbracket f \approx g \rrbracket$ but $p \nVdash \llbracket \neg X \rrbracket \equiv \llbracket \neg X' \rrbracket$—we derive a contradiction. Say $p \nVdash \llbracket \neg X \rrbracket \supset \llbracket \neg X' \rrbracket$—the other direction is similar. Thus $p \Vdash \llbracket \neg X \rrbracket$ but $p \nVdash \llbracket \neg X' \rrbracket$. Since $p \nVdash \llbracket \neg X' \rrbracket$, for some p' with $p\mathcal{R}p'$, $p' \Vdash \llbracket X' \rrbracket$. But since $p \Vdash \llbracket \neg X \rrbracket$, also $p' \Vdash \llbracket \neg X \rrbracket$, and so $p' \nVdash \llbracket X \rrbracket$. Since $p' \Vdash \llbracket f \approx g \rrbracket$, this contradicts the induction hypothesis. ∎

Corollary 2.8. *Suppose $\varphi(x)$ is a formula of $\mathcal{L}_\mathcal{C}$ with only x free, and with no occurrences of y. The following is $S4$ valid in $\mathcal{M}$:*

$$\llbracket (\forall x)(\forall y)[(x \approx y) \supset (\varphi(x) \equiv \varphi(y))] \rrbracket.$$

Proof Exercise 2.2. ∎

We now know that $\approx$ behaves the way equality should in the sense of formal logic. The next thing we need is that it behaves correctly in the set-theoretic sense.

Lemma 2.9. *If $f, g \in R_\alpha^\mathcal{G}$ and $p \Vdash (\forall x)\llbracket x \in f \equiv x \in g \rrbracket$ then $p \Vdash \llbracket f \approx_\alpha g \rrbracket$.*

Proof By transfinite induction on α. The 0 and limit ordinal cases are trivial—we give the successor case in full. Suppose the result is known for members of $R_\alpha^\mathcal{G}$ (and any $p \in \mathcal{G}$), and we now have $f, g \in R_{\alpha+1}^\mathcal{G}$ and $p \Vdash (\forall x)\llbracket x \in f \equiv x \in g \rrbracket$, but $p \nVdash \llbracket f \approx_{\alpha+1} g \rrbracket$—we derive a contradiction.

Since $\llbracket f \approx_{\alpha+1} g \rrbracket$ is not true at p either $p \nVdash \llbracket (\forall x)(x \, \varepsilon \, f \supset (\exists y)(y \, \varepsilon \, g \wedge x \approx_\alpha y)) \rrbracket$, or $p \nVdash \llbracket (\forall x)(x \, \varepsilon \, g \supset (\exists y)(y \, \varepsilon \, f \wedge x \approx_\alpha y)) \rrbracket$. Say the first of these is the case—the second is similarly dealt with. Then for some a and some p' with $p\mathcal{R}p'$, $p' \Vdash [\![a \,\varepsilon\, f]\!]$ and $p' \Vdash (\forall y)[\![y \,\varepsilon\, g \supset \neg(a \approx_\alpha y)]\!]$. Note that since $[\![a \,\varepsilon\, f]\!]$ is true at p', the $\mathcal{G}$-rank of a is less than that of f, and so $a \in R_\alpha^\mathcal{G}$.

Since $p' \Vdash [\![a \,\varepsilon\, f]\!]$ then $p' \Vdash [\![a \in f]\!]$. Also $p \Vdash (\forall x)[\![x \in f \equiv x \in g]\!]$ and consequently $p' \Vdash [\![a \in f \equiv a \in g]\!]$. It follows that $p' \Vdash [\![a \in g]\!]$. Then by definition, for some b, $p' \Vdash [\![a \approx b]\!]$ and $p' \Vdash [\![b \,\varepsilon\, g]\!]$. This implies the $\mathcal{G}$-rank of b is less than that of g and hence $b \in R_\alpha^\mathcal{G}$.

Now trivially, $p' \Vdash [\![(\forall x)(x \in a \equiv x \in a)]\!]$. But we also know that $[\![a \approx b]\!]$ is true at p' so by Theorem 2.7, $p' \Vdash [\![(\forall x)(x \in a \equiv x \in b)]\!]$. Since $a,b \in R_\alpha^\mathcal{G}$, by the induction hypothesis, $p' \Vdash [\![a \approx_\alpha b]\!]$. But $p' \Vdash (\forall y)[\![y \,\varepsilon\, g \supset \neg(a \approx_\alpha y)]\!]$ and it follows that $p' \Vdash [\![\neg(a \approx_\alpha b)]\!]$, and this is a contradiction. ∎

This immediately gives us the following.

Theorem 2.10. *If $p \Vdash (\forall x)[\![x \in f \equiv x \in g]\!]$ then $p \Vdash [\![f \approx g]\!]$.*

Corollary 2.11. *The following is $S4$ valid in $\mathcal{M}$:*

$$[\![(\forall x)(\forall y)\,[(\forall z)(z \in x \equiv z \in y) \supset (\forall z)(x \in z \equiv y \in z)]]\!].$$

Proof If this were not true at p, for some a, b, $p \not\Vdash [\![(\forall z)(z \in a \equiv z \in b) \supset (\forall z)(a \in z \equiv b \in z)]\!]$. But then for some p' with $p\mathcal{R}p'$, $p' \Vdash [\![(\forall z)(z \in a \equiv z \in b)]\!]$ and $p' \Vdash [\![\neg(\forall z)(a \in z \equiv b \in z)]\!]$. By Theorem 2.10, $p' \Vdash [\![a \approx b]\!]$. But then by Theorem 2.7 we would have that $p' \Vdash [\![\neg(\forall z)(b \in z \equiv b \in z)]\!]$, and this is impossible. ∎

Remark This corollary says the translate of the axiom of extensionality is valid in $\mathcal{M}$. Thus we have a start on showing we have an $S4$-ZF model.

We conclude the section with a useful result that says part (2), the inductive part, of Definition 1.4 "lifts."

Proposition 2.12. *For any $p \in \mathcal{G}$ and any $f, g \in \mathcal{D}^\mathcal{G}$, $p \Vdash [\![f \approx g]\!]$ iff*

$$p \Vdash [\![(\forall x)[(x \,\varepsilon\, f) \supset (\exists y)(y \,\varepsilon\, g \wedge y \approx x)] \wedge$$
$$(\forall x)[(x \,\varepsilon\, g) \supset (\exists y)(y \,\varepsilon\, f \wedge y \approx x)]]\!].$$

Proof

(1) Suppose $p \Vdash [\![f \approx g]\!]$. As a first step we show that for any p' such that $p\mathcal{R}p'$, and for any $a \in \mathcal{D}^\mathcal{G}$, if $p' \Vdash [\![a \,\varepsilon\, f]\!]$ then $p' \Vdash [\![(\exists y)(y \,\varepsilon\, g \wedge y \approx a)]\!]$. So, suppose $p' \Vdash [\![a \,\varepsilon\, f]\!]$, and hence $[\![a \in f]\!]$ is true at p'. It follows from Lemma 2.6 that $[\![a \in g]\!]$ is true at p', and by the remark following Definition 1.7, this is equivalent to $[\![(\exists y)(y \approx a \wedge y \,\varepsilon\, g)]\!]$ being true at p'.

Now, still assuming that $[\![f \approx g]\!]$ is true at p, suppose $[\![(\forall x)[x \,\varepsilon\, f \supset (\exists y)(y \,\varepsilon\, g \wedge y \approx x)]]\!]$ were not true at p. Then for some p' with $p\mathcal{R}p'$, and for some a, both $[\![a \,\varepsilon\, f]\!]$ and $[\![\neg(\exists y)(y \,\varepsilon\, g \wedge y \approx a)]\!]$ are true at p', and this contradicts what was shown above. Thus $[\![(\forall x)[x \,\varepsilon\, f \supset (\exists y)(y \,\varepsilon\, g \wedge y \approx x)]]\!]$ is true at p, and a similar argument establishes $[\![(\forall x)[x \,\varepsilon\, g \supset (\exists y)(y \,\varepsilon\, f \wedge y \approx x)]]\!]$ at p.

(?) Suppose $[\![(\forall x)[x \,\varepsilon\, f \supset (\exists y)(y \,\varepsilon\, g \wedge y \approx x)]]\!]$ and $[\![(\forall x)[x \,\varepsilon\, g \supset (\exists y)(y \,\varepsilon\, f \wedge y \approx x)]]\!]$ are both true at p, but $[\![f \approx g]\!]$ is not—we derive a contradiction. By Theorem 2.10, $p \not\Vdash (\forall x)[\![x \in f \equiv x \in g]\!]$, so for some a, $p \not\Vdash [\![a \in f \equiv a \in g]\!]$, say $p \not\Vdash [\![a \in f \supset a \in g]\!]$ (the other direction is similar). Then for some p' with $p\mathcal{R}p'$, $[\![a \in f]\!]$ and $[\![\neg(a \in g)]\!]$ are both true at p'. Since $[\![a \in f]\!]$ is true at p', for some b, both $[\![b \approx a]\!]$ and $[\![b \,\varepsilon\, f]\!]$ are true at p'.

Now $[\![(\forall x)[x \,\varepsilon\, f \supset (\exists y)(y \,\varepsilon\, g \wedge y \approx x)]]\!]$ is true at p, hence at p', and it follows that $p' \Vdash [\![(\exists y)(y \,\varepsilon\, g \wedge y \approx b)]\!]$ is true at p'. Then for some p'' and for some c, $p'' \Vdash [\![c \,\varepsilon\, g]\!]$ and $p'' \Vdash [\![c \approx b]\!]$. Since $[\![b \approx a]\!]$ is true at p', it is true at p'' and it follows that $p'' \Vdash [\![c \approx a]\!]$. Since $p'' \Vdash [\![c \,\varepsilon\, g]\!]$, by definition $p'' \Vdash [\![a \in g]\!]$, and this is impossible since $[\![\neg(a \in g)]\!]$ is true at p' and hence also at p''. ∎

EXERCISES

Exercise 2.1. Show that if $f \in R_\alpha^{\mathcal{G}}$ then $p \Vdash [\![f \approx_\alpha f]\!]$.

Exercise 2.2. Prove Corollary 2.8.

§3 The well founded sets are present

Let $\mathcal{M} = (\mathcal{G}, \mathcal{R}, \mathcal{D}^{\mathcal{G}}, \Vdash)$ be an $S4$-regular model over M. Earlier we observed that if $\mathcal{G}$ consisted of a single world, $\mathcal{D}^{\mathcal{G}}$ was isomorphic to the class of well founded sets in M (which is M itself since we are assuming the axiom of well foundedness is true in M). Now we show something considerably stronger. No matter whether $\mathcal{G}$ consists of a single world or not, each set in M has a "copy" in $\mathcal{D}^{\mathcal{G}}$, and that "copy" behaves like it in a very precise sense. One outcome of this will be the validity of translates of two more of the ZF axioms.

Definition 3.1. To each (well founded) set x in M we associate a member $\hat{x}$ of $\mathcal{D}^{\mathcal{G}}$ as follows: $\hat{x} = \{\langle p, \hat{y} \rangle \mid p \in \mathcal{G} \wedge y \in x\}$.

Remark This is really an inductive definition, based on the rank of x. We leave it to you to check the easy fact that if x has rank α, then $\hat{x}$ has $\mathcal{G}$-rank α. Also the definition makes $\hat{x}$ into a "constant function," in the sense that for *any* $p \in \mathcal{G}$, $\hat{x}''(\{p\}) = \{\hat{y} \mid y \in x\}$.

Now in what sense does $\hat{x}$ behave like x? The definition essentially says $p \Vdash (\hat{x} \,\varepsilon\, \hat{y})$ iff $x \in y$. Also if $p \Vdash (a \,\varepsilon\, \hat{y})$ then a must be $\hat{x}$ for some well founded set $x \in y$. But stronger things than this can be said. In particular, Δ_0 properties are preserved. We will show this shortly. (Recall the definition of Δ_0, from §2 of Chapter 12, essentially says that in a Δ_0 formula all quantifiers are bounded.)

Lemma 3.2. *If $p \Vdash (\hat{x} \approx \hat{y})$ then x and y are the same regular set.*

Proof This is left as Exercise 3.1. ∎

Proposition 3.3. *Let $\varphi(x_1, \ldots, x_n)$ be a Δ_0 formula in the language $\mathcal{L}_C$, with free variables among $x_1, \ldots, x_n$. Let $s_1, \ldots, s_n$ be (well founded) sets in M. Then, $\varphi(s_1, \ldots, s_n)$ is true in M if and only if $\varphi(s_1, \ldots, s_n)$ is true in V iff $[\![\varphi(\widehat{s_1}, \ldots, \widehat{s_n})]\!]$ is true at some world of $\mathcal{G}$ if and only if $[\![\varphi(\widehat{s_1}, \ldots, \widehat{s_n})]\!]$ is true at every world of $\mathcal{G}$.*

Proof The first equivalence is by the absoluteness of Δ_0 formulas. We concentrate on the rest.

Without loss of generality we can assume φ has all negations at the atomic level, and that it otherwise contains only $\wedge$, $\vee$, $(\forall x \in y)$ and $(\exists x \in y)$, since every Δ_0 formula is classically equivalent to one in this form. Now the proof is by induction on the complexity of φ, and we begin with the atomic case, where φ is $(x \in y)$.

Let s and t be (well founded) sets in M. If $s \in t$ then $(\hat{s} \, \varepsilon \, \hat{t})$ is true at every p, but so is $[\![\hat{s} \approx \hat{s}]\!]$, and it follows by definition that $(\hat{s} \in \hat{t})$ (or equivalently, $[\![\hat{s} \in \hat{t}]\!]$) is true at every p.

Conversely, suppose that for some p we have $p \Vdash [\![\hat{s} \in \hat{t}]\!]$. Then for some a, $p \Vdash [\![a \approx \hat{s}]\!]$ and $p \Vdash [\![a \varepsilon \hat{t}]\!]$. But then $p \Vdash (a \varepsilon \hat{t})$, so a is $\hat{x}$ for some set $x \in t$. Since $p \Vdash [\![\hat{x} \approx \hat{s}]\!]$, by the lemma above, x and s are the same set, and so $s \in t$. This concludes the atomic case.

The negation case is rather easy. If $\neg(s \in t)$ is true in M, $(s \in t)$ is false. Then by what was just shown, for no p can we have $p \Vdash [\![\hat{s} \in \hat{t}]\!]$. Now, if there were some world p such that $p \not\Vdash [\![\neg(\hat{s} \in \hat{t})]\!]$, there would be some p' with $p\mathcal{R}p'$ such that $p' \Vdash [\![\hat{s} \in \hat{t}]\!]$, and we just showed this cannot happen. Consequently $[\![\neg(\hat{s} \in \hat{t})]\!]$ is true at every p.

Conversely, if $p \Vdash [\![\neg(\hat{s} \in \hat{t})]\!]$ it follows that $p \not\Vdash [\![\hat{s} \in \hat{t}]\!]$. But then $s \in t$ cannot be true, since if it were $(\hat{s} \in \hat{t})$ would be true at every p. Consequently $\neg(s \in t)$ is true.

The conjunction case is simple, since $[\![X \wedge Y]\!] \equiv [\![X]\!] \wedge [\![Y]\!]$. The disjunction case is a little more interesting. Suppose that $\varphi(x_1, \ldots, x_n)$ is of the form $\psi_1(x_1, \ldots, x_n) \vee \psi_2(x_1, \ldots, x_n)$, and that the result is known for ψ_1 and for ψ_2.

If $\psi_1(s_1, \ldots, s_n) \vee \psi_2(s_1, \ldots, s_n)$ is true in M, one of the disjuncts is true, say the first, $\psi_1(s_1, \ldots, s_n)$. By the induction hypothesis, $[\![\psi_1(\widehat{s_1}, \ldots, \widehat{s_n})]\!]$ is true at every p, but this is enough to ensure the truth at every p of $[\![\psi_1(\widehat{s_1}, \ldots, \widehat{s_n}) \vee \psi_2(\widehat{s_1}, \ldots, \widehat{s_n})]\!]$.

Conversely, if $[\![\psi_1(\widehat{s_1}, \ldots, \widehat{s_n}) \vee \psi_2(\widehat{s_1}, \ldots, \widehat{s_n})]\!]$ is true at p, there is some p' with $p\mathcal{R}p'$, such that either $[\![\psi_1(\widehat{s_1}, \ldots, \widehat{s_n})]\!]$ or $[\![\psi_2(\widehat{s_1}, \ldots, \widehat{s_n})]\!]$ is true at p'; say the first is true there. Then by the induction hypothesis, $\psi_1(s_1, \ldots, s_n)$ is true in M, hence so is $\psi_1(s_1, \ldots, s_n) \vee \psi_2(s_1, \ldots, s_n)$.

This leaves the two bounded quantifier cases, which we leave to you as Exercise 3.2. ∎

Corollary 3.4. *Let $\varphi(x_1,\ldots,x_n)$ be a Σ formula in the language $\mathcal{L}_C$, with $s_1,\ldots,s_n \in M$. If $\varphi(s_1,\ldots,s_n)$ is true in M then $\lVert \varphi(\widehat{s_1},\ldots,\widehat{s_n}) \rVert$ is valid in the $S4$-regular model $\mathcal{M}$.*

Proof Without loss of generality, we can assume all unbounded existential quantifiers occur first, so that $\varphi(x_1,\ldots,x_n) = (\exists y_1)\cdots(\exists y_k)\psi(y_1,\ldots,y_k,x_1,\ldots,x_n)$. Then, if $\varphi(s_1,\ldots,s_n)$ is true in M, for some $t_1,\ldots,t_k$ in M, $\psi(t_1,\ldots,t_k,s_1,\ldots,s_n)$ is true in M. Then by the previous proposition, $\lVert \psi(\hat{t}_1,\ldots,\hat{t}_k,\hat{s}_1,\ldots,\hat{s}_n) \rVert$ is valid in $\mathcal{M}$, and from this the validity of $\lVert (\exists y_1)\cdots(\exists y_k)\psi(\hat{y}_1,\ldots,\hat{y}_k,\hat{s}_1,\ldots,\hat{s}_n) \rVert$ follows directly. ∎

These results let us verify the validity of the translates of two more ZF axioms.

Corollary 3.5. *Let N be the empty set axiom and I be the axiom of infinity. Both $\lVert N \rVert$ and $\lVert I \rVert$ are true at every $p \in \mathcal{G}$.*

Proof It was shown in Chapter 12, §2 that the property, x *is empty*, is Δ_0, so the empty set axiom is Σ, $(\exists x)(x \text{ is empty})$. Since this is true in M, its translate is valid in $\mathcal{M}$.

ω can be characterized by saying it is a limit ordinal but no member of it is a limit ordinal. In Chapter 12 it was shown that being a limit ordinal is Δ_0, and it follows easily that there is a Δ_0 characterization of ω. (The well foundedness of M plays a role here—it was used in giving a Δ_0 characterization of ordinals.) Now the validity of the translate of the axiom of infinity is shown the same way as that of the empty set axiom was. ∎

EXERCISES

Exercise 3.1. Prove Lemma 3.2. Hint, transfinite induction on the definition of $\approx$ is involved.

Exercise 3.2. Complete the proof of Proposition 3.3 by giving the two bounded quantifier cases.

§4 Four more axioms

Once again, let $\mathcal{M} = (\mathcal{G},\mathcal{R},\mathcal{D}^\mathcal{G},\Vdash)$ be an $S4$-regular model over M. In this section we show translates of the axioms of unordered pair, power set, union, and well foundedness are valid in $\mathcal{M}$. The proofs are all rather straightforward and are loosely related to similar results about R_Ω that were shown in Chapter 7. We begin with the union axiom, which is one of the simplest to verify.

Theorem 4.1. *For any $f \in \mathcal{D}^\mathcal{G}$ there is some $g \in \mathcal{D}^\mathcal{G}$ such that*

$$\lVert (\forall x)(x \in g \equiv x \in \cup f) \rVert$$

is valid in $\mathcal{M}$, where we write $x \in \cup f$ as an abbreviation for $(\exists y)(y \in f \wedge x \in y)$.

Proof Say $f \in R^\mathcal{G}_{\alpha+2}$. Define a set g as follows:

$$\langle p, a \rangle \in g \Leftrightarrow p \in \mathcal{G} \text{ and } a \in R^\mathcal{G}_\alpha \text{ and } p \Vdash \lVert a \in \cup f \rVert.$$

Obviously $g \in R^{\mathcal{G}}_{\alpha+1}$, and equally obviously, for $a \in R^{\mathcal{G}}_\alpha$,

$$p \Vdash (a \, \varepsilon \, g) \Leftrightarrow p \Vdash \lVert a \in \cup f \rVert.$$

There are a few other easy items we will need. First, if $p \Vdash a \, \varepsilon \, g$, it follows that $p \Vdash \lVert a \, \varepsilon \, g \rVert$. The reasoning is simple. Suppose $p \Vdash a \, \varepsilon \, g$. Then $p \Vdash \lVert a \in \cup f \rVert$, and so $\lVert a \in \cup f \rVert$ is true at every p' that is accessible from p. But then $p' \Vdash a \, \varepsilon \, g$ for every world accessible from p, so $p \Vdash \Box(a \, \varepsilon \, g)$. From this, trivially, $p \Vdash \Box \Diamond (a \, \varepsilon \, g)$, i.e., $p \Vdash \lVert a \, \varepsilon \, g \rVert$. The second item we need is: if $p \Vdash \lVert a \, \varepsilon \, g \rVert$ then for some accessible p', $p' \Vdash a \, \varepsilon \, g$. Again the argument is simple. If $p \Vdash \Box \Diamond (a \, \varepsilon \, g)$, by reflexivity, $p \Vdash \Diamond (a \, \varepsilon \, g)$, and the conclusion is immediate.

Now we proceed with the heart of the argument. Suppose that for some p, $\lVert (\forall x)(x \in g \equiv x \in \cup f) \rVert$ is not true at p; we derive a contradiction.

(1) Suppose for some a, $p \not\Vdash \lVert a \in g \supset a \in \cup f \rVert$. Then there is some p' with $p \mathcal{R} p'$, such that $p' \Vdash \lVert a \in g \rVert$ and $p' \Vdash \lVert \neg (a \in \cup f) \rVert$. By definition, for some a', $p' \Vdash \lVert a \approx a' \rVert$ and $p' \Vdash \lVert a' \, \varepsilon \, g \rVert$. But then, for some p'' with $p' \mathcal{R} p''$, $p'' \Vdash a' \, \varepsilon \, g$, so by the definition of g, $p'' \Vdash \lVert a' \in \cup f \rVert$. Then by substitutivity, $p'' \Vdash \lVert a \in \cup f \rVert$, and this is impossible since $\lVert \neg (a \in \cup f) \rVert$ is true at p' and hence at p''.

(2) Suppose for some a, $p \not\Vdash \lVert a \in \cup f \supset a \in g \rVert$. Then for some p' with $p \mathcal{R} p'$, $p' \Vdash \lVert a \in \cup f \rVert$ and $p' \Vdash \lVert \neg (a \in g) \rVert$. Since $p' \Vdash \lVert a \in \cup f \rVert$, $p' \Vdash \lVert (\exists y)(y \in f \wedge a \in y) \rVert$, and so for some p'' with $p' \mathcal{R} p''$, and for some $b \in \mathcal{G}$, $p'' \Vdash \lVert b \in f \rVert$ and $p'' \Vdash \lVert a \in b \rVert$. Then further, there exists $b' \in \mathcal{D}^\mathcal{G}$ such that $p'' \Vdash \lVert b \approx b' \rVert$ and $p'' \Vdash \lVert b' \, \varepsilon \, f \rVert$. But then b' must be of lower $\mathcal{G}$-rank than that of f, hence $b' \in R^{\mathcal{G}}_{\alpha+1}$. And by substitutivity, $p'' \Vdash \lVert a \in b' \rVert$ so again, for some $a' \in \mathcal{D}^\mathcal{G}$, $p'' \Vdash \lVert a \approx a' \rVert$ and $p'' \Vdash \lVert a' \, \varepsilon \, b' \rVert$. Then a' must be of lower $\mathcal{G}$-rank than b', and it follows that $a' \in R^\mathcal{G}_\alpha$. Now we have $p'' \Vdash \lVert b' \in f \rVert \wedge \lVert a' \in b' \rVert$, and it follows that $p'' \Vdash \lVert (\exists y)(y \in f \wedge a' \in y) \rVert$, that is, $p'' \Vdash \lVert a' \in \cup f \rVert$, where $a' \in R^\mathcal{G}_\alpha$. Then by definition of g, $p'' \Vdash a' \, \varepsilon \, g$, and so $p'' \Vdash \lVert a' \, \varepsilon \, g \rVert$. But since $p'' \Vdash \lVert a \approx a' \rVert$, then $p'' \Vdash \lVert a \in g \rVert$, and this is impossible since $\lVert \neg (a \in g) \rVert$ is true at p' and hence also at p''. ∎

The theorem above is not quite what is needed, but that follows easily now. (For the remaining axioms we will not formally present this extra step.)

Corollary 4.2. $\lVert (\forall f)(\exists g)(\forall x)(x \in g \equiv x \in \cup f) \rVert$ *is valid in* $\mathcal{M}$.

Proof If this were not the case, for some $p \in \mathcal{G}$,

$$p \not\Vdash \lVert (\forall f)(\exists g)(\forall x)(x \in g \equiv x \in \cup f) \rVert$$

and hence for some f,

$$p \not\Vdash \lVert (\exists g)(\forall x)(x \in g \equiv x \in \cup f) \rVert.$$

§4. FOUR MORE AXIOMS

But then for some p' with $p\mathcal{R}p'$ and for some g,

$$p' \not\Vdash \left[\!\left[(\forall x)(x \in g \equiv x \in \cup f)\right]\!\right]$$

and this contradicts the previous theorem. ∎

The pairing axiom is treated in a way that is similar to the axiom of unions, and we leave it to you.

Theorem 4.3. *For every $f, g \in \mathcal{D}^\mathcal{G}$ there is an $h \in \mathcal{D}^\mathcal{G}$ such that*

$$\left[\!\left[(\forall x)(x \in h \equiv x \in \{f, g\})\right]\!\right]$$

is valid in $\mathcal{M}$*, where* $x \in \{f, g\}$ *abbreviates* $(x \approx f \vee x \approx g)$.

Proof Exercise 4.1. ∎

Next we turn to the power set axiom. We start off with a preliminary result that contains the basic technical work. Roughly it says that subsets of f are no more complicated than f itself (or at least are equivalent to things meeting this condition).

Lemma 4.4. *Suppose $f \in R^\mathcal{G}_{\alpha+1}$ and $\left[\!\left[a \subseteq f\right]\!\right]$ is true at p (that is, $p \Vdash \left[\!\left[(\forall y)(y \in a \supset y \in f)\right]\!\right]$). Then for some $b \in R^\mathcal{G}_{\alpha+1}$, $\left[\!\left[a \approx b\right]\!\right]$ is true at p.*

Proof Suppose $f \in R^\mathcal{G}_{\alpha+1}$ and $p \Vdash \left[\!\left[a \subseteq f\right]\!\right]$. Define a set b as follows:

$$\langle q, d \rangle \in b \Leftrightarrow q \in \mathcal{G} \text{ and } d \in R^\mathcal{G}_\alpha \text{ and } q \Vdash (\exists w)\left[\!\left[w \approx d \wedge w \varepsilon a\right]\!\right].$$

Obviously $b \in R^\mathcal{G}_{\alpha+1}$ and, for each $q \in \mathcal{G}$,

$$q \Vdash (d \varepsilon b) \Leftrightarrow d \in R^\mathcal{G}_\alpha \text{ and } q \Vdash \left[\!\left[c \approx d\right]\!\right] \text{ and } q \Vdash \left[\!\left[c \varepsilon a\right]\!\right] \text{ for some } c.$$

In addition, if $q \Vdash d \varepsilon b$ then $q \Vdash \left[\!\left[d \varepsilon b\right]\!\right]$, and the argument for this is basically the same as in the proof of Theorem 4.1. Now, we will show $\left[\!\left[a \approx b\right]\!\right]$ is true at p.

Say $\left[\!\left[a \approx b\right]\!\right]$ is not true at p. Then by Proposition 2.12, either $p \not\Vdash (\forall x)\left[\!\left[x \varepsilon a \supset (\exists y)(y \varepsilon b \wedge x \approx y)\right]\!\right]$ or $p \not\Vdash (\forall x)\left[\!\left[x \varepsilon b \supset (\exists y)(y \varepsilon a \wedge x \approx y)\right]\!\right]$, say the first (the argument is similar either way). Then for some c and for some p' with $p\mathcal{R}p'$, $p' \Vdash \left[\!\left[c \varepsilon a\right]\!\right]$ and $p' \Vdash (\forall y)\left[\!\left[y \varepsilon b \supset \neg(c \approx y)\right]\!\right]$.

Since $p' \Vdash \left[\!\left[c \varepsilon a\right]\!\right]$ then $p' \Vdash \left[\!\left[c \in a\right]\!\right]$. Also $\left[\!\left[a \subseteq f\right]\!\right]$, or $\left[\!\left[(\forall y)(y \in a \supset y \in f)\right]\!\right]$, is true at p and hence at p', so it follows that $p' \Vdash \left[\!\left[c \in f\right]\!\right]$. Then for some d, $p' \Vdash \left[\!\left[c \approx d\right]\!\right]$ and $p' \Vdash \left[\!\left[d \varepsilon f\right]\!\right]$. But then d must be of lower $\mathcal{G}$-rank than f, and consequently $d \in R^\mathcal{G}_\alpha$. Also $\left[\!\left[c \varepsilon a\right]\!\right]$ is true at p', so by the definition of b, $p' \Vdash d \varepsilon b$, and hence $p' \Vdash \left[\!\left[d \varepsilon b\right]\!\right]$. Finally, we have that $p' \Vdash (\forall y)\left[\!\left[y \varepsilon b \supset \neg(c \approx y)\right]\!\right]$ and it follows that $p' \Vdash \left[\!\left[\neg(c \approx d)\right]\!\right]$, and this is our contradiction. ∎

Now the validity of the translate of the power set axiom is easy to show. It is a direct consequence of the following.

Theorem 4.5. *For any $f \in \mathcal{D}^{\mathcal{G}}$ there is some $g \in \mathcal{D}^{\mathcal{G}}$ such that*

$$\left\lVert (\forall x)(x \in g \equiv x \subseteq f) \right\rVert$$

is valid in M, *where $x \subseteq f$ abbreviates $(\forall y)(y \in x \supset y \in f)$.*

Proof Say $f \in R^{\mathcal{G}}_{\alpha+1}$. Define $g \in \mathcal{D}^{\mathcal{G}}$ by:

$$\langle p, a \rangle \in g \Leftrightarrow a \in R^{\mathcal{G}}_{\alpha+1} \text{ and } p \Vdash \left\lVert a \subseteq f \right\rVert.$$

Then $g \in R^{\mathcal{G}}_{\alpha+2}$. And further, as in the previous proofs, if $p \Vdash a \varepsilon g$ then $p \Vdash \left\lVert a \varepsilon g \right\rVert$ as well. Now we will show that if $\left\lVert (\forall x)(x \in g \equiv x \subseteq f) \right\rVert$ is not true at some p a contradiction results. The argument has two parts.

(1) Suppose for some a, $p \nVdash \left\lVert a \in g \supset a \subseteq f \right\rVert$. Then for some p' with $p\mathcal{R}p'$, $p' \Vdash \left\lVert a \in g \right\rVert$ and $p' \Vdash \left\lVert \neg(a \subseteq f) \right\rVert$. Now, since $p' \Vdash \left\lVert a \in g \right\rVert$, for some h, $p' \Vdash \left\lVert h \approx a \right\rVert$ and $p' \Vdash \left\lVert h \varepsilon g \right\rVert$. But then for some p'' with $p'\mathcal{R}p''$, $p'' \Vdash h \varepsilon g$, so by definition of g, $p'' \Vdash \left\lVert h \subseteq f \right\rVert$. Since $\left\lVert h \approx a \right\rVert$ is true at p' and hence at p'', by substitutivity, $p'' \Vdash \left\lVert a \subseteq f \right\rVert$, and this is impossible since $\left\lVert \neg(a \subseteq f) \right\rVert$ is true at p' and hence also at p''.

(2) Suppose for some a, $p \nVdash \left\lVert a \subseteq f \supset a \in g \right\rVert$. Then for some p' with $p\mathcal{R}p'$, $p' \Vdash \left\lVert a \subseteq f \right\rVert$ and $p' \Vdash \left\lVert \neg(a \in g) \right\rVert$.

Since $p' \Vdash \left\lVert a \subseteq f \right\rVert$ and $f \in R^{\mathcal{G}}_{\alpha+1}$, by the lemma above, for some $b \in R^{\mathcal{G}}_{\alpha+1}$, $p' \Vdash \left\lVert a \approx b \right\rVert$. Then by substitutivity, $p' \Vdash \left\lVert b \subseteq f \right\rVert$ so by definition of g, $\langle p', b \rangle \in g$, and $p \Vdash (b \varepsilon g)$. Since $p' \Vdash \left\lVert a \approx b \right\rVert$, we have $p' \Vdash \left\lVert a \in g \right\rVert$, which is also impossible. ∎

Finally we show the validity of the translate of the axiom of well foundedness. The heart of the argument is contained in the following.

Lemma 4.6. *If $p \Vdash \left\lVert b \in a \right\rVert$ then for some p' with $p\mathcal{R}p'$ and for some c, $p' \Vdash \left\lVert c \in a \right\rVert$ and $p' \Vdash (\forall z) \left\lVert (z \in a) \supset \neg(z \in c) \right\rVert$.*

Proof Assume $p \Vdash \left\lVert b \in a \right\rVert$. Let A consist of those x such that $\left\lVert x \in a \right\rVert$ is true at some possible world accessible from p. By our assumption $A \neq \emptyset$, since p is accessible from itself. (In fact, A is a proper class over M, though this is not important for what follows). Let c be a member of A of least $\mathcal{G}$-rank. Since $c \in A$, there is some p' with $p\mathcal{R}p'$ such that $p' \Vdash \left\lVert c \in a \right\rVert$. We show this choice of c and of p' satisfies the conclusions of the lemma.

First, $p' \Vdash \left\lVert c \in a \right\rVert$, by definition.

Next, suppose $p' \nVdash (\forall z) \left\lVert (z \in a) \supset \neg(z \in c) \right\rVert$; say $p' \nVdash \left\lVert (d \in a) \supset \neg(d \in c) \right\rVert$. We derive a contradiction.

By our supposition, there must be some p'' with $p'\mathcal{R}p''$ such that $p'' \Vdash [\![d \in a]\!]$ and $p'' \Vdash [\![d \in c]\!]$. Since $p'' \Vdash [\![d \in c]\!]$, for some d', $p'' \Vdash [\![d \approx d']\!]$ and $p'' \Vdash [\![d' \varepsilon c]\!]$. Then, as usual, d' must be of lower $\mathcal{G}$-rank than c. But by substitutivity, $p'' \Vdash [\![d' \in a]\!]$, so $d' \in A$, and c was a member of A of lowest $\mathcal{G}$-rank. This is our contradiction. ∎

And now the validity of the axiom translate follows easily. We leave the verification to you as an exercise.

Theorem 4.7. *For any $a \in \mathcal{D}^{\mathcal{G}}$ the formula*

$$[\![(\exists y)(y \in a) \supset (\exists y)(y \in a \wedge (\forall z)(z \in a \supset \neg(z \in y)))]\!]$$

is valid in $\mathcal{M}$.

Proof Exercise 4.3. ∎

EXERCISES

Exercise 4.1. Prove Theorem 4.3.

Exercise 4.2. Show that the validity of

$$[\![(\forall y)(\forall z)(\exists w)(\forall x)(x \in w \equiv x \in \{y, z\})]\!]$$

follows from Theorem 4.3.

Exercise 4.3. Prove Theorem 4.7.

§5 The definability of forcing

Once more, let $\mathcal{M} = (\mathcal{G}, \mathcal{R}, \mathcal{D}^{\mathcal{G}}, \Vdash)$ be an $S4$-regular model over M. We have one more family of ZF axioms to verify: the axioms of substitution. In order to show the translate of it is valid in $\mathcal{M}$ we will use the axiom itself in M. To do this in M we need to have a first-order formula to work with and, since forcing is the key notion we deal with, this means we need a first-order characterization of the forcing relation itself. It is to this that the present section is devoted. Essentially we just work our way through the early sections of this chapter, rewriting in a first-order language what was formerly said in English.

Proposition 5.1. *There is a formula in the language $\mathcal{L}_C$, with $\mathcal{G}$ as a parameter, that defines the class $\mathcal{D}^{\mathcal{G}}$. (We write this formula as $\mathcal{D}(x)$.)*

Proof Exercise 5.1. ∎

For the following lemma we find some temporary terminology useful. We say the three place relation $\mathcal{A}$ *embodies* $\approx_\alpha$ if

$$\langle p, f, g \rangle \in \mathcal{A} \Leftrightarrow p \Vdash [\![f \approx_\alpha g]\!].$$

Lemma 5.2. *There is a formula in the language $\mathcal{L}_C$, Next_Equal(x, y, z, w) (with $\mathcal{G}$ and $\mathcal{R}$ as parameters), such that if the M-set $\mathcal{A}$ embodies $\approx_\alpha$ and*

$$\mathcal{A}' = \{\langle y, z, w \rangle \mid \text{Next\_Equal}(\mathcal{A}, y, z, w) \text{ is true over } M\}$$

then $\mathcal{A}'$ embodies $\approx_{\alpha+1}$.

Proof We need a way of capturing in a first-order formula the contents of part (2) of Definition 1.4, that is, we need to express:

$$p \Vdash [\![(\forall x)[(x \,\varepsilon\, f) \supset (\exists y)(y \,\varepsilon\, g \wedge y \approx_\alpha x)] \wedge$$
$$(\forall x)[(x \,\varepsilon\, g) \supset (\exists y)(y \,\varepsilon\, f \wedge y \approx_\alpha x)]]\!].$$

Now, this has the form of a conjunction—we begin with one of the conjuncts, loosely characterizing subset. Taking the first conjunct of the formula above, and translating away the mapping $[\![\ ]\!]$ in favor of $\square$ and $\lozenge$, we find we need a first-order formula to express:

$$p \Vdash (\forall x)\square\lozenge[\square\lozenge(x \,\varepsilon\, f) \supset \square\lozenge(\exists y)(\square\lozenge(y \,\varepsilon\, g) \wedge (x \approx_\alpha y))].$$

(Recall that $(x \approx_\alpha y)$ and $[\![x \approx_\alpha y]\!]$ are equivalent.) And to express this, we simply write out the truth conditions for the modal operators, quantifiers, etc. The result is tedious but straightforward (in fact, it might be better for you to write your own version, instead of trying to read ours). To make things simpler, we use the following abbreviations.

$$\begin{array}{rcl}
(\forall x \in \mathcal{D})\varphi & \text{for} & (\forall x)(\mathcal{D}(x) \supset \varphi) \\
(\exists x \in \mathcal{D})\varphi & \text{for} & (\exists x)(\mathcal{D}(x) \wedge \varphi) \\
(\forall q \overleftarrow{\mathcal{R}} p)\varphi & \text{for} & (\forall q)(\mathcal{R}(p,q) \supset \varphi) \\
(\exists q \overleftarrow{\mathcal{R}} p)\varphi & \text{for} & (\exists q)(\mathcal{R}(p,q) \wedge \varphi)
\end{array}$$

Now, let $F(\mathcal{A}, p, f, g)$ be the formula:

$$(\forall x \in \mathcal{D})(\forall q \overleftarrow{\mathcal{R}} p)(\exists q' \overleftarrow{\mathcal{R}} q)\{(\forall r \overleftarrow{\mathcal{R}} q')(\exists r' \overleftarrow{\mathcal{R}} r)\langle r', x\rangle \in f \supset$$
$$(\forall r \overleftarrow{\mathcal{R}} q')(\exists r' \overleftarrow{\mathcal{R}} r)(\exists y \in \mathcal{D})[(\forall s \overleftarrow{\mathcal{R}} r')(\exists s' \overleftarrow{\mathcal{R}} s)\langle s', y\rangle \in g \wedge \langle s', x, y\rangle \in \mathcal{A}]\}.$$

And finally, just take for $\mathsf{Next\_Equal}(\mathcal{A}, p, f, g)$ the formula:

$$F(\mathcal{A}, p, f, g) \wedge F(\mathcal{A}, p, g, f).$$

∎

Next we use this lemma to show the equality relation, $\approx$, is definable over M.

Proposition 5.3. *There is a formula in the language* $\mathcal{L}_C$, $\mathsf{Equal}(x, y, z)$, *such that*

$$p \Vdash [\![f \approx g]\!] \Leftrightarrow \mathsf{Equal}(p, f, g) \text{ is true over } M.$$

Proof The idea is simple: we just write a formula saying $\approx$ holds if $\approx_\alpha$ holds for some ordinal α. The rest is details. Now, *using straightforward abbreviations*, take the formula Equal(p, f, g) to be

$(\exists \alpha)(\exists s)\{$Ordinal$(\alpha) \wedge$
 Sequence_With_Length$(s, \alpha) \wedge$
 $(\forall \beta \leq \alpha)$
 $[(\beta = 0 \supset s_\beta = \emptyset) \wedge$
 $(\forall y, z, w)(\langle y, z, w \rangle \in s_{\beta+1} \equiv$ Next_Equal$(s_\beta, y, z, w)) \wedge$
 (Limit_Ordinal$(\beta) \supset s_\beta = \cup_{\gamma < \beta} s_\gamma$)
 $] \wedge$
 $\langle p, f, g \rangle \in s_\alpha$
 $\}$

■

Finally, we can state and easily prove the definability result we have been after.

Theorem 5.4. *Let $\varphi(x_1, \ldots, x_n)$ be a formula in the language $\mathcal{L}_C$, with $x_1, \ldots, x_n$ free. There is a formula $F_\varphi(z, x_1, \ldots, x_n)$ such that*

$$p \Vdash [\![\varphi(f_1, \ldots, f_n)]\!] \text{ iff } F_\varphi(p, f_1, \ldots, f_n) \text{ is true over } M.$$

Proof By induction on the complexity of φ. In the atomic case, $\varphi(x, y)$ is $x \in y$, in which case we take $F_\varphi(z, x, y)$ to be the formula:

$$(\exists w)[\text{Equal}(z, w, x) \wedge \langle z, w \rangle \in y]$$

where we have used the formula constructed in the previous proposition.

Next, suppose $\varphi = (\psi_1 \wedge \psi_2)$, and the result is already known for ψ_1 and ψ_2. Then take F_φ to be the following:

$$F_{\psi_1} \wedge F_{\psi_2}$$

which will work because $[\![\psi_1 \wedge \psi_2]\!] \equiv [\![\psi_1]\!] \wedge [\![\psi_2]\!]$.

If $\varphi(x_1, \ldots, x_n) = \neg \psi(x_1, \ldots, x_n)$ things are a little more complicated. Recall that $[\![\neg\psi(x_1, \ldots, x_n)]\!]$ is $\Box \Diamond \neg [\![\psi(x_1, \ldots, x_n)]\!]$. Now, using the notation introduced for the proof of Lemma 5.2, take $F_\varphi(x_1, \ldots, x_n)$ to be the following:

$$(\forall p' \overleftarrow{\mathcal{R}} p)(\exists p'' \overleftarrow{\mathcal{R}} p') \neg F_\psi(z, x_1, \ldots, x_n).$$

The other cases are treated in a similar way, and we leave them to you. ■

EXERCISES

Exercise 5.1. Prove Proposition 5.1.

Exercise 5.2. Complete the proof of Theorem 5.4.

§6 The substitution axiom schema

It will make our work a little easier if, instead of showing the axioms of substitution are valid in our $S4$ models, we show the validity of the axioms of *collection*. In §6 of Chapter 8 this principle was stated as the following.

A$_8^*$ For any relation R and set A, if for every $a \in A$ there is some x such that $R(a,x)$, then there is a set B such that for every $a \in A$ there is some $b \in B$ such that $R(a,b)$.

And it was shown to be equivalent to substitution, provided we assumed the universe was well founded. (We are assuming this about M, and the translate of the axiom of well foundedness is valid in our $S4$-model.) Just as with substitution, when stated using first-order formalism, it becomes the following collection of formulas. For each formula $\varphi(x,y)$, with only x and y free:

$$(\forall a)[(\forall x \in a)(\exists y)\varphi(x,y) \supset (\exists b)(\forall x \in a)(\exists y \in b)\varphi(x,y)].$$

It is our intention to show the translate of each of these is valid in any $S4$-regular model $\mathcal{M}$. To do this we will, of course, use the fact that substitution (or collection) holds in M.

To begin with, we discuss a point concerning the ε relation that is simple, but has not come up previously. Suppose $g \in \mathcal{D}^{\mathcal{G}}$ and $\langle p, f \rangle \in g$. Then by definition, $p \Vdash f \varepsilon g$, but it may or may not be the case that $p \Vdash \llbracket f \varepsilon g \rrbracket$, since this depends not only on p but on members of $\mathcal{G}$ that are accessible from p. Now suppose we use g to define a new member g^* of $\mathcal{D}^{\mathcal{G}}$, as follows:

$$g^* = \{\langle q, f \rangle \mid \langle p, f \rangle \in g \text{ for some } p \in \mathcal{G} \text{ such that } p\mathcal{R}q\}.$$

It is easy to check that g^* will again be a member of $\mathcal{D}^{\mathcal{G}}$. Suppose $p \Vdash f \varepsilon g$, so that $\langle p, f \rangle \in g$. Let q be any member of $\mathcal{G}$ such that $p\mathcal{R}q$. By definition, $\langle q, f \rangle \in g^*$, so $q \Vdash f \varepsilon g^*$. Since q is arbitrary, it follows that $p \Vdash \Box(f \varepsilon g^*)$, and hence $p \Vdash \Box\Diamond(f \varepsilon g^*)$, that is, $p \Vdash \llbracket f \varepsilon g^* \rrbracket$. This observation will play a useful role below, so we present it in summarized form.

Summary Suppose g^* is constructed from g as above. If $p \Vdash f \varepsilon g$ then $p \Vdash \llbracket f \varepsilon g^* \rrbracket$.

Now we are ready to give the chief result on which the validity of the collection schema translate rests.

Proposition 6.1. *Suppose $p \in \mathcal{G}$, $f \in \mathcal{D}^{\mathcal{G}}$, and $p \Vdash \llbracket (\forall x)[x \in f \supset (\exists y)\varphi(x,y)] \rrbracket$, for φ a formula of $\mathcal{L}_C$. Then for some $g^* \in \mathcal{D}^{\mathcal{G}}$, $p \Vdash \llbracket (\forall x)[x \in f \supset (\exists y)(y \in g^* \wedge \varphi(x,y))] \rrbracket$.*

Proof Consider the following relation between $u, v \in \mathcal{D}^{\mathcal{G}}$:

$$\begin{aligned} u &= \langle q, a \rangle \\ \text{and} \quad v &= \langle q', b \rangle \text{ with } q\mathcal{R}q' \\ \text{and} \quad q' &\Vdash \llbracket \varphi(a,b) \rrbracket. \end{aligned}$$

§6. THE SUBSTITUTION AXIOM SCHEMA

Using definability of the forcing relation (Theorem 5.4), definability of $\mathcal{D}^{\mathcal{G}}$, (Proposition 5.1), and the assumption that $\mathcal{G}$ and $\mathcal{R}$ are sets in M, it is routine to show this relation is definable over M by a formula, say $\psi(u, v)$. We assume this has been done, and do not give details.

Now let

$$F = \{\langle q, a \rangle \mid q \in \mathcal{G}, q \Vdash [\![a \, \varepsilon \, f]\!] \text{ and } q \Vdash [\![(\forall x)[x \in f \supset (\exists y)\varphi(x, y)]]\!]\}.$$

F is a set in M—the use of ε instead of $\in$ is critical here. In fact, although we will not need it, one can verify that $F \in \mathcal{D}^{\mathcal{G}}$. Setting things up for an application of a collection axiom, we claim that for every $u \in F$ there is some v such that $\psi(u, v)$ is true over M. The argument goes as follows.

Suppose $\langle q, a \rangle \in F$. Then $q \Vdash [\![a \, \varepsilon \, f]\!]$, and also $q \Vdash [\![(\forall x)[x \in f \supset (\exists y)\varphi(x, y)]]\!]$ and hence $[\![a \in f \supset (\exists y)\varphi(a, y)]\!]$ is true at q. Since $q \Vdash [\![a \, \varepsilon \, f]\!]$ then $q \Vdash [\![a \in f]\!]$, and consequently $q \Vdash [\![(\exists y)\varphi(a, y)]\!]$. Then for some q' with $q \mathcal{R} q'$, and for some b, $q' \Vdash [\![\varphi(a, b)]\!]$. But this says that $\psi(\langle q, a \rangle, \langle q', b \rangle)$ is true over M. We have established our claim that for every $u \in F$ there is some v such that $\psi(u, v)$ is true over M.

We now apply collection in M. There is some set h in M such that

$$(\forall u \in F)(\exists v \in h)\psi(u, v).$$

Still, h might have some extraneous items in it, so let g be the subset of h such that $v \in g$ iff $\psi(u, v)$ for some $u \in F$. Then g is a set in M such that

$$(\forall u \in F)(\exists v \in g)\psi(u, v) \land (\forall v \in g)(\exists u \in F)\psi(u, v).$$

Since each $v \in g$ makes $\psi(u, v)$ true for some u, by the defining conditions of ψ, members of g are in $\mathcal{D}^{\mathcal{G}}$. Since g is a *set* in M, it follows that $g \in \mathcal{D}^{\mathcal{G}}$. Hence also $g^* \in \mathcal{D}^{\mathcal{G}}$.

Now with this background out of the way, we can verify the claim made in the proposition. Assume $p \Vdash [\![(\forall x)[x \in f \supset (\exists y)\varphi(x, y)]]\!]$; we will show that

$$p \Vdash [\![(\forall x)[x \in f \supset (\exists y)(y \in g^* \land \varphi(x, y))]]\!].$$

Well, suppose not. Then for some a, $p \nVdash [\![a \in f \supset (\exists y)(y \in g^* \land \varphi(a, y))]\!]$, and so for some p' with $p \mathcal{R} p'$, $p' \Vdash [\![a \in f]\!]$ and $p' \Vdash (\forall y)[\![y \in g^* \supset \neg \varphi(a, y)]\!]$.

Since $p' \Vdash [\![a \in f]\!]$, for some a', $p' \Vdash [\![a \approx a']\!]$ and $p' \Vdash [\![a' \, \varepsilon \, f]\!]$. Then by definition, $\langle p', a' \rangle \in F$, so for some $\langle p'', b \rangle \in g$, $\psi(\langle p', a' \rangle, \langle p'', b \rangle)$. That is, $p' \mathcal{R} p''$ and $p'' \Vdash [\![\varphi(a', b)]\!]$, so by substitutivity of equality, $p'' \Vdash [\![\varphi(a, b)]\!]$. Since $\langle p'', b \rangle \in g$, $p'' \Vdash b \, \varepsilon \, g$, and so $p'' \Vdash [\![b \, \varepsilon \, g^*]\!]$, and consequently $p'' \Vdash [\![b \in g^*]\!]$. Now $p' \Vdash (\forall y)[\![y \in g^* \supset \neg \varphi(a, y)]\!]$, so $[\![b \in g^* \supset \neg \varphi(a, b)]\!]$ is true at p', and hence at p''. It follows that $p'' \Vdash [\![\neg \varphi(a, b)]\!]$, contradicting the fact that $p'' \Vdash [\![\varphi(a, b)]\!]$. This contradiction concludes the proof. ∎

Now the validity in all $S4$-regular models over M of the translate of the collection schema follows easily.

EXERCISES

Exercise 6.1. Write out the formula $\psi(u,v)$ used in the proof of Proposition 6.1.

Exercise 6.2. Give a full proof that the set g, used in the proof of Proposition 6.1, is a member of $\mathcal{D}^{\mathcal{G}}$.

§7 The axiom of choice

Suppose we assume that our underlying first-order model, M, is a model in which the axiom of choice is true. Using this we can show the translate of the axiom of choice is valid in every $S4$-regular model over M as well. Of the many equivalent forms of the axiom of choice, we show the validity of one that says every collection of non-empty, non-overlapping sets has a choice set. That is, if members of f are non-empty, and non-overlapping, then there is a set s that contains exactly one member from each $a \in f$.

For this section let $\mathcal{M} = (\mathcal{G}, \mathcal{R}, \mathcal{D}^{\mathcal{G}}, \Vdash)$ be an $S4$-regular model over M. In addition, assume f is some fixed member of $R^{\mathcal{G}}_{\alpha+1}$, so that f is of $\mathcal{G}$-rank α or less. Using the axiom of choice in M, the set $R^{\mathcal{G}}_{\alpha}$ can be well ordered—let $<$ be a well ordering of it, also fixed for this section. Not surprisingly, we are going to use the well ordering of $R^{\mathcal{G}}_{\alpha}$ to choose members from subsets of $R^{\mathcal{G}}_{\alpha}$, and we will do this relative to each possible world $p \in \mathcal{G}$. But it is not enough to work with what p "knows" to be in a subset a of $R^{\mathcal{G}}_{\alpha}$—we must also take into account what p "thinks might be" in a as well. This vague and somewhat silly idea leads to the following formal notion.

Definition 7.1. For each $p \in \mathcal{G}$ and $a \in \mathcal{D}^{\mathcal{G}}$, let $m(p,a)$ be the least $b \in R^{\mathcal{G}}_{\alpha}$ (least in the well ordering of $R^{\mathcal{G}}_{\alpha}$) such that $p \Vdash \Diamond \llbracket b \in a \rrbracket$, if this exists, and otherwise $m(p,a)$ is undefined.

Note that if $p \Vdash \Diamond \llbracket b \in a \rrbracket$ for some $b \in R^{\mathcal{G}}_{\alpha}$, then $m(p,a)$ exists, and $m(p,a) \leq b$ (in the well ordering of $R^{\mathcal{G}}_{\alpha}$). Also if $p \Vdash \llbracket b \in a \rrbracket$ for some $b \in R^{\mathcal{G}}_{\alpha}$ this is enough for $m(p,a)$ to exist, since $\llbracket b \in a \rrbracket \supset \Diamond \llbracket b \in a \rrbracket$ is valid in any $S4$ model. On the other hand, if $m(p,a)$ exists, then $p \Vdash \Diamond \llbracket m(p,a) \in a \rrbracket$, by definition.

Lemma 7.2. *If* $p \Vdash \llbracket a \approx b \rrbracket$ *then* $m(p,a)$ *exists if and only if* $m(p,b)$ *exists, and if both exist,* $m(p,a) = m(p,b)$.

Proof Using substitutivity of equality, $p \Vdash \Diamond \llbracket c \in a \rrbracket$ if and only if $p \Vdash \Diamond \llbracket c \in b \rrbracket$, from which the lemma follows immediately. ∎

Lemma 7.3. *Suppose* $p\mathcal{R}q$ *and* $m(q,a)$ *is defined. Then* $m(p,a)$ *is also defined, and* $m(p,a) \leq m(q,a)$.

Proof If $m(q,a)$ exists, $q \Vdash \Diamond \llbracket m(q,a) \in a \rrbracket$. Since $p\mathcal{R}q$ then $p \Vdash \Diamond \llbracket m(q,a) \in a \rrbracket$ as well. And from this it follows that $m(p,a)$ exists, and $m(p,a) \leq m(q,a)$. ∎

§7. THE AXIOM OF CHOICE   245

Definition 7.4. Suppose $m(p,a)$ is defined. We say the pair $\langle p,a \rangle$ *stabilizes* at q provided $p\mathcal{R}q$ and, for every r with $q\mathcal{R}r$, $m(r,a)$ is defined, $m(r,a) = m(p,a)$, and $r \Vdash \llbracket m(r,a) \in a \rrbracket$.

Lemma 7.5. *If $m(p,a)$ is defined, there is some q at which $\langle p,a \rangle$ stabilizes.*

Proof Since $m(p,a)$ is defined, $p \Vdash \Diamond \llbracket m(p,a) \in a \rrbracket$, so for some q, $p\mathcal{R}q$ and $q \Vdash \llbracket m(p,a) \in a \rrbracket$. We will show $\langle p,a \rangle$ stabilizes at q.

Let r be such that $q\mathcal{R}r$. Then $r \Vdash \llbracket m(p,a) \in a \rrbracket$, and as observed earlier, this is enough to ensure $m(r,a)$ exists and that $m(r,a) \leq m(p,a)$. By Lemma 7.3, $m(p,a) \leq m(r,a)$ as well, so $m(p,a) = m(r,a)$. Finally, since $r \Vdash \llbracket m(p,a) \in a \rrbracket$ and $m(p,a) = m(r,a)$, then $r \Vdash \llbracket m(r,a) \in a \rrbracket$. ∎

Now assume p_0 is some member of $\mathcal{G}$, fixed for the time being.

Definition 7.6.

$$s = \{\langle q, m(q,a) \rangle \mid p_0 \mathcal{R} q, a \in R_\alpha^\mathcal{G}, q \Vdash \llbracket a \in f \rrbracket, \text{ and } m(q,a) \text{ exists}\}.$$

Notice that, directly from the definition, $s \in R_{\alpha+1}^\mathcal{G}$. We will show that, under the right circumstances, s acts like a choice set for f.

Lemma 7.7. *Suppose $p_0\mathcal{R}p$ and $\llbracket a \varepsilon f \wedge b \varepsilon a \rrbracket$ is true at p. Then $m(p,a)$ is defined, $\langle p,a \rangle$ stabilizes at q for some q, and $\llbracket m(p,a) \in a \wedge m(p,a) \in s \rrbracket$ is true at q.*

Proof Since $\llbracket a \varepsilon f \wedge b \varepsilon a \rrbracket$ is true at p, the $\mathcal{G}$-rank of a is less than that of f, and the $\mathcal{G}$-rank of b is less than that of a. Since $f \in R_{\alpha+1}^\mathcal{G}$, it follows that both a and b are in $R_\alpha^\mathcal{G}$.

Since $\llbracket b \in a \rrbracket$ is true at p, $m(p,a)$ exists, and we can apply Lemma 7.5: there is some q at which $\langle p,a \rangle$ stabilizes. We will show that both $\llbracket m(p,a) \in a \rrbracket$ and $\llbracket m(p,a) \in s \rrbracket$ are true at q.

Let r be an arbitrary member of $\mathcal{G}$ such that $q\mathcal{R}r$. Since $\llbracket a \in f \rrbracket$ is true at p it is true at r as well. Since $\langle p,a \rangle$ stabilizes at q, $m(r,a)$ is defined, and so $\langle r, m(r,a) \rangle \in s$. But since $m(r,a) = m(p,a)$, $\langle r, m(p,a) \rangle \in s$ for every r with $q\mathcal{R}r$, and so $r \Vdash m(p,a) \varepsilon s$ for each such r. Then $q \Vdash \Box(m(p,a) \varepsilon s)$, hence trivially, $q \Vdash \Box\Diamond(m(p,a) \varepsilon s)$, or $q \Vdash \llbracket m(p,a) \varepsilon s \rrbracket$. Thus $q \Vdash \llbracket m(p,a) \in s \rrbracket$. But also, since $\langle p,a \rangle$ stabilizes at q and $q\mathcal{R}q$, $q \Vdash \llbracket m(q,a) \in a \rrbracket$, and $m(q,a) = m(p,a)$, so $q \Vdash \llbracket m(p,a) \in a \rrbracket$. ∎

Now assume the following two items connecting p_0 and f:

A1 $p_0 \Vdash \llbracket (\forall a)(a \in f \supset (\exists x)(x \in a)) \rrbracket$

A2 $p_0 \Vdash \llbracket (\forall a)(\forall b)(\forall x)((a \in f \wedge b \in f \wedge x \in a \wedge x \in b) \supset a \approx b) \rrbracket$.

Item **A1** says p_0 thinks f consists of non-empty sets, and item **A2** says p_0 thinks the members of f don't overlap. We will show this is enough to ensure that p_0 thinks s acts like a choice set for f.

Proposition 7.8 (*s chooses something*).

$$p_0 \Vdash [\![(\forall a)(a \in f \supset (\exists x)(x \in a \land x \in s))]\!].$$

Proof Suppose not. Then for some p with $p_0 \mathcal{R} p$ and for some a, $p \Vdash [\![a \in f]\!]$ and $p \Vdash [\![\neg(\exists x)(x \in a \land x \in s)]\!]$.

Since $p \Vdash [\![a \in f]\!]$, for some a', both $[\![a \approx a']\!]$ and $[\![a' \,\varepsilon\, f]\!]$ are true at p. By assumption **A1**, $[\![a' \in f \supset (\exists x)(x \in a')]\!]$ is true at p_0, and hence at p. Since $p \Vdash [\![a' \in f]\!]$ then $p \Vdash [\![(\exists x)(x \in a')]\!]$. But then, for some b and for some p' with $p\mathcal{R} p'$, $p' \Vdash [\![b \in a']\!]$. So again, for some b', both $[\![b \approx b']\!]$ and $[\![b' \,\varepsilon\, a']\!]$ are true at p'. Thus $[\![a' \,\varepsilon\, f \land b' \,\varepsilon\, a']\!]$ is true at p'.

Now apply Lemma 7.7: for some q with $p'\mathcal{R} q$, $[\![m(p',a') \in a' \land m(p',a') \in s]\!]$ is true at q, from which it follows that $[\![(\exists x)(x \in a' \land x \in s)]\!]$ is true at q, and by substitutivity of equality, so is $[\![(\exists x)(x \in a \land x \in s)]\!]$. But this contradicts the fact that $[\![\neg(\exists x)(x \in a \land x \in s)]\!]$ is true at p, hence also true at q.

This contradiction concludes the proof. ∎

Now we turn to showing s chooses exactly one item from each member of f, at p_0.

Lemma 7.9. *If $p_0\mathcal{R} p$ and $p \Vdash [\![a\,\varepsilon\, f \land x\,\varepsilon\, a \land x\,\varepsilon\, s]\!]$ then $x = m(p,a)$.*

Proof Assume $p \Vdash [\![a\,\varepsilon\, f \land x\,\varepsilon\, a \land x\,\varepsilon\, s]\!]$. We show $m(p,a) \leq x$ and $x \leq m(p,a)$ in the well ordering of $R_\alpha^\mathcal{G}$. The first of these is easy. We are assuming that $f \in R_{\alpha+1}^\mathcal{G}$, so since $[\![a\,\varepsilon\, f]\!]$ and $[\![x\,\varepsilon\, a]\!]$ are both true at p, it follows that both a and x are in $R_\alpha^\mathcal{G}$. Since $p \Vdash [\![x \in a]\!]$ then $p \Vdash \Diamond [\![x \in a]\!]$ and so $m(p,a)$ exists and $m(p,a) \leq x$. Now we turn to the harder half.

Using Lemma 7.5 there is some p' at which $\langle p,a \rangle$ stabilizes. Since $[\![x\,\varepsilon\, s]\!]$ is true at p it is also true at p', so there is some p'' at which $x\,\varepsilon\, s$ is true. Then $\langle p'',x\rangle \in s$, and hence by the definition of s, $x = m(p'',b)$ for some b, where $b \in R_\alpha^\mathcal{G}$ and $p'' \Vdash [\![b \in f]\!]$. Also by definition of m, $\Diamond[\![m(p'',b) \in b]\!]$ is true at p'', so for some q, $p''\mathcal{R} q$ and $[\![m(p'',b) \in b]\!]$, or $[\![x \in b]\!]$, is true at q.

Since $p\mathcal{R} q$ and both $[\![a \in f]\!]$ and $[\![x \in a]\!]$ are true at p, they are true at q as well, as is $[\![x \in b]\!]$, as we have just seen. And since $[\![b \in f]\!]$ is true at p'', it too is true at q. But then, since $p_0\mathcal{R} p\mathcal{R} q$, by **A2**, $[\![(\forall a)(\forall b)(\forall x)((a \in f \land b \in f \land x \in a \land x \in b) \supset a \approx b)]\!]$ is true at q, and consequently $q \Vdash [\![a \approx b]\!]$. Also, since $\langle p,a\rangle$ stabilizes at p', and $p'\mathcal{R} q$, $m(q,a)$ exists, and $m(p,a) = m(q,a)$. Now, applying Lemma 7.2 we conclude that $m(p,a) = m(q,a) = m(q,b)$.

Finally, since $p''\mathcal{R}q$, by Lemma 7.3 $m(p'',b) \leq m(q,b) = m(p,a)$. But also $x = m(p'',b)$, so $x \leq m(p,a)$, and this concludes the proof. ∎

Proposition 7.10 (*s* **chooses uniquely**).

$$p_0 \Vdash [\![(\forall a)(\forall x)(\forall y)[(a \in f \wedge x \in a \wedge x \in s \wedge y \in a \wedge y \in s) \supset x \approx y]]\!].$$

Proof Suppose not. Then for some p with $p_0\mathcal{R}p$, and for some a, x, and y, $p \Vdash [\![a \in f \wedge x \in a \wedge x \in s \wedge y \in a \wedge y \in s]\!]$, and $p \Vdash [\![\neg(x \approx y)]\!]$. Then, as usual, there are a', x', and y' so that $[\![a \approx a' \wedge x \approx x' \wedge y \approx y']\!]$ and $[\![a' \varepsilon f \wedge x' \varepsilon a' \wedge x' \varepsilon s \wedge y' \varepsilon a \wedge y' \varepsilon s]\!]$ are true at p. Now by Lemma 7.9 we have both $x' = m(p,a')$ and $y' = m(p,a')$. By reflexivity, $p \Vdash [\![x' \approx y']\!]$, and so $p \Vdash [\![x \approx y]\!]$, and this is a contradiction. ∎

Now the following theorem is an easy consequence of Proposition 7.8 and Proposition 7.10.

Theorem 7.11. *Let* AC *be the axiom of choice. Then* $[\![\text{AC}]\!]$ *is valid in the S4-regular model* $\mathcal{M}$, *provided it is true in* M.

§8 Where we stand now

After much work, we now have quite a variety of $S4$-ZF models. If M is any well founded first-order universe, any $S4$-regular model over M is an $S4$-ZF model. In addition if M satisfies the axiom of choice, its translate will be valid in any $S4$-regular model over M.

In more detail, we now know the following. Suppose M is a first-order model in which the axiom of choice is true. Let $\langle \mathcal{G}, \mathcal{R} \rangle$ be any $S4$ frame that is a set in M. Let us construct, as domain, $\mathcal{D}^{\mathcal{G}}$—a proper class in M. And suppose we define truth at worlds, $\Vdash$, as in Definition 1.7 (which admittedly is rather complicated). If we do all these things, producing the $S4$-regular model $\mathcal{M} = \langle \mathcal{G}, \mathcal{R}, \mathcal{D}^{\mathcal{G}}, \Vdash \rangle$, then we have a structure in which for every axiom X of ZF and for the axiom of choice, $[\![X]\!]$ is valid. It follows that if X is any *theorem* of ZF + choice, $[\![X]\!]$ is valid in $\mathcal{M}$. And we have all this for *any* choice of $\langle \mathcal{G}, \mathcal{R} \rangle$. At an extreme, if we take $\mathcal{G}$ to be a trivial one-world set, $\mathcal{M}$ is (isomorphically) just M. But we will see that by making more complicated choices, we can produce models in which translates of various interesting assertions about sets—such as the continuum hypothesis—can be invalidated. It is in this way that we shall establish independence results that match the consistency results established in Part II.

CHAPTER 18

THE AXIOM OF CONSTRUCTIBILITY IS INDEPENDENT

§1 Introduction

Gödel proved the axiom of constructibility implies the generalized continuum hypothesis and the axiom of choice, and in turn the axiom of constructibility is consistent with the axioms of set theory, thus establishing its, and their, relative consistency. In this chapter we give Paul Cohen's proof that the axiom of constructibility is independent of the axioms of set theory—it cannot be proved even if we allow the use of the axiom of choice and the generalized continuum hypothesis. Since nobody claims the axiom of constructibility is an intuitively plausible truth about sets, its independence, while interesting, is not terribly important. But the proof will serve as a warm-up for the proof of the independence of the continuum hypothesis in the next chapter.

For this chapter, take M to be the transitive subclass L of V. Then M is a well founded first-order universe in which the axiom of choice and the generalized continuum hypothesis are true. Further, all members of M are constructible.

The general plan of the chapter is this. Since we are interested in the constructible sets, we begin by showing the notion of constructibility is more-or-less well behaved in $S4$-regular models. This requires a prior consideration of the properties of ordinals in $S4$-regular models. Then we give a simple $S4$-regular model over M, due to Cohen, in which the translate of the axiom of constructibility is not valid. We already know the translate of the axiom of choice is valid in $S4$-regular models over M, since we are assuming the axiom of choice holds in M, and a little more work on the behavior of cardinals in $S4$-regular models lets us establish that the translate of the generalized continuum hypothesis is valid as well, thus establishing the full independence result promised above.

§2 Ordinals are well behaved

We show that in $S4$-regular models, the notion of being an ordinal is almost as well behaved as in a classical model. We know each member x of M (which consists of well founded sets) has a representative $\hat{x}$ in each $S4$-regular model over M—see §3 of Chapter 17. If α is an ordinal in M, we will show that $\hat{\alpha}$ behaves like an ordinal in any $S4$-regular model $\mathcal{M} = (\mathcal{G}, \mathcal{R}, \mathcal{D}^{\mathcal{G}}, \Vdash)$ over M. Further, we will show that if something behaves like an ordinal in $\mathcal{M}$, then it is possible that it is $\hat{\alpha}$ for some ordinal α. (We will make good sense of this shortly.)

To start, we need a definition of ordinal that is as simple as we can manage. According to Theorem 5.3 of Chapter 8, the ordinals of a well founded model are the sets that are transitive and ∈-connected. This means we can use the Δ_0 characterization of ordinal from §2 of Chapter 12.

$$\text{Ordinal}(x) \Leftrightarrow (\forall y \in x)(\forall z \in y)(z \in x) \land (\forall y \in x)(\forall z \in x)(y \in z \lor z \in y \lor y = z).$$

(We have used $y = z$ as an abbreviation for $(\forall w)(w \in y \equiv w \in z)$.) Now an application of Proposition 3.3 of Chapter 17 gives us the following.

Theorem 2.1. *If α is an ordinal of M, then $[\![\text{Ordinal}(\hat{\alpha})]\!]$ is valid in M (and conversely).*

We still must settle the status of members of $\mathcal{D}^{\mathcal{G}}$ satisfying $[\![\text{Ordinal}(x)]\!]$ that are not of the form $\hat{a}$ for some ordinal $a \in M$? After a straightforward lemma, we find a simple answer to this. The lemma essentially says that the collection of items that are thought to equal an ordinal in $R_\beta^{\mathcal{G}}$, by any forcing condition, is an M-set, provided β is an ordinal in M.

Lemma 2.2. *Let β be some fixed ordinal in M, and take S to be $\{\gamma \in \text{On} \cap M \mid q \Vdash [\![\hat{\gamma} \approx g]\!]$ for some $q \in \mathcal{G}$ and some $g \in R_\beta^{\mathcal{G}}\}$. Then S is an M-set.*

Proof First, let q_0 be a fixed member of $\mathcal{G}$. Define a function F on a subset of $R_\beta^{\mathcal{G}}$ as follows. $F(g) = \gamma$ provided γ is an ordinal and $q_0 \Vdash [\![\hat{\gamma} \approx g]\!]$, and otherwise $F(g)$ is undefined. F is really a function since if $q_0 \Vdash [\![\hat{\gamma} \approx g]\!]$ and $q_0 \Vdash [\![\hat{\delta} \approx g]\!]$ then $q_0 \Vdash [\![\hat{\gamma} \approx \hat{\delta}]\!]$ and it follows that $\gamma = \delta$ (Lemma 3.2 of Chapter 17). Now the domain of F, being a subclass of $R_\beta^{\mathcal{G}}$, is an M-set, hence so is the range, using the axiom of substitution. Call the range $S(q_0)$ for convenience. It is easy to see that $S = \cup_{q_0 \in \mathcal{G}} S(q_0)$ and since $\mathcal{G}$ is an M-set, so is S. ∎

Theorem 2.3. *Suppose $p \Vdash [\![\text{Ordinal}(f)]\!]$ for some $f \in \mathcal{D}^{\mathcal{G}}$. Then for some ordinal $\alpha \in M$, $p \Vdash \Diamond [\![f \approx \hat{\alpha}]\!]$.*

Proof It is a theorem of set theory that any two ordinals are comparable, that is

$$(\forall x)(\forall y)((\text{Ordinal}(x) \land \text{Ordinal}(y)) \supset (x \in y \lor y \in x \lor x \approx y)).$$

Now suppose f is of $\mathcal{G}$-rank β, so that $f \in R_{\beta+1}^{\mathcal{G}}$. Let $S = \{\gamma \in \text{On} \cap M \mid q \Vdash [\![\hat{\gamma} \approx g]\!]$ for some $q \in \mathcal{G}$ and some $g \in R_\beta^{\mathcal{G}}\}$. By the lemma S is an M-set, so we can choose an ordinal $\delta \in M$ such that $\delta \notin S$. Now, M has been shown to be an $S4$-ZF model, so every theorem of set theory has its translate valid in M. It follows that $p \Vdash [\![(\text{Ordinal}(f) \land \text{Ordinal}(\hat{\delta})) \supset (f \in \hat{\delta} \lor \hat{\delta} \in f \lor f \approx \hat{\delta})]\!]$. We are assuming that $p \Vdash [\![\text{Ordinal}(f)]\!]$, and we know $p \Vdash [\![\text{Ordinal}(\hat{\delta})]\!]$, so $p \Vdash [\![f \in \hat{\delta} \lor \hat{\delta} \in f \lor f \approx \hat{\delta}]\!]$.

Now, for some q, $p\mathcal{R}q$ and either $q \Vdash [\![f \in \hat{\delta}]\!]$ or $q \Vdash [\![\hat{\delta} \in f]\!]$ or $q \Vdash [\![f \approx \hat{\delta}]\!]$. Suppose we have that $q \Vdash [\![\hat{\delta} \in f]\!]$. Then for some g, $q \Vdash [\![\hat{\delta} \approx g \land g \varepsilon f]\!]$. Since $f \in R_{\beta+1}^{\mathcal{G}}$, then $g \in R_\beta^{\mathcal{G}}$, and since $q \Vdash [\![\hat{\delta} \approx g]\!]$ it follows that $\delta \in S$, which contradicts our choice of δ. Consequently either $q \Vdash [\![f \in \hat{\delta}]\!]$ or $q \Vdash [\![f \approx \hat{\delta}]\!]$. Trivially, if $q \Vdash [\![f \approx \hat{\delta}]\!]$ then $p \Vdash \Diamond [\![f \approx \hat{\delta}]\!]$ for some ordinal $\delta \in M$ and we are done.

Finally, if $q \Vdash [\![f \in \hat{\delta}]\!]$ then for some h, $q \Vdash [\![f \approx h \wedge h \varepsilon \hat{\delta}]\!]$. But from the definition of $\hat{\delta}$, h must be $\hat{\alpha}$ for some $\alpha \in \delta$. Of course this implies that α is an ordinal. But then $q \Vdash [\![f \approx \hat{\alpha}]\!]$, and so $p \Vdash \Diamond [\![f \approx \hat{\alpha}]\!]$ for some ordinal α. ∎

§3 Constructible sets are well behaved too

Of course each constructible set in M has its representative in every $S4$-regular model $\mathcal{M} = (\mathcal{G}, \mathcal{R}, \mathcal{D}^{\mathcal{G}}, \Vdash)$ over M. We show these representatives have properties similar to those sets that represent ordinals, considered in the previous section.

First, there is a Σ formula with two free variables, expressing in M that y is the xth term in the sequence $L_0, L_1, \ldots, L_\alpha, L_{\alpha+1}, \ldots$, of members of the constructible hierarchy. In §4 of Chapter 14 this Σ formula was denoted $\mathcal{M}(x, y)$, but we do not want to use this notation here because we are using $\mathcal{M}$ to denote our $S4$-modal model. Instead we now denote this Σ formula by:

$$y \text{ is } L_x.$$

Then, as was also noted in Chapter 14 the class of constructible sets is defined over V by the Σ formula:

$$\mathsf{Const}(x) = (\exists y)(\exists z)(z \text{ is } L_y \wedge x \in z).$$

Further, this same formula defines the class of constructible sets over L as well (Theorem 4.5 of Chapter 14), and we are assuming $M = L$. Since we are dealing with Σ formulas, Corollary 3.4 of Chapter 17 immediately gives us the following.

Theorem 3.1. *If c is a member (i.e. a constructible set) in M then $[\![\mathsf{Const}(\hat{c})]\!]$ is valid in $\mathcal{M}$. Likewise the formula $[\![\widehat{L_\alpha} \text{ is } L_{\hat{\alpha}}]\!]$ is valid in $\mathcal{M}$ for each ordinal α of M.*

We also have an analog to Theorem 2.3 of the previous section.

Theorem 3.2. *Suppose $p \Vdash [\![\mathsf{Const}(f)]\!]$ for some $f \in \mathcal{D}^{\mathcal{G}}$. Then for some constructible set $c \in M$, $p \Vdash \Diamond [\![f \approx \hat{c}]\!]$.*

Proof Assume $p \Vdash [\![\mathsf{Const}(f)]\!]$, so $p \Vdash [\![(\exists y)(\exists z)(z \text{ is } L_y \wedge f \in z)]\!]$. Then for some p' with $p\mathcal{R}p'$, and for some $a, b \in \mathcal{D}^{\mathcal{G}}$, $p' \Vdash [\![b \text{ is } L_a \wedge f \in b]\!]$. Since

$$(\forall x)(\forall y)(x \text{ is } L_y \supset \mathsf{Ordinal}(y))$$

is a theorem of set theory, and $\mathcal{M}$ is an $S4$-ZF model, it follows that $p' \Vdash [\![\mathsf{Ordinal}(a)]\!]$. Consequently, by Theorem 2.3, there is an ordinal α of M such that $p' \Vdash \Diamond [\![a \approx \hat{\alpha}]\!]$, and so for some $q \in \mathcal{D}^{\mathcal{G}}$ with $p'\mathcal{R}q$, $q \Vdash [\![a \approx \hat{\alpha}]\!]$. Then also $q \Vdash [\![b \text{ is } L_{\hat{\alpha}} \wedge f \in b]\!]$. By the previous theorem, $q \Vdash [\![\widehat{L_\alpha} \text{ is } L_{\hat{\alpha}}]\!]$. And it is also the case that

$$(\forall x)(\forall y)(\forall z)(x \text{ is } L_z \wedge y \text{ is } L_z \supset x = y)$$

is a theorem of set theory, and so

$$\llbracket b \text{ is } L_{\hat{\alpha}} \wedge \widehat{L_\alpha} \text{ is } L_{\hat{\alpha}} \supset b \approx \widehat{L_\alpha} \rrbracket$$

is valid in $\mathcal{M}$. Consequently $q \Vdash \llbracket b \approx \widehat{L_\alpha} \rrbracket$, and so $q \Vdash \llbracket f \in \widehat{L_\alpha} \rrbracket$. But then, for some $g \in \mathcal{D}^\mathcal{G}$, $q \Vdash \llbracket f \approx g \wedge g \, \varepsilon \, \widehat{L_\alpha} \rrbracket$, and it follows from the definition of $\widehat{L_\alpha}$ that g must be $\hat{c}$ for some $c \in L_\alpha$. Thus $q \Vdash \llbracket f \approx \hat{c} \rrbracket$ for $c \in L_\alpha$, and so $p \Vdash \Diamond \llbracket f \approx \hat{c} \rrbracket$ for a set c in M. ∎

§4 A real $S4$ model, at last

So far we have been proving things about all $S4$-regular models, but we have given no concrete examples (except for the trivial single-world one). Now, finally, we produce a specific example, one that establishes the independence of the axiom of constructibility. The model is due to Paul Cohen.

The idea behind the model is really simple. We want to create an f_0 that will be different than any constructible set, and we do this by giving each possible world only a finite amount of information about f_0, so that we have enough freedom to "steer f_0 away from" any given constructible set. In fact, we will even arrange to make f_0 a set of *integers*, so that independence will be established in a very strong sense: one cannot even prove that every set of integers is constructible.

Model specifics A *forcing condition* is a pair $\langle P, N \rangle$ where P and N are finite, disjoint sets of integers. The idea is, P contains the positive, and N the negative information about our candidate for a non-constructible set—P says what is known to be present, N says what is known to be excluded. $P \cap N = \emptyset$ because our information should be non-contradictory. We do not require that $P \cup N = \omega$—in fact it is forbidden, since P and N must be finite—so no forcing condition contains complete information.

Let $\mathcal{G}$ be the set of all forcing conditions. And for $p_1 = \langle P_1, N_1 \rangle$ and $p_2 = \langle P_2, N_2 \rangle$, take $p_1 \mathcal{R} p_2$ to mean $P_1 \subseteq P_2$ and $N_1 \subseteq N_2$. Notice that $\mathcal{R}$ is reflexive and transitive.

This is enough to determine an $S4$-regular model $\mathcal{M} = (\mathcal{G}, \mathcal{R}, \mathcal{D}^\mathcal{G}, \Vdash)$ over M completely.

Now we create $f_0 \in R^\mathcal{G}_{\omega+1} - R^\mathcal{G}_\omega$ as follows:

$$f_0 = \{\langle p, \hat{n} \rangle \mid p = \langle P, N \rangle \text{ and } n \in P\}.$$

By definition, $\langle P, N \rangle \Vdash (\hat{n} \, \varepsilon \, f_0)$ iff $n \in P$. Recall that if $\langle P_1, N_1 \rangle \mathcal{R} \langle P_2, N_2 \rangle$ then $P_1 \subseteq P_2$. It follows that if $\langle P, N \rangle \Vdash (\hat{n} \, \varepsilon \, f_0)$, then $\langle P, N \rangle \Vdash \Box(\hat{n} \, \varepsilon \, f_0)$, and so $\langle P, N \rangle \Vdash \llbracket \hat{n} \, \varepsilon \, f_0 \rrbracket$. Likewise if $n \in N$, since P and N must be disjoint, $n \notin P$, so $\langle P, N \rangle \nVdash (n \, \varepsilon \, f_0)$. And again, from the way that $\mathcal{R}$ is defined, it follows easily that if $n \in N$, then $\langle P, N \rangle \Vdash \llbracket \neg(\hat{n} \, \varepsilon \, f_0) \rrbracket$.

Now, recall there is a Σ formula $\mathsf{Const}(x)$ characterizing the constructible sets.

Theorem 4.1. *In the model $\mathcal{M}$ the formula $\llbracket \neg \mathsf{Const}(f_0) \rrbracket$ is valid.*

Proof Suppose not, say $p \not\Vdash \llbracket \neg\mathsf{Const}(f_0) \rrbracket$. Then $q \Vdash \llbracket \mathsf{Const}(f_0) \rrbracket$ for some q with $p\mathcal{R}q$. And then, by Theorem 3.2, for some constructible set c in M, $q \Vdash \Diamond \llbracket f_0 \approx \hat{c} \rrbracket$, so for some r with $q\mathcal{R}r$, $\llbracket f_0 \approx \hat{c} \rrbracket$ is true at r. But we will show this is impossible. Say $r = \langle P, N \rangle$. Now there are two cases, each easy.

Case 1) $c \subseteq P$. Since P and N are finite, we can choose an integer n such that $n \notin P \cup N$ (so $n \notin c$ as well). Let $P' = P \cup \{n\}$, and let $r' = \langle P', N \rangle$. Then $r\mathcal{R}r'$, and $r' \Vdash \llbracket \hat{n} \in f_0 \rrbracket$. But since $n \not\subset c$, $r' \Vdash \llbracket \neg(\hat{n} \in \hat{c}) \rrbracket$, so in this case we could not have had $r \Vdash \llbracket f_0 \approx \hat{c} \rrbracket$.

Case 2) $c \not\subseteq P$. Let n be an integer in c but not in P, let $N' = N \cup \{n\}$, and let $r' = \langle P, N' \rangle$. Again $r\mathcal{R}r'$. And this time $r' \Vdash \llbracket \neg(\hat{n} \in f_0) \rrbracket$ but $r' \Vdash \llbracket \hat{n} \in \hat{c} \rrbracket$, so again we can not have had $r \Vdash \llbracket f_0 \approx \hat{c} \rrbracket$. ∎

We have now shown that the axiom of constructibility is not provable from the axioms of ZF together with the axiom of choice. After some technical work, we will show the generalized continuum hypothesis has a translate that is valid in $\mathcal{M}$, which yields that the axiom of constructibility is not provable, even if GCH is added to ZF.

EXERCISES

Exercise 4.1. Prove that if $n \in N$, then $\langle P, N \rangle \Vdash \llbracket \neg(\hat{n} \, \varepsilon \, f_0) \rrbracket$.

§5 Cardinals are sometimes well behaved

In order to show the continuum hypothesis has a valid translate in the model constructed in the previous section, we need to know something about how cardinals behave in this model. The situation is not quite as simple as it was with ordinals and constructible sets, but at least the reason for the complications is clear. An ordinal is a cardinal if there does not exist a 1-1 mapping between it and a smaller ordinal. This means that a formula characterizing cardinals will have a negated existential quantifier, and so will not be a Σ formula. Then our general results about Σ formulas are simply not applicable. Special pleading is needed.

It is straightforward to produce a Δ_0 formula expressing that x is a function from y to z, or that x is a 1-1, onto function from y to z—this was done in Chapter 12. Let $x : y \longrightarrow z$ and $x : y \xrightarrow[\text{onto}]{\text{1-1}} z$ be these Δ_0 formulas. Then, one formula characterizing cardinals is the following.

$$\mathsf{Cardinal}(\alpha) \Leftrightarrow \mathsf{Ordinal}(\alpha) \wedge (\forall \beta \in \alpha)\neg(\exists f)(f : \beta \xrightarrow[\text{onto}]{\text{1-1}} \alpha).$$

While this is not a Σ formula, it is easy to see that its negation is, and this gives us the following easy result.

Proposition 5.1. *Let $\mathcal{M}$ be any $S4$-regular model and let α be an ordinal of M. If the formula $\llbracket \mathsf{Cardinal}(\hat{\alpha}) \rrbracket$ is true at some possible world of $\mathcal{M}$, then α is, in fact, a cardinal of M.*

Proof If α is not a cardinal, then $\neg\mathsf{Cardinal}(\alpha)$ is true in M. Since this is a Σ formula, by Corollary 3.4 of Chapter 17, $\llbracket \neg\mathsf{Cardinal}(\hat{\alpha}) \rrbracket$ is valid in $\mathcal{M}$, so $\llbracket \mathsf{Cardinal}(\hat{\alpha}) \rrbracket$ could not be true at any world. ∎

The other direction is more important to us, and it does not hold for all $S4$-regular models. Cohen identified a nice class for which it does hold, and we discuss that now.

Definition 5.2. Let $\langle \mathcal{G}, \mathcal{R} \rangle$ be a frame. Two members $p, q \in \mathcal{G}$ are called *compatible* if they have a common extension, that is, for some $r \in \mathcal{G}$, $p\mathcal{R}r$ and $q\mathcal{R}r$. They are called *incompatible* if they are not compatible.

The frame $\langle \mathcal{G}, \mathcal{R} \rangle$ is said to meet the *countable chain condition* (abbreviated c.c.c.) if every collection of pairwise incompatible members of $\mathcal{G}$ is at most countable.

There is a connection between the countable chain condition and chains, but it plays no role here. For us, c.c.c. is simply standard, convenient terminology.

Suppose that at some world p of an $S4$-regular model the assertion that f is a function from a to b is true. Can this be "lifted" to simulate that function in M? Well, not exactly. Loosely the problem is this. If $[\![c \in a]\!]$ is true at p then $[\![(\exists x)(x \in b \wedge x \text{ is } f(c))]\!]$ will also be true (assuming obvious formula abbreviations). But as we have seen, this implies that for some d and for some q with $p\mathcal{R}q$, $[\![d \in b \wedge d \text{ is } f(c)]\!]$ will be true at q. We had to move from p to q to instantiate the existential quantifier. Now, there may be several different possible worlds we could move to, in each of which that existential quantifier is instantiated, but instantiated differently from world to world. Which instantiation should we choose for our simulation of f in M? There is no natural way of choosing, so we make the simulation of f map members of the set that corresponds to a in M to *subsets* of the correspondent of b—we simply throw in all the various existential quantifier instantiations. The following theorem says that, under the right circumstances, this is not too bad.

Lemma 5.3. *Suppose $(\mathcal{G}, \mathcal{R}, \mathcal{D}^{\mathcal{G}}, \Vdash)$ is an $S4$-regular model over M, with $\langle \mathcal{G}, \mathcal{R} \rangle$ satisfying c.c.c. Suppose also that $a, b \in M$, $p \in \mathcal{G}$, and $[\![f : \hat{a} \longrightarrow \hat{b}]\!]$ is true at p. Define a function $F : a \longrightarrow \mathcal{P}(b)$ by*

$$F(c) = \{d \in b \mid p \Vdash \Diamond[\![\hat{d} \text{ is } f(\hat{c})]\!]\}.$$

Then for each $c \in a$, $F(c)$ is at most countable.

Proof We are, of course, assuming that x is $f(y)$ abbreviates the obvious Δ_0 formula, which we do not write out. All that we need of it is the classical validity of the following: $[x \text{ is } f(y) \wedge x' \text{ is } f(y)] \supset x = x'$.

Suppose $c \in a$ and $d_1, d_2 \in F(c)$. Then both $\Diamond[\![\hat{d}_1 \text{ is } f(\hat{c})]\!]$ and $\Diamond[\![\hat{d}_2 \text{ is } f(\hat{c})]\!]$ are true at p, so for some q_1 and q_2, $p\mathcal{R}q_1$ and $p\mathcal{R}q_2$, and $q_1 \Vdash [\![\hat{d}_1 \text{ is } f(\hat{c})]\!]$ and $q_2 \Vdash [\![\hat{d}_2 \text{ is } f(\hat{c})]\!]$. If q_1 and q_2 were compatible, for some r, $q_1\mathcal{R}r$, $q_2\mathcal{R}r$, and then both the formulas $[\![\hat{d}_1 \text{ is } f(\hat{c})]\!]$ and $[\![\hat{d}_2 \text{ is } f(\hat{c})]\!]$ would be true at r. Then, since $[x \text{ is } f(y) \wedge x' \text{ is } f(y)] \supset x = x'$ is classically valid, by the usual $S4$ embedding results, it follows that $r \Vdash [\![\hat{d}_1 \approx \hat{d}_2]\!]$, and so $d_1 = d_2$.

Now, to each $d \in F(c)$ associate some q so that $p\mathcal{R}q$ and $q \Vdash [\![\hat{d} \text{ is } f(\hat{c})]\!]$ (the axiom of choice is needed here to pick one q for each d). It follows, from what was proved above, that distinct members of $F(c)$ have incompatible members of $\mathcal{G}$ associated with

§5. CARDINALS ARE SOMETIMES WELL BEHAVED

them. Since we are assuming the c.c.c. condition, the set of associates, and hence $F(c)$ must be at most countable. ∎

Lemma 5.4. *Again suppose $(\mathcal{G}, \mathcal{R}, \mathcal{D}^{\mathcal{G}}, \Vdash)$ is an S4-regular model over M such that $\langle \mathcal{G}, \mathcal{R} \rangle$ satisfies c.c.c. And suppose α and β are infinite ordinals with $\beta < \alpha$, and α is a cardinal. Then for no $p \in \mathcal{G}$ do we have $p \Vdash \llbracket f : \hat{\beta} \xrightarrow[onto]{1\text{-}1} \hat{\alpha} \rrbracket$.*

Proof Suppose the hypothesis, and also that $p \Vdash \llbracket f : \hat{\beta} \xrightarrow[onto]{1\text{-}1} \hat{\alpha} \rrbracket$. We derive a contradiction.

Define a function $F : \beta \longrightarrow \mathcal{P}(\alpha)$ as in the previous lemma, by

$$F(\gamma) = \left\{ \delta \in \alpha \mid p \Vdash \Diamond \llbracket \hat{\delta} \text{ is } f(\hat{\gamma}) \rrbracket \right\}.$$

Let δ be an arbitrary member of α. Since $\llbracket f : \hat{\beta} \xrightarrow[onto]{1\text{-}1} \hat{\alpha} \rrbracket$ is true at p, so is $\llbracket (\hat{\delta} \in \hat{\alpha} \supset (\exists y)(y \in \hat{\beta} \land \hat{\delta} \text{ is } f(y))) \rrbracket$. $\llbracket \hat{\delta} \in \hat{\alpha} \rrbracket$ is true at p, hence so is $\llbracket (\exists y)(y \in \hat{\beta} \land \hat{\delta} \text{ is } f(y)) \rrbracket$. Then it is a simple exercise to verify that for some $\gamma \in \beta$, $p \Vdash \Diamond \llbracket \hat{\delta} \text{ is } f(\hat{\gamma}) \rrbracket$ and consequently $\delta \in F(\gamma)$. Briefly, each member of α is in $F(\gamma)$ for some $\gamma \in \beta$.

Now, let A be $\{F(\gamma) \mid \gamma \in \beta\}$. We have just verified that $\cup A = \alpha$. Further, $\overline{\overline{\beta}} \geq \overline{\overline{A}}$ (since F maps β onto A, but is not necessarily 1-1). Then $\aleph_0 \cdot \overline{\overline{\beta}} \geq \aleph_0 \cdot \overline{\overline{A}}$, and since β is an infinite ordinal, $\aleph_0 \cdot \overline{\overline{\beta}} = \overline{\overline{\beta}}$, hence $\overline{\overline{\beta}} \geq \aleph_0 \cdot \overline{\overline{A}}$. But also, by the previous lemma, $\aleph_0$ is an upper bound on size for members of A, consequently $\aleph_0 \cdot \overline{\overline{A}} \geq \overline{\overline{\cup A}} = \overline{\overline{\alpha}} = \alpha$. Consequently $\overline{\overline{\beta}} \geq \alpha$, which is impossible since $\beta < \alpha$ and α is a cardinal. ∎

Now the main result of this section.

Theorem 5.5. *Suppose $(\mathcal{G}, \mathcal{R}, \mathcal{D}^{\mathcal{G}}, \Vdash)$ is an S4-regular model over M with $\langle \mathcal{G}, \mathcal{R} \rangle$ satisfying c.c.c., and let α be an ordinal of M. Then α is a cardinal of M iff $\llbracket \text{Cardinal}(\hat{\alpha}) \rrbracket$ is true at some member of $\mathcal{G}$ iff $\llbracket \text{Cardinal}(\hat{\alpha}) \rrbracket$ is true at every member of $\mathcal{G}$.*

Proof If $\llbracket \text{Cardinal}(\hat{\alpha}) \rrbracket$ is true at some member of $\mathcal{G}$ then α is a cardinal, by Proposition 5.1.

In the other direction, suppose α is a cardinal—we show $\llbracket \text{Cardinal}(\hat{\alpha}) \rrbracket$ is true at every member of $\mathcal{G}$.

If α is a finite cardinal things are easily handled. In this case α is a natural number. There is a Δ_0 characterization of natural numbers, say $N(x)$, so $N(\hat{\alpha})$ is true at every member of $\mathcal{G}$. Also $(\forall x)(N(x) \supset \text{Cardinal}(x))$ is a theorem of ZF, so $\llbracket (\forall x)(N(x) \supset \text{Cardinal}(x)) \rrbracket$ is true at every member of $\mathcal{G}$ as well. Then it follows that $\llbracket \text{Cardinal}(\hat{\alpha}) \rrbracket$ is true at all members of $\mathcal{G}$ too.

Now suppose α is an infinite cardinal. If $\llbracket \text{Cardinal}(\hat{\alpha}) \rrbracket$ were not true at every member of $\mathcal{G}$ there would be some world at which $\llbracket \neg \text{Cardinal}(\hat{\alpha}) \rrbracket$ were true. Since α is an ordinal, and ordinals are well behaved, there would then be a world p such that $p \Vdash \llbracket (\exists x)(x \in \hat{\alpha} \land (\exists f)(f : x \xrightarrow[onto]{1\text{-}1} \hat{\alpha})) \rrbracket$. Then for some p' with $p\mathcal{R}p'$, and some b,

$p' \Vdash \llbracket b \in \hat{\alpha} \wedge (\exists f)(f : b \xrightarrow[\text{onto}]{1\text{-}1} \hat{\alpha}) \rrbracket$. Since $p' \Vdash \llbracket b \in \hat{\alpha} \rrbracket$, in the usual way, for some $\beta \in \alpha$, $p' \Vdash \llbracket b \approx \hat{\beta} \rrbracket$. Consequently $p' \Vdash \llbracket (\exists f)(f : \hat{\beta} \xrightarrow[\text{onto}]{1\text{-}1} \hat{\alpha}) \rrbracket$. But then, of course, for some p'' with $p'Rp''$, and for some f, $p'' \Vdash \llbracket f : \hat{\beta} \xrightarrow[\text{onto}]{1\text{-}1} \hat{\alpha} \rrbracket$, and this contradicts the previous lemma. Hence $\llbracket \mathsf{Cardinal}(\hat{\alpha}) \rrbracket$ must be true at every member of $\mathcal{G}$. ∎

EXERCISES

Exercise 5.1. In the proof of Lemma 5.4, at a certain point we concluded that $\aleph_0 \cdot \overline{\overline{A}} \geq \overline{\overline{UA}}$. Give an argument for why this is correct.

§6 The status of the generalized continuum hypothesis

We have shown the axiom of constructibility is independent of ZF, even allowing the axiom of well foundedness and the axiom of choice. Now we will show the axiom of constructibility is independent of ZF together with the generalized continuum hypothesis as well. Given what has been done in this chapter thus far, all that is left is to show the translate of the generalized continuum hypothesis is valid in the $S4$-regular model constructed in §4.

Let $\mathcal{M}$ be the model $(\mathcal{G}, \mathcal{R}, \mathcal{D}^{\mathcal{G}}, \Vdash)$ constructed in §4. The chief fact we need is quite simple: the set $\mathcal{G}$ of possible worlds is *countable* (since the collection of finite subsets of a countable set is countable). Then the countable chain condition, from the previous section, is trivially satisfied, so cardinals are well behaved in $\mathcal{M}$. Now, the basic technical item is the following. In it we use various fairly obvious formula abbreviations.

Lemma 6.1. *Suppose α and β are infinite cardinals, with $\alpha \leq \beta$, and, for some $p \in \mathcal{G}$, both $\llbracket (\forall x)(x \in s \equiv x \subseteq \hat{\alpha}) \rrbracket$ and $\llbracket f : \hat{\beta} \xrightarrow[\text{onto}]{1\text{-}1} s \rrbracket$ are true at p. Then $\overline{\overline{\beta}} \leq \overline{\overline{\mathcal{P}(\alpha)}}$.*

Proof We define a mapping $F : \beta \longrightarrow \mathcal{P}(\mathcal{G} \times \alpha)$ as follows. For $\gamma \in \beta$:

$$F(\gamma) = \left\{ \langle q, \delta \rangle \mid pRq, \delta \in \alpha, q \Vdash \llbracket \hat{\delta} \in f(\hat{\gamma}) \rrbracket \right\}.$$

We will show that F is 1-1, after which we have the conclusion by the following easy argument. Since F is 1-1, $\overline{\overline{\beta}} \leq \overline{\overline{\mathcal{P}(\mathcal{G} \times \alpha)}}$. But $\overline{\overline{\mathcal{G}}} = \aleph_0$ in our model, and α is infinite, so $\overline{\overline{\mathcal{G} \times \alpha}} = \overline{\overline{\alpha}}$. It follows that $\overline{\overline{\mathcal{P}(\mathcal{G} \times \alpha)}} = \overline{\overline{\mathcal{P}(\alpha)}}$, so $\overline{\overline{\beta}} \leq \overline{\overline{\mathcal{P}(\alpha)}}$ as promised.

Now we turn to the argument that F is 1-1. Suppose $\gamma_1, \gamma_2 \in \beta$ and $\gamma_1 \neq \gamma_2$—we will show $F(\gamma_1) \neq F(\gamma_2)$.

Since $\gamma_1, \gamma_2 \in \beta$, both $\llbracket \hat{\gamma}_1 \in \hat{\beta} \rrbracket$ and $\llbracket \hat{\gamma}_2 \in \hat{\beta} \rrbracket$ are valid in the $S4$ model. And since $p \Vdash \llbracket f : \hat{\beta} \xrightarrow[\text{onto}]{1\text{-}1} s \rrbracket$ it follows that $p \Vdash \llbracket (\forall x)(x \in \hat{\beta} \supset (\exists y)(y \text{ is } f(x) \wedge y \in s)) \rrbracket$. Then both $\llbracket (\exists y)(y \text{ is } f(\hat{\gamma}_1) \wedge y \in s) \rrbracket$ and $\llbracket (\exists y)(y \text{ is } f(\hat{\gamma}_2) \wedge y \in s) \rrbracket$ are true at p. It follows that there is some p' with pRp', and there are $a, b \in \mathcal{D}^{\mathcal{G}}$ such that both $\llbracket a \text{ is } f(\hat{\gamma}_1) \wedge a \in s \rrbracket$ and $\llbracket b \text{ is } f(\hat{\gamma}_2) \wedge b \in s \rrbracket$ are true at p'.

Again, since $p \Vdash \llbracket f : \hat{\beta} \xrightarrow[\text{onto}]{1\text{-}1} s \rrbracket$ it follows that

$$p \Vdash \llbracket (\forall x)(\forall y)(\forall a)(\forall b)[(a \text{ is } f(x) \wedge b \text{ is } f(y) \wedge \neg(x \approx y)) \supset \neg(a \approx b)] \rrbracket.$$

§6. THE STATUS OF THE GENERALIZED CONTINUUM HYPOTHESIS   257

Also since $\gamma_1 \neq \gamma_2$, it follows that $[\![\neg(\hat{\gamma}_1 \approx \hat{\gamma}_2)]\!]$ is valid in the $S4$ model. Consequently $p' \Vdash [\![\neg(a \approx b)]\!]$.

Since $[\![\neg(a \approx b)]\!]$ is true at p', so is $[\![(\exists x)(x \in a \wedge \neg(x \in b)) \vee (\exists x)(x \in b \wedge \neg(x \in a))]\!]$. Then for some p'' with $p'\mathcal{R}p''$, and for some c, either $p'' \Vdash [\![c \in a \wedge \neg(c \in b)]\!]$ or $p'' \Vdash [\![c \in b \wedge \neg(c \in a)]\!]$. Say the first alternative is the case.

We have that $p'' \Vdash [\![c \in a]\!]$. We also have that $p' \Vdash [\![a \subset s]\!]$ and $p \Vdash [\![(\forall x)(x \in s \equiv x \subseteq \hat{\alpha})]\!]$. It follows that $p'' \Vdash [\![a \subseteq \hat{\alpha}]\!]$ and consequently $p'' \Vdash [\![c \in \hat{\alpha}]\!]$. Then for some d, $p'' \Vdash [\![c \approx d \wedge d\varepsilon\hat{\alpha}]\!]$. Since $[\![d\varepsilon\hat{\alpha}]\!]$ is true at p'', d must be $\hat{\delta}$ for some $\delta \in \alpha$. Thus we have all of $[\![\hat{\delta} \in a]\!]$, $[\![\neg(\hat{\delta} \in b)]\!]$, $[\![a \text{ is } f(\hat{\gamma}_1)]\!]$ and $[\![b \text{ is } f(\hat{\gamma}_2)]\!]$ true at p'', so $p'' \Vdash [\![\hat{\delta} \in f(\hat{\gamma}_1)]\!]$ but $p'' \nVdash [\![\hat{\delta} \in f(\hat{\gamma}_2)]\!]$. Then by definition, $\langle p'', \delta \rangle \in F(\gamma_1)$ but $\langle p'', \delta \rangle \notin F(\gamma_2)$, so $F(\gamma_1) \neq F(\gamma_2)$.

This establishes that F is 1-1 and concludes the proof of the lemma. ∎

Theorem 6.2. *The translate of the generalized continuum hypothesis is valid in* $\mathcal{M}$.

Proof If the translate of the generalized continuum hypothesis were not valid in $\mathcal{M}$, then at some world $p \in \mathcal{G}$, the following would true:

$[\![(\exists \alpha)(\exists \gamma)(\exists \beta)(\exists f)(\exists s)$
$[\text{Cardinal}(\alpha) \wedge \text{Cardinal}(\gamma) \wedge \text{Cardinal}(\beta) \wedge$
$\hat{\omega} \in \alpha \wedge \alpha \in \gamma \wedge \gamma \in \beta \wedge$
$(\forall x)(x \in s \equiv x \subseteq \alpha) \wedge f : \beta \xrightarrow[\text{onto}]{\text{1-1}} s]]\!]$.

In the usual way, it follows that there is some p' with $p\mathcal{R}p'$, with all existential quantifiers instantiated at p'. Since $(\forall \alpha)(\text{Cardinal}(\alpha) \supset \text{Ordinal}(\alpha))$, by applying Theorem 2.3, there is some p'' with $p'\mathcal{R}p''$, and there are ordinals α, γ and β, such that at p'' the following is true:

$[\![\text{Cardinal}(\hat{\alpha}) \wedge \text{Cardinal}(\hat{\gamma}) \wedge \text{Cardinal}(\hat{\beta})$
$\wedge \hat{\omega} \in \hat{\alpha} \wedge \hat{\alpha} \in \hat{\gamma} \wedge \hat{\gamma} \in \hat{\beta}$
$\wedge (\forall x)(x \in s \equiv x \subseteq \hat{\alpha}) \wedge (f : \hat{\beta} \xrightarrow[\text{onto}]{\text{1-1}} s)]\!]$.

Since c.c.c. is true of $\mathcal{M}$, by Theorem 5.5, α, β, and γ are cardinals of M. Now by the lemma above, $\overline{\overline{\beta}} \leq \overline{\overline{\mathcal{P}(\alpha)}}$. Since $\alpha \in \gamma \in \beta$, all of which are cardinals, the generalized continuum hypothesis fails in M. However, in this chapter, $M = L$, and the generalized continuum hypothesis holds in L. ∎

CHAPTER 19

INDEPENDENCE OF THE CONTINUUM HYPOTHESIS

§1 Power politics

In this short chapter we show the continuum hypothesis (hence also the generalized continuum hypothesis) cannot be proved in ZF plus choice plus well foundedness by producing an S4-regular model in which the size of the continuum—the power set of ω—is at least $\aleph_2$. Once again the construction is due to Cohen. Furthermore, it will be clear that the construction can easily be generalized to produce models in which the size of the continuum is at least $\aleph_3$, or, $\aleph_{100}$, or as big as we want. The chapter is short because of all the preliminary work that has been done. At this point we can just give the model, prove a few things about it, and we're done. Our work takes place in M as usual, a transitive well founded first-order universe in which the axiom of choice is true. In the last few sections we assume the generalized continuum hypothesis holds in M as well, which means taking M to be L will serve.

A word to justify the title of this section. To produce an S4-regular model in which the power set of ω has some specific size, there are two approaches we might try. We could play with the notion of size—of cardinality—or we could play with the notion of power set. It is the second approach that succeeds—the model we produce will satisfy the countable chain condition, and so its cardinals will be the same as the cardinals of M (except, of course, that we are in a modal instead of a classical model). It is the notion of power set that will be manipulated—we will arrange things so that ω has lots of subsets. (The technique is very similar to the one used in the previous chapter to produce a non-constructible set.) The seemingly simple power set operation turns out to be one of the least understood operations of set theory.

§2 The model

In the previous chapter we used as forcing conditions disjoint pairs $\langle P, N \rangle$ of finite sets of integers, and this allowed us to produce a single member of the model that was not constructible. This time we want to make sure ω has $\aleph_2$ subsets in our model, so we will produce $\aleph_2$ non-constructible sets, but we use the same technique for each one of them.

One word on notation first, however. Cardinals are initial ordinals—those not the same size as any smaller ordinal. The cardinal $\aleph_2$ is thus some ordinal. Ordinals are always well behaved in modal models, but this is not true for cardinals. We will, for the most part, be using $\aleph_2$ in its role as ordinal, rather than as cardinal. To emphasize this, we generally denote it as ω_2 in this chapter, rather than as $\aleph_2$. Thus ω_2 is a particular ordinal that, in M, is the second cardinal after ω. It remains to be determined whether this ordinal, or rather its modal counterpart, still plays the same role in the $S4$-regular model we construct.

Model specifics A *forcing condition* is a function p with domain ω_2 such that:

(1) for each $\alpha \in \omega_2$, $p(\alpha) = \langle P_\alpha, N_\alpha \rangle$, where P_α and N_α are finite, disjoint sets of integers;

(2) $p(\alpha) = \langle \emptyset, \emptyset \rangle$ for all but a finite number of $\alpha \in \omega_2$.

$\mathcal{G}$ is the set of all forcing conditions. For p and q in $\mathcal{G}$, set $p\mathcal{R}q$ provided, for each $\alpha \in \omega_2$, if $p_\alpha = \langle P_\alpha, N_\alpha \rangle$ and $q_\alpha = \langle P'_\alpha, N'_\alpha \rangle$, then $P_\alpha \subseteq P'_\alpha$ and $N_\alpha \subseteq N'_\alpha$.

It is easy to see that $\mathcal{R}$ is reflexive and transitive, and that $\langle \mathcal{G}, \mathcal{R} \rangle$ is an M-set. Let $\mathcal{M} = \langle \mathcal{G}, \mathcal{R}, \mathcal{D}^{\mathcal{G}}, \Vdash \rangle$ be the resulting $S4$-regular model over M.

Now, for each $\alpha \in \omega_2$, create a member $f_\alpha \in R^{\mathcal{G}}_{\omega+1} - R^{\mathcal{G}}_\omega$ as follows:

$$f_\alpha = \{\langle p, \hat{n} \rangle \mid p \in \mathcal{G}, p(\alpha) = \langle P_\alpha, N_\alpha \rangle \text{ and } n \in P_\alpha\}.$$

As expected, if $p(\alpha) = \langle P_\alpha, N_\alpha \rangle$ then $p \Vdash (\hat{n} \,\varepsilon\, f_\alpha)$ iff $n \in P_\alpha$. Note too that if $p \Vdash \hat{n} \,\varepsilon\, f_\alpha$ then $p \Vdash \llbracket \hat{n} \,\varepsilon\, f_\alpha \rrbracket$. Also if $p(\alpha) = \langle P_\alpha, N_\alpha \rangle$ and $n \in N_\alpha$, then $p \Vdash \llbracket \neg(\hat{n} \,\varepsilon\, f_\alpha) \rrbracket$. We leave it to you to verify these items.

The goal is to show that as far as the $S4$-model $\mathcal{M}$ is concerned, if $\alpha \neq \beta$ then f_α and f_β are distinct, each is a subset of ω, and there are ω_2 of them. We do this in the next few sections.

EXERCISES

Exercise 2.1. For each $\alpha \in \omega_2$ show f_α is non-constructible in the model $\mathcal{M}$. That is, show the validity of $\llbracket \neg \mathsf{Const}(f_\alpha) \rrbracket$.

§3 Cardinals stay cardinals

We show the countable chain condition holds for the frame $\langle \mathcal{G}, \mathcal{R} \rangle$, essentially a result of Marczewski (1947). Then by Theorem 5.5 of Chapter 18, the cardinals of $\mathcal{M}$ are well-behaved—that is, they are the same as the "real" cardinals. For convenience, we introduce some temporary terminology.

Suppose $p \in \mathcal{G}$, and $p(\alpha) = \langle P_\alpha, N_\alpha \rangle$. We say $p(\alpha)$ is *non-trivial* if $P_\alpha \cup N_\alpha \neq \emptyset$. If $n \in P_\alpha \cup N_\alpha$ we say p *commits on* $\langle n, \alpha \rangle$—positively if $n \in P_\alpha$, negatively if $n \in N_\alpha$. If also $q \in \mathcal{G}$ we say p and q *differ on* $\langle n, \alpha \rangle$ if $q(\alpha) = \langle P'_\alpha, N'_\alpha \rangle$, and either $n \in P_\alpha \cap N'_\alpha$ or $n \in N_\alpha \cap P'_\alpha$. It is easy to see that if p and q are incompatible—that is, if p and q have no common extension—they must differ on some $\langle n, \alpha \rangle$. Likewise if $p\mathcal{R}q$ then q must commit on $\langle n, \alpha \rangle$ whenever p does, and must commit the same way—positively if p commits positively, negatively if p commits negatively.

Proposition 3.1. $\langle \mathcal{G}, \mathcal{R} \rangle$ *satisfies c.c.c.*

§3. CARDINALS STAY CARDINALS

Proof Suppose $\mathcal{B} \subseteq \mathcal{G}$, members of $\mathcal{B}$ are pairwise incompatible, and $\mathcal{B}$ is uncountable. We derive a contradiction.

For each $p \in \mathcal{G}$, for only a finite number of α is $p(\alpha)$ non-trivial, and if $p(\alpha) = \langle P_\alpha, N_\alpha \rangle$ is non-trivial, P_α and N_α are both finite. Consequently we can introduce a measure for the complexity of p as follows:

$$\|p(\alpha)\| = \overline{\overline{P_\alpha}} + \overline{\overline{N_\alpha}} \text{ where } p(\alpha) = \langle P_\alpha, N_\alpha \rangle$$

$$\|p\| = \sum_{\alpha \in \omega_2} \|p(\alpha)\|.$$

For each n, let $\mathcal{B}_n = \{p \in \mathcal{B} \mid \|p\| < n\}$. Then $\mathcal{B} = \cup_{n \in \omega} \mathcal{B}_n$. Since $\mathcal{B}$ is uncountable, for some n, $\mathcal{B}_n$ must be uncountable. Fix such an n: then $\mathcal{B}_n$ is uncountable, its members are pairwise incompatible, and if $p \in \mathcal{B}_n$, then $\|p\| < n$.

Next, call p *good* if there are uncountably many $q \in \mathcal{B}_n$ such that $p\mathcal{R}q$ (if p is good p cannot be in $\mathcal{B}$ since members of $\mathcal{B}$ are mutually incompatible). If p is the function that is identically $\langle \emptyset, \emptyset \rangle$, p is good (since $\mathcal{B}_n$ is uncountable and $p\mathcal{R}q$ for any q). Also for every $p \in \mathcal{G}$ such that $\|p\| \geq n$, p is not good (since if $p\mathcal{R}q$ then $\|p\| \leq \|q\|$). Thus there are good members of $\mathcal{G}$, and there is an upper bound on their complexity. Now fix a good $p_0 \in \mathcal{G}$ with $\|p_0\|$ the largest possible.

Finally, let $\mathcal{B}^0 = \{q \in \mathcal{B}_n \mid p_0\mathcal{R}q\}$—an uncountable set since p_0 is good. Arbitrarily choose some $q_0 \in \mathcal{B}^0$. Then $p_0\mathcal{R}q_0$. If q is another member of $\mathcal{B}^0$, q_0 and q are incompatible, so they must differ on some item, and it cannot be an item on which p_0 commits, since both $p_0\mathcal{R}q$ and $p_0\mathcal{R}q_0$. Let $D = \{\langle n_1, \alpha_1 \rangle, \ldots, \langle n_t, \alpha_t \rangle\}$ be the items on which q_0 commits but p_0 does not—a finite set. Then q_0 and q must differ on some member of D. More generally, q_0 must differ with every $q \in \mathcal{B}^0$ on some member of D and, since D is finite and $\mathcal{B}^0$ is uncountable, there must be some $\langle n_i, \alpha_i \rangle \in D$ on which q_0 differs from uncountably many members of $\mathcal{B}^0$—say q_0 commits positively on $\langle n_i, \alpha_i \rangle$, the other possibility is treated similarly. Now, p_0 does not commit on this item, since it was taken from D. Let p' be the same as p_0 on all members of ω_2 except for α_i, and if $p_0(\alpha_i) = \langle P_{\alpha_i}, N_{\alpha_i} \rangle$, let $p'(\alpha_i) = \langle P_{\alpha_i}, N_{\alpha_i} \cup \{n_i\} \rangle$. p' is again a member of $\mathcal{G}$ (since p_0 did not commit on $\langle n_i, \alpha_i \rangle$). But also, if q differs from q_0 on $\langle n_i, \alpha_i \rangle$ it is easy to see that $p'\mathcal{R}q$, consequently there are uncountably many members of $\mathcal{B}^0$ accessible from p'. But then p' is also good, and this is a contradiction on our choice of p_0, since $\|p'\| = \|p_0\| + 1$. ∎

Remark All the $S4$-regular models we have presented so far have satisfied the c.c.c., and so the notion of cardinal was well behaved. It should be pointed out that this need not be the case. There is a technique for *collapsing cardinals* which can produce quite different results. Here is a simple example.

Temporarily, take a forcing condition to be a 1-1 function p such that:

(1) Dom(p) is a finite subset of ω,

(2) Ran(p) $\subseteq \omega_2$.

Let $\mathcal{G}_0$ be the set of these forcing conditions, and for $p, q \in \mathcal{G}_0$, let $p\mathcal{R}_0 q$ mean q extends p as a function. Then $\langle \mathcal{G}_0, \mathcal{R}_0 \rangle$ is a frame that is an M-set, so the theory developed so far applies, and we have an $S4$-regular model based on this frame.

Now, for each integer n and each $\alpha \in \omega_2$, $\langle n, \alpha \rangle$ is a regular set, so $\widehat{\langle n, \alpha \rangle}$ is in the domain of our $S4$-regular model. Let f be the member of the domain determined by: $p \Vdash \widehat{\langle n, \alpha \rangle} \varepsilon f$ iff $n \in \text{Dom}(p)$ and $p(n) = \alpha$. We leave it to you to show that $\llbracket f : \hat{\omega} \xrightarrow[\text{onto}]{1\text{-}1} \widehat{\omega_2} \rrbracket$ is valid in the model. This implies that $\widehat{\omega_2}$, the member of the model domain corresponding directly to the cardinal $\aleph_2$ of M, does not play the role of being the third infinite cardinal in the model. In fact, in the $S4$ model $\widehat{\omega_2}$ has the same cardinality as $\hat{\omega}$.

EXERCISES

Exercise 3.1. Carry out the proof suggested in the remark above, and prove that $\llbracket f : \hat{\omega} \xrightarrow[\text{onto}]{1\text{-}1} \widehat{\omega_2} \rrbracket$ is valid in the $S4$-regular model described.

§4 CH is independent

Now we continue from where we left off in §2. The following result concerning the members f_α of $\mathcal{D}^{\mathcal{G}}$ is fundamental.

Proposition 4.1. *If $\alpha \neq \beta$ then $\llbracket \neg (f_\alpha \approx f_\beta) \rrbracket$ is valid in $\mathcal{M}$.*

Proof Suppose $\alpha \neq \beta$ but $\llbracket \neg (f_\alpha \approx f_\beta) \rrbracket$ is not valid. Then there is some world at which $\llbracket f_\alpha \approx f_\beta \rrbracket$ is true, say at world p. Also say $p(\alpha) = \langle P_\alpha, N_\alpha \rangle$ and $p(\beta) = \langle P_\beta, N_\beta \rangle$. Choose a number n that is not in $P_\alpha \cup N_\alpha \cup P_\beta \cup N_\beta$, and let p' be like p except that $p'(\alpha) = \langle P_\alpha \cup \{n\}, N_\alpha \rangle$ and $p'(\beta) = \langle P_\beta, N_\beta \cup \{n\} \rangle$ (this is possible, of course, since $\alpha \neq \beta$). Then $p\mathcal{R}p'$, $p' \Vdash \llbracket \hat{n} \in f_\alpha \rrbracket$, and $p' \Vdash \llbracket \neg (\hat{n} \in f_\beta) \rrbracket$. This contradicts the assumption that $\llbracket f_\alpha \approx f_\beta \rrbracket$ is true at p, and hence also at p'. ∎

Now we are almost done. We want to show there is a member of $\mathcal{D}^{\mathcal{G}}$ that acts, in $\mathcal{M}$, like a 1-1 correspondence between ω_2 and a subset of $\mathcal{P}(\omega)$. To make it easier to do this we introduce a few simple operations, intended to represent unordered and ordered pairing respectively.

Definition 4.2. Two mappings from $\mathcal{D}^{\mathcal{G}}$ to itself are defined as follows.
$\text{Up}(x, y) = \mathcal{G} \times \{x, y\}$
$\text{Op}(x, y) = \text{Up}(\text{Up}(x, x), \text{Up}(x, y))$.

If $f, g \in R_\alpha^{\mathcal{G}}$, then $\text{Up}(f, g) \in R_{\alpha+1}^{\mathcal{G}}$ and $\text{Op}(f, g) \in R_{\alpha+2}^{\mathcal{G}}$. Now, each $f_\alpha \in R_{\omega+1}^{\mathcal{G}}$, and $\hat{\alpha} \in R_{\alpha+1}^{\mathcal{G}}$. Consequently, for $\alpha \in \omega_2$, $\text{Up}(\hat{\alpha}, f_\alpha) \in R_{\omega_2+2}^{\mathcal{G}}$, and so the following is in $R_{\omega_2+3}^{\mathcal{G}}$:

$$F = \{\text{Op}(\hat{\alpha}, f_\alpha) \mid \alpha \in \omega_2\}.$$

Also, let $S = \mathcal{G} \times \{f_\alpha \mid \alpha \in \omega_2\}$. Then $S \in R_{\omega+2}^{\mathcal{G}}$, so both S and F are in $\mathcal{D}^{\mathcal{G}}$.

Proposition 4.3. *The following is valid in* $\mathcal{M}$:

$$\left\lVert (F : \widehat{\omega_2} \xrightarrow[\text{onto}]{1\text{-}1} S) \land (\forall x)(x \in S \supset x \subseteq \hat{\omega}) \right\rVert.$$

By now the verification of this proposition is a routine exercise, and we omit the details. Since $\mathcal{M}$ meets the countable chain condition, the things it thinks are cardinals are the counterparts of the things M thinks are cardinals. Then an easy consequence of the proposition above is the following, whose proof we also leave to you.

Corollary 4.4. *The following is valid in* $\mathcal{M}$ *(where* a is ω *abbreviates a* Σ *definition of* ω*):*

$$\left\lVert (\exists a)(\exists b)(\exists c)(\exists S)(\exists F) \right. \\ [a \text{ is } \omega \land \mathsf{Cardinal}(b) \land \mathsf{Cardinal}(c) \\ \land a \in b \land b \in c \\ \land (\forall x)(x \in S \supset x \subseteq a) \\ \left. \land (F : c \xrightarrow[\text{onto}]{1\text{-}1} S)] \right\rVert.$$

Loosely, this says that in $\mathcal{M}$ the translate of a statement asserting $\overline{\overline{\mathcal{P}(\omega)}} \geq \aleph_2$ is valid. Consequently $\lVert \neg(CH) \rVert$ is also valid in $\mathcal{M}$.

EXERCISES

Exercise 4.1. Suppose $f, g \in \mathcal{D}^{\mathcal{G}}$. Show the validity in $\mathcal{M}$ of

$$\left\lVert (\forall x)[x \in \mathsf{Up}(f, g) \equiv (x \approx f \lor x \approx g)] \right\rVert.$$

Exercise 4.2. Prove Proposition 4.3 in detail.

§5 Cleaning it up

We established the independence of the continuum hypothesis by producing a modal model that thinks the power set of ω has at least $\aleph_2$ members. Anything more is extra, but it would be nice to know just what size $\mathcal{P}(\omega)$ has in the modal model. In this section we show it is exactly $\aleph_2$, but to show this, we need to assume the generalized continuum hypothesis is true in M. It is consistent to assume this, since M could have been taken to be L, consequently the work in this section amounts to showing the consistency of ZF plus well foundedness plus choice with the assertion that the continuum is of cardinality $\aleph_2$. The role of the generalized continuum hypothesis centers in the following simple calculation.

Lemma 5.1. *Assuming the generalized continuum hypothesis,* $\aleph_2^{\aleph_0} = \aleph_2$.

Proof

$$\begin{aligned} \aleph_2^{\aleph_0} &= (2^{\aleph_1})^{\aleph_0} && \text{by GCH} \\ &= 2^{\aleph_1 \cdot \aleph_0} && \text{by Exercise 8.4 of Chapter 9} \\ &= 2^{\aleph_1} && \text{by Corollary 8.7 of Chapter 9} \\ &= \aleph_2 && \text{by GCH.} \end{aligned}$$

∎

According to Lemma 4.4 of Chapter 17, since $\hat{\omega} \in R^{\mathcal{G}}_{\omega+1}$, if $[\![a \subseteq \hat{\omega}]\!]$ is true at a world p, then for some $b \in R^{\mathcal{G}}_{\omega+1}$, $[\![a \approx b]\!]$ is true at p. To say $b \in R^{\mathcal{G}}_{\omega+1}$ is to say $b \subseteq \mathcal{G} \times R^{\mathcal{G}}_\omega$. Now, this result can be improved, using an argument along the same lines (something we leave to you) to establish that if $[\![a \subseteq \hat{\omega}]\!]$ is true at p then $[\![a \approx b]\!]$ is true at p for some $b \subseteq \mathcal{G} \times \hat{\omega}$. Consequently, to investigate the size of the power set of $\hat{\omega}$ in the modal model, we begin by investigating the power set of $\mathcal{G} \times \hat{\omega}$ in M.

Let $C = \{a \mid a \subseteq \mathcal{G} \times \hat{\omega}\}$. From the point of view of the modal model M defined in §2, there is redundancy in C, which the following is intended to eliminate. Call two members a and b of C *equivalent* if $[\![a \approx b]\!]$ is valid in M. This is an equivalence relation — let C_0 be a subset of C containing exactly one member from each equivalence class (we use the axiom of choice in M here, of course). Our immediate goal is an upper bound on the size of C_0 in M.

For each $a \in C_0$, let $\varphi_a : \omega \to \mathcal{P}(\mathcal{G})$ be defined by:

$$\varphi_a(n) = \{p \in \mathcal{G} \mid p \Vdash [\![\hat{n} \in a]\!]\}.$$

And let C_1 be the set of such functions. We show there is a 1-1 correspondence between C_0 and C_1, so that it is enough to bound the size of C_1. And to show this, since we already have a mapping from C_0 onto C_1 (namely, map a to φ_a), it is enough to show this mapping is 1-1. The following does this.

Lemma 5.2. *For $a, b \in C_0$, if $a \neq b$ then $\varphi_a \neq \varphi_b$.*

Proof If $a \neq b$, $[\![a \approx b]\!]$ is not valid by definition of C_0, so for some $p \in \mathcal{G}$, $p \not\Vdash [\![a \approx b]\!]$. It follows that for some p' with $p\mathcal{R}p'$, $p' \Vdash [\![\neg(a \approx b)]\!]$, and consequently for some c and some p'' with $p'\mathcal{R}p''$, either $p'' \Vdash [\![c \in a \wedge \neg(c \in b)]\!]$ or $p'' \Vdash [\![c \in b \wedge \neg(c \in a)]\!]$ — say the first, since the argument is similar either way.

Since $p'' \Vdash [\![c \in a]\!]$, for some d, $p'' \Vdash [\![c \approx d \wedge d \varepsilon a]\!]$. Since $a \subseteq \mathcal{G} \times \hat{\omega}$, it follows that $d = \hat{n}$ for some integer n. But then $p'' \in \varphi_a(n)$ and clearly $p'' \notin \varphi_b(n)$, so $\varphi_a \neq \varphi_b$. ∎

Each φ_a maps ω to $\mathcal{P}(\mathcal{G})$. Now, $\overline{\overline{\mathcal{G}}}$ is $\aleph_2$ (Exercise 5.2), so assuming the generalized continuum hypothesis, $\overline{\overline{\mathcal{P}(\mathcal{G})}} = \aleph_3$. Then an obvious upper bound on the cardinality of C_1 is $\overline{\overline{\mathcal{P}(\mathcal{G})^\omega}} = \aleph_3^{\aleph_0} = \aleph_3$ (the last step is by a proof essentially identical to that of Lemma 5.1). But we can do better than this by showing that there is a subset T of $\mathcal{P}(\mathcal{G})$, with $\overline{\overline{T}} \leq \aleph_2$, so that each φ_a actually maps ω to T. Then a better upper bound on $\overline{\overline{C_1}}$ is $\overline{\overline{T^\omega}}$, and this is $\leq \aleph_2$ by Lemma 5.1. So the final step in bounding C_0, or equivalently, C_1, is to produce an appropriate set T.

If a set is in the range of φ_a, it is $\{p \mid p \Vdash [\![\hat{n} \in a]\!]\}$. Let X be an arbitrary formula, and let us say a subset A of $\mathcal{G}$ is the *X set* if $A = \{p \in \mathcal{G} \mid p \Vdash [\![X]\!]\}$. Then members of the range of φ_a are X sets, where X is of the form $\hat{n} \in a$. Let T be the collection of all X sets—then each φ_a maps ω into T. We proceed to show $\overline{\overline{T}} \leq \aleph_2$.

Suppose $A \in T$. By a *core* of A we mean a maximal subset of A whose members are pairwise incompatible. Using Zorn's lemma, cores exist. They are not unique, but

§5. CLEANING IT UP

it is the case that if $A_1 \neq A_2$, no core for A_1 can also be a core for A_2—this is the content of the following.

Lemma 5.3. *Suppose A_1 is the X_1 set, A_2 is the X_2 set, and C is a core for both A_1 and A_2. Then $[\![X_1 \equiv X_2]\!]$ is valid in $\mathcal{M}$, and it follows that $A_1 = A_2$.*

Proof If $[\![X_1 \equiv X_2]\!]$ is not valid, either there is some world at which $[\![X_1 \wedge \neg X_2]\!]$ is true, or there is some world at which $[\![X_2 \wedge \neg X_1]\!]$ is true—say $p \Vdash [\![X_1 \wedge \neg X_2]\!]$, the other case is similar.

Since $p \Vdash [\![X_1]\!]$, $p \in A_1$. If we had that $p \in C$, then $p \in A_2$, so $p \Vdash [\![X_2]\!]$, which is impossible. Hence $p \notin C$. If p were incompatible with every member of C, p would be in C by maximality. Consequently for some $q \in C$, p and q are compatible. Then for some $r \in \mathcal{G}$, $p\mathcal{R}r$ and $q\mathcal{R}r$. Since $[\![\neg X_2]\!]$ is true at p, it is also true at r. But since $q \in C$, $q \in A_2$, so $[\![X_2]\!]$ is true at q, hence also at r, and this is impossible. ∎

Thus there is a many-one mapping from the set of all cores of members of T to the set T itself, so an upper bound on the size of the set of all T-cores would also be an upper bound on the size of T.

Lemma 5.4. *Assume the generalized continuum hypothesis holds in $\mathcal{M}$. Then the cardinality of the set of cores of members of T is $\leq \aleph_2$.*

Proof Every core is a subset of $\mathcal{G}$ whose members are pairwise incompatible. Since $\langle \mathcal{G}, \mathcal{R} \rangle$ satisfies c.c.c., each core must be countable. Therefore the cardinality of all cores is bounded by the cardinality of all countable subsets of $\mathcal{G}$, that is, the cardinality is $\leq \overline{\overline{\mathcal{G}^\omega}}$. And this is $\aleph_2$ by Lemma 5.1 again. ∎

We summarize things thus far.

Proposition 5.5. $\overline{\overline{C_0}} \leq \aleph_2$.

We are finished investigating $\mathcal{P}(\mathcal{G} \times \hat{\omega})$ and its subset C_0. According to the proposition above, there is a 1-1 function $f_0 : C_0 \to \omega_2$ (recall, ω_2 is the initial ordinal that plays the role of $\aleph_2$ in M). Now we define a member of $\mathcal{D}^{\mathcal{G}}$ as follows.

$$f = \{\mathsf{Op}(a, \hat{\alpha}) \mid a \in C_0 \text{ and } f_0(a) = \alpha\}.$$

Proposition 5.6. *The following is valid in $\mathcal{M}$:*

$$[\![(f : \mathcal{P}(\hat{\omega}) \to \widehat{\omega_2}) \wedge \text{1-1}(f)]\!]$$

(where we have used abbreviations for the obvious formulas).

The proof of this is left to you. Combined with results of the previous section we have that, as far as $\mathcal{M}$ is concerned, the cardinality of $\mathcal{P}(\omega)$ is exactly $\aleph_2$.

EXERCISES

Exercise 5.1. Following the proof of Lemma 4.4 of Chapter 17, show that if $p \Vdash \llbracket a \subseteq \hat{\omega} \rrbracket$ then $p \Vdash \llbracket a \approx b \rrbracket$ for some $b \subseteq \mathcal{G} \times \hat{\omega}$.

Exercise 5.2. Prove $\overline{\overline{\mathcal{G}}} = \aleph_2$.

Exercise 5.3. Prove Proposition 5.6.

§6 Wrapping it up

It should be clear that by using the methods of this chapter we can produce modal models in which the cardinality of the continuum is as big as we wish, at least $\aleph_{100}$, or $\aleph_\omega$, for instance. But when do the methods of §5 apply—which cardinals can we hit exactly? There is an old theorem of König that says the cardinality of the continuum cannot be the sum of countably many smaller cardinals. According to this, for example, the cardinality of the continuum cannot be $\aleph_\omega$ since this is a countable sum of smaller cardinals: $\aleph_0 + \aleph_1 + \aleph_2 + \cdots$. König's theorem does not rule out $\aleph_{100}$, and in fact the methods of §5 and earlier sections can be used to show it is consistent to assume the continuum has size $\aleph_{100}$. More generally, every cardinal not explicitly ruled out by König's theorem can be attained in some modal model, so König's result is best possible.

CHAPTER 20

INDEPENDENCE OF THE AXIOM OF CHOICE

§1 A little history

Since the translate of the axiom of choice is valid in every $S4$-ZF model we have seen so far, it is not obvious how to prove its independence. The technique we use has an interesting history. As we noted back in Chapter 4, Bertrand Russell illustrated the role of the axiom of choice by using socks and shoes. If we have a countable set of pairs of shoes, we can choose one shoe from each pair by applying a *rule*—choose the left shoe, for instance. But if we have a countable set of pairs of socks no such rule is available, and the possibility of making a choice of one sock from each pair must be specially postulated. The key difference between the two examples lies in their *symmetries*. Socks are interchangeable, shoes are not—socks have a symmetry that shoes lack.

Fraenkel (1922) used this idea of symmetry to prove the independence of the axiom of choice. He produced a model with so much symmetry that there were items indistinguishable in the theory, and consequently choice functions could not exist for every collection. There was a basic difficulty, however. One cannot simply impose symmetry requirements on sets—they have internal structure and intimate relationships with other sets that make it very difficult to modify some without affecting all. Fraenkel's solution was to work with a version of set theory admitting *urelements*—items that can be members of sets, but are not sets themselves. (In a sense they are duals of proper classes, which can have sets as members, but are not sets themselves.) Urelements are, in fact, needed if one wishes to apply set theory to the everyday world. After all, chairs and tables and beer mugs are not sets, but one can form sets of them. Symmetries can be imposed on the collection of urelements with considerable freedom, and this was the basis of Fraenkel's independence proof. But urelements are not needed in pure mathematics, and it remained at least possible that without urelements around, the axiom of choice might turn out to be provable.

Cohen showed that Fraenkel's argument could be adapted to the setting of forcing with very few changes, and this yielded the full independence of the axiom of choice, even without urelements. What we present in this chapter is a modified version of Scott and Solovay's modification of Cohen's modification of Fraenkel's proof. We will produce a modal model with a countable family of sets whose members have so much symmetry that no choice function for the family can exist — a version of Russell's socks.

§2 Automorphism groups

Mathematically, symmetries are represented by *permutation groups*. In our case, permutations of the set of possible worlds are enough to consider since, as we will see, these induce mappings of model domains. For this section, $\mathcal{M} = (\mathcal{G}, \mathcal{R}, \mathcal{D}^{\mathcal{G}}, \mathcal{V})$ is an $S4$-regular model over M, where M is a transitive subclass of V that is a well founded first-order universe in which the axiom of choice is true.

Definition 2.1. An *automorphism* of $\langle \mathcal{G}, \mathcal{R} \rangle$ is a mapping $\theta : \mathcal{G} \xrightarrow[\text{onto}]{\text{1-1}} \mathcal{G}$ such that $p\mathcal{R}q$ if and only if $\theta(p)\mathcal{R}\theta(q)$. An *automorphism group* of $\langle \mathcal{G}, \mathcal{R} \rangle$ is a set of automorphisms of $\langle \mathcal{G}, \mathcal{R} \rangle$ that contains the identity map, is closed under function composition, and contains with each member its inverse.

Remark We will be speaking about groups throughout this chapter, but what we use from group theory is quite minimal—little more than the definition. For automorphisms, we will write $\theta_1 \theta_2$ to denote composition, and we note that this operation of composition is associative, so we will leave parentheses out when more than two items are involved.

The action of an automorphism on $\langle \mathcal{G}, \mathcal{R} \rangle$ can be extended (recursively) to the members of the domain of $\mathcal{M}$.

Definition 2.2. For $f \in \mathcal{D}^{\mathcal{G}}$ and an automorphism θ of $\langle \mathcal{G}, \mathcal{R} \rangle$:

$$\theta(f) = \{\langle \theta(p), \theta(g) \rangle \mid \langle p, g \rangle \in f\}.$$

Let us get a feeling for how this works. $R_0^{\mathcal{G}} = \emptyset$, so $\mathcal{G} \times R_0^{\mathcal{G}} = \emptyset$ and thus $R_1^{\mathcal{G}}$ has only $\emptyset$ as member. From the definition, $\theta(\emptyset) = \emptyset$. Next, the members of $R_2^{\mathcal{G}}$ are each of the form $S \times \{\emptyset\}$, where $S \subseteq \mathcal{G}$, so $\theta(S \times \{\emptyset\}) = \{\langle \theta(s), \theta(\emptyset) \rangle \mid s \in S\} = \{\langle \theta(s), \emptyset \rangle \mid s \in S\}$. And so on. Automorphisms map $\mathcal{G}$ 1-1 and onto itself—the following says the same is true of their extensions to $\mathcal{D}^{\mathcal{G}}$.

Proposition 2.3. *If θ is an automorphism of $\langle \mathcal{G}, \mathcal{R} \rangle$ then $\theta : \mathcal{D}^{\mathcal{G}} \xrightarrow[\text{onto}]{\text{1-1}} \mathcal{D}^{\mathcal{G}}$. Likewise, if e is the identity automorphism of $\langle \mathcal{G}, \mathcal{R} \rangle$, then e is the identity map on $\mathcal{D}^{\mathcal{G}}$, and similarly for inverses.*

Proof This is left to you as Exercise 2.1 ∎

Definition 2.4. Let $\mathfrak{G}$ be an automorphism group of $\langle \mathcal{G}, \mathcal{R} \rangle$. We say a member $f \in \mathcal{D}^{\mathcal{G}}$ is $\mathfrak{G}$-*invariant* if $\llbracket f \approx \theta(f) \rrbracket$ is valid in $\mathcal{M}$, for each $\theta \in \mathfrak{G}$.

Working with automorphism groups is a start in imposing symmetry conditions on $\mathcal{M}$—restrict the model to those members of $\mathcal{D}^{\mathcal{G}}$ that are invariant with respect to an automorphism group. But this is not yet flexible enough for our purposes. We need something more dynamic—a *sequence* of automorphism groups, which we can think of as finer and finer approximations to some limit. However a completely arbitrary sequence of automorphism groups won't do—some structure connecting members of the sequence is needed.

§2. AUTOMORPHISM GROUPS

Notation If θ_1 and θ_2 are automorphisms, and H is a set of automorphisms, by $\theta_1 H \theta_2$ is meant $\{\theta_1 h \theta_2 \mid h \in H\}$. (Recall that we have associativity, so parentheses are not needed.)

Definition 2.5. Let θ be an automorphism and H be a set of automorphisms. We say θ *normalizes* H if $\theta^{-1} H \theta = H$.

Definition 2.6. An *automorphism sequence* of $\langle \mathcal{G}, \mathcal{R} \rangle$ is a sequence $\mathfrak{F} = \mathfrak{G}_0, \mathfrak{G}_1, \mathfrak{G}_2, \ldots$ of automorphism groups of $\langle \mathcal{G}, \mathcal{R} \rangle$ such that:

(1) $\mathfrak{G}_{n+1} \subseteq \mathfrak{G}_n$, for each n;

(2) for every $\theta \in \mathfrak{G}_0$, there is some n so that θ normalizes each of $\mathfrak{G}_n, \mathfrak{G}_{n+1}, \mathfrak{G}_{n+2}, \ldots$

Since $\mathfrak{G}_{n+1} \subseteq \mathfrak{G}_n$, and both are groups, $\mathfrak{G}_{n+1}$ is a *subgroup* of $\mathfrak{G}_n$, though we will make no special use of this fact. We use the terminology that θ normalizes *almost all* the members of $\mathfrak{F}$, meaning that θ normalizes all of $\mathfrak{G}_n, \mathfrak{G}_{n+1}, \mathfrak{G}_{n+2}, \ldots$ for some n. The requirement of normalization in the definition of automorphism sequence may seem somewhat peculiar. It is needed for the proof of Theorem 3.3. Now we come to the central notion of this chapter.

Definition 2.7. Let $\mathfrak{F}$ be an automorphism sequence of $\langle \mathcal{G}, \mathcal{R} \rangle$. We say a member $f \in \mathcal{D}^{\mathcal{G}}$ is $\mathfrak{F}$-*invariant* if:

(1) for some $\mathfrak{G}$ in the sequence $\mathfrak{F}$, f is $\mathfrak{G}$-invariant;

(2) for each $\langle p, g \rangle \in f$, g is $\mathfrak{F}$-invariant.

Notice that if f is $\mathfrak{G}$-invariant for some $\mathfrak{G}$ in the sequence $\mathfrak{F}$, it is also $\mathfrak{G}'$-invariant for all later members of the sequence $\mathfrak{F}$ as well, since later members are subsets of $\mathfrak{G}$. We could informally describe this as: $\mathcal{M}$ eventually can't distinguish between f and images of f. Now the essence of the recursive definition above is quite simple. f is $\mathfrak{F}$-invariant if $\mathcal{M}$ eventually can't distinguish between f and images of f, and the same is true for items any world thinks are members of f, or members of members of f, and so on. The following proposition is a simple consequence of this definition, and will turn out to be quite useful.

Proposition 2.8. *Suppose f is $\mathfrak{F}$-invariant, and $p \Vdash \llbracket g \, \varepsilon \, f \rrbracket$. Then g is also $\mathfrak{F}$-invariant.*

Proof Since $p \Vdash \llbracket g \, \varepsilon \, f \rrbracket$, $p \Vdash \square \Diamond (g \, \varepsilon \, f)$, so $p \Vdash \Diamond (g \, \varepsilon \, f)$ and for some p' with $p \mathcal{R} p'$, $p' \Vdash (g \, \varepsilon \, f)$. Then $\langle p', g \rangle \in f$, so g is $\mathfrak{F}$-invariant. ∎

The following has a simple proof, via an obvious transfinite induction on rank.

Proposition 2.9. *If a is an M-set, $\theta(\hat{a}) = \hat{a}$ for every automorphism θ. Consequently the counterparts of M-sets are $\mathfrak{F}$-invariant for every automorphism sequence $\mathfrak{F}$.*

Now a few general notions, after which we can outline the rest of the chapter.

Definition 2.10. Let $\mathfrak{F}$ be an automorphism sequence. Set $\mathcal{D}^{\mathcal{G}}_{\mathfrak{F}}$ to be the subset of $\mathcal{D}^{\mathcal{G}}$ consisting of those members that are $\mathfrak{F}$-invariant. Set $\Vdash_{\mathfrak{F}}$ to be the forcing relation determined by restricting $\mathcal{V}$ to atomic formulas over $\mathcal{D}_{\mathfrak{F}}^{\mathcal{G}}$. Finally, let $\mathcal{M}_{\mathfrak{F}}$ be the $S4$ model $(\mathcal{G}, \mathcal{R}, \mathcal{D}_{\mathfrak{F}}^{\mathcal{G}}, \Vdash_{\mathfrak{F}})$.

There is some ambiguity in the definition of $\Vdash_{\mathfrak{F}}$ above, which we must clarify. The forcing relation of $\mathcal{M}_{\mathfrak{F}}$ is determined at the atomic level by restricting $\Vdash$ to atomic formulas involving only constants from $\mathcal{D}_{\mathfrak{F}}^{\mathcal{G}}$. The ambiguity lies in our not specifying the language—do we allow only $\in$, or also $\approx$? What about ε, and $\approx_\alpha$? Basically we are interested in $\mathcal{L}_C$, the language of set theory that has only $\in$ as a relation symbol. In defining $\mathcal{M}_{\mathfrak{F}}$ we use the modal counterpart of $\mathcal{L}_C$, which we called $\mathcal{L}_C^*$—thus atomic formulas contain $\in$ and no other relation symbols. The role of ε, $\approx_\alpha$, and $\approx$ in $\mathcal{M}$ was secondary all along. They aided in the definition of forcing for formulas involving $\in$ in $\mathcal{M}$. They still play that role, of course, but they have no direct function in $\mathcal{M}_{\mathfrak{F}}$.

Now, what is the plan for this chapter? First we will show that, for any automorphism sequence $\mathfrak{F}$ that is an M-set, $\mathcal{M}_{\mathfrak{F}}$ is always an $S4$-ZF model. And second, we will produce a particular $S4$-regular model $\mathcal{M}$ over M, and an automorphism sequence $\mathfrak{F}$ in M, so that in $\mathcal{M}_{\mathfrak{F}}$ the translate of the axiom of choice fails. In $\mathcal{M}_{\mathfrak{F}}$ we will have an analog of Russell's socks—there will be a countable family T, each member of which is a countable family whose members are created in essentially the same way the non-constructible set of Chapter 18 was. Then we will create an automorphism sequence $\mathfrak{F}$ that imposes so much symmetry that in $\mathcal{M}_{\mathfrak{F}}$ no choice function for T is possible.

EXERCISES

Exercise 2.1. Automorphisms are required to be 1-1 and onto on $\mathcal{G}$. Show they are 1-1 and onto on $\mathcal{D}^{\mathcal{G}}$ as well.

Exercise 2.2. Prove that $\theta(\hat{a}) = \hat{a}$ for every regular set a and any automorphism θ of $\langle \mathcal{G}, \mathcal{R} \rangle$. Show it follows that every member of $\mathcal{G}$ of the form $\hat{a}$ is $\mathfrak{F}$-invariant for every automorphism sequence $\mathfrak{F}$.

§3 Automorphisms preserve truth

For this section let $\mathcal{M} = (\mathcal{G}, \mathcal{R}, \mathcal{D}^{\mathcal{G}}, \Vdash)$ be an $S4$-regular model over M. Automorphisms of $\langle \mathcal{G}, \mathcal{R} \rangle$ act on both $\mathcal{G}$ and $\mathcal{D}^{\mathcal{G}}$. Their action can be extended to formulas as well.

Definition 3.1. Let X be a formula with constants from $\mathcal{D}^{\mathcal{G}}$, and let θ be an automorphism of $\langle \mathcal{G}, \mathcal{R} \rangle$. By $\theta(X)$ is meant the formula that results when each constant f in X is replaced with its image $\theta(f)$ under the automorphism.

Now we show the following unsurprising, but basic, result.

Theorem 3.2. *For any $p \in \mathcal{G}$, any closed formula X of $\mathcal{L}_M$ with constants from $\mathcal{D}^{\mathcal{G}}$, and any automorphism θ, $p \Vdash X$ if and only if $\theta(p) \Vdash \theta(X)$.*

Proof The proof, while not complicated, has several parts to it which we treat one by one. While we are primarily interested in formulas whose only relation symbol is $\in$, the behavior of $\in$ was defined in terms of $\approx$ and ε, and these must be dealt with as well. We begin with a "build-up" result.

§3. AUTOMORPHISMS PRESERVE TRUTH

(1) Let $\mathcal{F}$ be a set of formulas with free variables, and suppose Theorem 3.2 holds for every instance of a formula in $\mathcal{F}$. Then Theorem 3.2 also holds for instances of formulas built up from members of $\mathcal{F}$ using propositional connectives, modal operators, and quantifiers.

Proof of (1). The argument is by induction on the complexity of formula instances, with the base case given. For the modal case, suppose the result is known for X; we show it for $\Box X$. In one direction, suppose $p \Vdash \Box X$. Let q be an arbitrary member of $\mathcal{G}$ such that $\theta(p)\mathcal{R}q$. Since $\theta : \mathcal{G} \xrightarrow[\text{onto}]{\text{1-1}} \mathcal{G}$, for some p', $\theta(p') = q$, and so $\theta(p)\mathcal{R}\theta(p')$. Since θ is an automorphism, $p\mathcal{R}p'$, so $p' \Vdash X$. By the induction hypothesis, $\theta(p') \Vdash \theta(X)$, or $q \Vdash \theta(X)$. Since q was arbitrary, $\theta(p) \Vdash \Box\theta(X)$, but $\Box\theta(X) = \theta(\Box X)$. We have shown that $p \Vdash \Box X$ implies $\theta(p) \Vdash \theta(\Box X)$. The other direction is similar, using θ^{-1}.

The propositional connective cases are trivial. Finally, the quantifier cases are easy consequences of the fact that $\theta : \mathcal{D}^\mathcal{G} \xrightarrow[\text{onto}]{\text{1-1}} \mathcal{D}^\mathcal{G}$ (Proposition 2.3).

(2) Theorem 3.2 holds when X is of the form $f \, \varepsilon \, g$.

Proof of (2). This is almost immediate from the definition, since $p \Vdash f \varepsilon g$ iff $\langle p, f \rangle \in g$ iff $\langle \theta(p), \theta(f) \rangle \in \theta(g)$ iff $\theta(p) \Vdash \theta(f) \varepsilon \theta(g)$.

(3) Theorem 3.2 holds if X is of the form $f \approx_\alpha g$ or $f \approx g$.

Proof of (3). The result for $f \approx g$ follows once that for $f \approx_\alpha g$ has been shown, and this is proved by induction on α. The cases where α is 0 or a limit ordinal are trivial. For the successor ordinal case, suppose the result is true for α. Then it follows for $\alpha+1$ since, by Definition 1.4 of Chapter 17, the truth of $f \approx_{\alpha+1} g$ at a world depends on the truth, at that world, of a formula built up from ε and $\approx_\alpha$, and we have the induction hypothesis and parts (1) and (2).

(4) Theorem 3.2 holds generally.

Proof of (4). Truth of formulas of the form $f \in g$ is determined by the behavior of formulas involving ε and $\approx$ (Definition 1.7 of Chapter 17), and we have parts (1), (2) and (3), so the result holds for all atomic formulas. Now an application of item (1) finishes the argument. ∎

Now we get to use the normalization clause of the definition of automorphism sequence.

Theorem 3.3. *Suppose $\mathfrak{F}$ is an automorphism sequence of $\langle \mathcal{G}, \mathcal{R} \rangle$ and $\theta \in \mathfrak{G}_0$. If f is $\mathfrak{F}$-invariant, so is $\theta(f)$.*

Proof The proof is an easy consequence (via transfinite induction on the $\mathcal{G}$-rank of f) of the following lemma. ∎

Lemma 3.4. *Suppose $\mathfrak{F}$ is the automorphism sequence $\mathfrak{G}_0$, $\mathfrak{G}_1$, $\mathfrak{G}_2$,..., and suppose θ is any member of $\mathfrak{G}_0$. If f is $\mathfrak{G}_n$-invariant then $\theta(f)$ is $\mathfrak{G}_k$-invariant, for some k.*

Proof f is $\mathfrak{G}_n$-invariant. Since $\theta \in \mathfrak{G}_0$, θ normalizes almost all the terms of the sequence $\mathfrak{F}$, so there must be a number $k \geq n$ such that θ normalizes $\mathfrak{G}_k$. Since $k \geq n$, f is $\mathfrak{G}_k$-invariant as well. We will show that $\theta(f)$ is $\mathfrak{G}_k$-invariant.

Let π be an arbitrary member of $\mathfrak{G}_k$. Since θ normalizes $\mathfrak{G}_k$, $\theta^{-1}\mathfrak{G}_k\theta = \mathfrak{G}_k$, and since $\pi \in \mathfrak{G}_k$, $\theta^{-1}\pi\theta \in \mathfrak{G}_k$. Since f is $\mathfrak{G}_k$-invariant, $[\![f \approx \theta^{-1}\pi\theta(f)]\!]$ is $\mathcal{M}$-valid. Let p be an arbitrary member of $\mathcal{G}$, so $p \Vdash [\![f \approx \theta^{-1}\pi\theta(f)]\!]$. By the theorem above, $\theta(p) \Vdash [\![\theta(f) \approx \pi\theta(f)]\!]$. Finally, since θ is an automorphism, and so onto, every member of $\mathcal{G}$ is of the form $\theta(p)$ for some p, and hence $[\![\theta(f) \approx \pi\theta(f)]\!]$ is valid in $\mathcal{M}$. Since π was an arbitrary member of $\mathfrak{G}_k$, $\theta(f)$ is $\mathfrak{G}_k$-invariant. ∎

Now, using this, we can transfer Theorem 3.2 from $\mathcal{M}$ to $\mathcal{M}_{\mathfrak{F}}$.

Theorem 3.5. *Let X be any closed formula of $\mathcal{L}_C^*$ (the language with only $\in$ as relation symbol) having all constants in $\mathcal{D}_{\mathfrak{F}}^{\mathcal{G}}$. For any $p \in \mathcal{G}$ and any $\theta \in \mathfrak{G}_0$, $p \Vdash_{\mathfrak{F}} X$ if and only if $\theta(p) \Vdash_{\mathfrak{F}} \theta(X)$.*

Proof First note that by the previous theorem, if X has all its constants in $\mathcal{D}_{\mathfrak{F}}^{\mathcal{G}}$, the same is true for $\theta(X)$, so the statement of the theorem at least makes sense. Now the result is true for atomic X in $\mathcal{L}_C^*$, since at the atomic level, $\Vdash_{\mathfrak{F}}$ and $\Vdash$ agree, and we have Theorem 3.2. The result then follows for non-atomic formulas by induction on degree, more-or-less as in part (1) of the proof of Theorem 3.2. We omit details. ∎

§4 Model and submodel

Again for this section let $\mathcal{M} = (\mathcal{G}, \mathcal{R}, \mathcal{D}^{\mathcal{G}}, \Vdash)$ be an $S4$-regular model over M and let $\mathfrak{F}$ be an automorphism sequence of $\langle \mathcal{G}, \mathcal{R} \rangle$, so that $\mathcal{M}_{\mathfrak{F}}$ is also an $S4$ model—a *submodel* of $\mathcal{M}$. We want to show $\mathcal{M}_{\mathfrak{F}}$ is itself an $S4$-ZF model, and some simple relationships between it and $\mathcal{M}$ will be useful for this. The two models agree at the atomic level on members of the common domain $\mathcal{D}_{\mathfrak{F}}^{\mathcal{G}}$, but beyond that there can be difficulties, since quantifiers in the two models have different ranges. *Bounded* quantifiers should be well-behaved, however. This is the content of the following.

Proposition 4.1. *Let $\varphi(x_1, \ldots, x_n)$ be a Δ_0 formula in the language $\mathcal{L}_C$, having only $\in$ as a relation symbol, with all free variables among $x_1, \ldots, x_n$. And let $s_1, \ldots, s_n$ be members of $\mathcal{D}_{\mathfrak{F}}^{\mathcal{G}}$. Then for all $p \in \mathcal{G}$,*

$$p \Vdash [\![\varphi(s_1, \ldots, s_n)]\!] \text{ if and only if } p \Vdash_{\mathfrak{F}} [\![\varphi(s_1, \ldots, s_n)]\!].$$

Proof The argument is by induction on the degree of φ. The atomic case is from the definition of $\Vdash_{\mathfrak{F}}$ (recall that $f \in g$ and $[\![f \in g]\!]$ are equivalent in $\mathcal{M}$). The propositional connective cases are easy. We consider the quantifier case in more detail.

Suppose $\varphi(x_1, \ldots, x_n)$ is $(\exists y)[y \in x_1 \wedge \psi(y, x_1, \ldots, x_n)]$, the result is known for ψ, $p \Vdash [\![\varphi(s_1, \ldots, s_n)]\!]$, where $s_1, \ldots, s_n \in \mathcal{D}_{\mathfrak{F}}^{\mathcal{G}}$, but $p \nVdash_{\mathfrak{F}} [\![\varphi(s_1, \ldots, s_n)]\!]$. We derive a contradiction. Incidentally, the argument the other way around is similar but simpler, and we omit it.

Since $p \nVdash_{\mathfrak{F}} [\![\varphi(s_1, \ldots, s_n)]\!]$, for some p', $p\mathcal{R}p'$ and $p' \Vdash_{\mathfrak{F}} [\![\neg\varphi(s_1, \ldots, s_n)]\!]$, or equivalently, $p' \Vdash_{\mathfrak{F}} [\![(\forall y)[y \in s_1 \supset \neg\psi(y, s_1, \ldots, s_n)]]\!]$.

Since $p \Vdash \llbracket \varphi(s_1,\ldots,s_n) \rrbracket$, also $p' \Vdash \llbracket \varphi(s_1,\ldots,s_n) \rrbracket$, that is, $p' \Vdash \llbracket (\exists y)[y \in s_1 \wedge \psi(y,s_1,\ldots,s_n)] \rrbracket$. Then for some p'' with $p'\mathcal{R}p''$, and for some $f \in \mathcal{D}^{\mathcal{G}}$, $p'' \Vdash \llbracket f \in s_1 \wedge \psi(f,s_1,\ldots,s_n) \rrbracket$. Since $p'' \Vdash \llbracket f \in s_1 \rrbracket$, for some g, $p'' \Vdash \llbracket g \approx f \wedge g \varepsilon s_1 \rrbracket$. Now $s_1 \in \mathcal{D}^{\mathcal{G}}_{\mathfrak{F}}$ so s_1 is $\mathfrak{F}$-invariant. Since $p'' \Vdash \llbracket g \varepsilon s_1 \rrbracket$, by Proposition 2.8, g is $\mathfrak{F}$-invariant as well, so $g \in \mathcal{D}^{\mathcal{G}}_{\mathfrak{F}}$. But then, by substitutivity, $p'' \Vdash \llbracket g \in s_1 \wedge \psi(g,s_1,\ldots,s_n) \rrbracket$. Now both g and s_1 are in $\mathcal{D}^{\mathcal{G}}_{\mathfrak{F}}$, and $p'' \Vdash \llbracket g \in s_1 \rrbracket$, so since the result is known at the atomic level, $p'' \Vdash_{\mathfrak{F}} \llbracket g \in s_1 \rrbracket$. But also, since $p'\mathcal{R}p''$, $p'' \Vdash_{\mathfrak{F}} \llbracket (\forall y)[y \in s_1 \supset \neg\psi(y,s_1,\ldots,s_n)] \rrbracket$, and it follows that $p'' \Vdash_{\mathfrak{F}} \llbracket \neg\psi(g,s_1,\ldots,s_n) \rrbracket$. But, $p'' \Vdash \llbracket \psi(g,s_1,\ldots,s_n) \rrbracket$ has also been established, and this contradicts the induction hypothesis. ∎

In Chapter 14, §1, we introduced the notion of Σ-formula. It will be convenient, for the rest of this chapter, to also use the dual notion of Π-formula, for which the universal quantifier replaces the existential one.

Definition 4.2 (Π-formulas). The class of Π-formulas is characterized by the following.

(1) Every Δ_0 formula is a Π-formula.

(2) If φ and ψ are Π-formulas, so are $\varphi \wedge \psi$ and $\varphi \vee \psi$.

(3) If φ is a Π-formula, so is $(\forall x)\varphi$.

(4) If φ is a Π-formula, so are $(\forall x \in y)\varphi$ and $(\exists x \in y)\varphi$.

Corollary 4.3. *Let $\varphi(x_1,\ldots,x_n)$ be a formula of $\mathcal{L}_C$, and $s_1,\ldots,s_n$ be members of $\mathcal{D}^{\mathcal{G}}_{\mathfrak{F}}$. If φ is a Σ-formula and $p \Vdash_{\mathfrak{F}} \llbracket \varphi(s_1,\ldots,s_n) \rrbracket$ then $p \Vdash \llbracket \varphi(s_1,\ldots,s_n) \rrbracket$. If φ is a Π-formula and $p \Vdash \llbracket \varphi(s_1,\ldots,s_n) \rrbracket$ then $p \Vdash_{\mathfrak{F}} \llbracket \varphi(s_1,\ldots,s_n) \rrbracket$.*

EXERCISES

Exercise 4.1. Show the following.

(a) If X is a Π-formula then $\neg X$ is equivalent to some Σ-formula.

(b) If X is a Σ-formula then $\neg X$ is equivalent to some Π-formula.

Exercise 4.2. Prove Corollary 4.3.

§5 Verifying the axioms

Once again, let $\mathcal{M} = (\mathcal{G}, \mathcal{R}, \mathcal{D}^{\mathcal{G}}, \Vdash)$ be an arbitrary $S4$-regular model over M and let $\mathfrak{F}$ be an automorphism sequence of $\langle \mathcal{G}, \mathcal{R} \rangle$. In this section we require that $\mathfrak{F}$ be an M-set. We then show that the axioms of set theory have valid translates in $\mathcal{M}_{\mathfrak{F}}$, so that it is an $S4$-ZF model. The requirement that $\mathfrak{F}$ be an M-set comes into play in showing the validity of the translate of the axiom of substitution.

Several of the axioms are easy to verify. In one form the axiom of extensionality reads:

$$(\forall x)(\forall y)(\forall z)[((\forall w \in x)(w \in y) \wedge (\forall w \in y)(w \in x) \wedge x \in z) \supset y \in z].$$

Since this is a Π formula whose translate is valid in M, by Corollary 4.3 the translate is valid in $M_{\mathfrak{F}}$ as well, so we have extensionality.

The null set axiom and the axiom of infinity are readily checked since both $\emptyset$ and ω are in M, so that $\hat{\emptyset}$ and $\hat{\omega}$ are in $\mathcal{D}^{\mathcal{G}}$, and hence in $\mathcal{D}^{\mathcal{G}}_{\mathfrak{F}}$ as well.

The axiom of well foundedness is a Π formula, so validity of its translate in $M_{\mathfrak{F}}$ follows immediately from its validity in M.

Thus we are left with the axioms of unordered pairs, power set, union, and substitution to verify. Proofs that translates of these axioms are valid in $M_{\mathfrak{F}}$ are modifications of earlier proofs that the translates are valid in M. We present the arguments for the power set axiom and the axiom of substitution in detail, and leave the other two as exercises. We begin with power set.

In order to show the power set axiom holds in $M_{\mathfrak{F}}$, we modify the proof of Theorem 4.5 from Chapter 17 showing it holds in M—it would be a good idea to review that proof before continuing. We begin with an altered version of Lemma 4.4 of Chapter 17 (it differs from the original in adding that f, a, and b are in $\mathcal{D}^{\mathcal{G}}_{\mathfrak{F}}$).

Lemma 5.1. Suppose $f \in R^{\mathcal{G}}_{\alpha+1} \cap \mathcal{D}^{\mathcal{G}}_{\mathfrak{F}}$, $a \in \mathcal{D}^{\mathcal{G}}_{\mathfrak{F}}$, and $p \Vdash [\![a \subseteq f]\!]$. Then for some $b \in R^{\mathcal{G}}_{\alpha+1} \cap \mathcal{D}^{\mathcal{G}}_{\mathfrak{F}}$, $p \Vdash [\![a \approx b]\!]$.

Proof Assume the hypothesis, so that both f and a are $\mathfrak{F}$-invariant, and $f \in R^{\mathcal{G}}_{\alpha+1}$. Then by Lemma 4.4 of Chapter 17, for some $b \in R^{\mathcal{G}}_{\alpha+1}$, $p \Vdash [\![a \approx b]\!]$, and from the proof of that lemma, $b = \{\langle q, d \rangle \mid d \in R^{\mathcal{G}}_{\alpha}$ and $q \Vdash [\![c \approx d \wedge c \varepsilon a]\!]$ for some $c\}$. Our modification of this is simple—for this proof let

$$b = \{\langle q, d \rangle \mid d \in R^{\mathcal{G}}_{\alpha}, d \text{ is } \mathfrak{F}\text{-invariant, and } q \Vdash [\![c \approx d \wedge c \varepsilon a]\!] \text{ for some } c\}.$$

Notice it is an easy consequence of the definition that if $\langle q, d \rangle \in b$ and qRq', then $\langle q', d \rangle \in b$. It follows easily that if $q \Vdash d \varepsilon b$ then $q \Vdash [\![d \varepsilon b]\!]$. Obviously $b \in R^{\mathcal{G}}_{\alpha+1}$. We will show that b is also $\mathfrak{F}$-invariant, and that $p \Vdash [\![a \approx b]\!]$, which will complete the argument.

(1) We begin by showing that b is $\mathfrak{F}$-invariant. By definition, if $\langle q, d \rangle \in b$, d is $\mathfrak{F}$-invariant, so we have part (2) of the definition. We must verify part (1). Since f is $\mathfrak{F}$-invariant, it is $\mathfrak{G}_f$-invariant for some automorphism group $\mathfrak{G}_f$ in the sequence $\mathfrak{F}$. Similarly since a is $\mathfrak{F}$-invariant, there is an automorphism group $\mathfrak{G}_a$ corresponding to a as well. Let $\mathfrak{G}$ be the latter of the two groups in the sequence $\mathfrak{F}$, so that both f and a are $\mathfrak{G}$-invariant. We will show that b is also $\mathfrak{G}$-invariant, thus establishing part (1) of the $\mathfrak{F}$-invariance definition.

So we need to show that $[\![b \approx \theta(b)]\!]$ is valid in M, for every $\theta \in \mathfrak{G}$. And this can be simplified. Suppose we show that $[\![b \subseteq \theta(b)]\!]$ is valid, for every $\theta \in \mathfrak{G}$. Since $\mathfrak{G}$

§5. VERIFYING THE AXIOMS   275

is closed under inverses, for each $\theta \in \mathfrak{G}$, $\theta^{-1} \in \mathfrak{G}$ and so $[\![b \subseteq \theta^{-1}(b)]\!]$ is valid. Then the validity of $[\![\theta(b) \subseteq b]\!]$ follows, using Theorem 3.2. So we show the validity of $[\![b \subseteq \theta(b)]\!]$, for arbitrary $\theta \in \mathfrak{G}$.

Suppose $[\![b \subseteq \theta(b)]\!]$ were not valid in $\mathcal{M}$. It follows that there is some world at which $[\![\neg(b \subseteq \theta(b))]\!]$, or equivalently $[\![(\exists x)(x \in b \wedge \neg(x \in \theta(b)))]\!]$, is true. Then further, there must be some $x \in \mathcal{D}^\mathcal{G}$, and some accessible world, at which $[\![x \in b \wedge \neg(x \in \theta(b))]\!]$ is true. As usual, since $[\![x \in b]\!]$ is true at this world, for some $x' \in \mathcal{D}^\mathcal{G}$, $[\![x \approx x' \wedge x' \varepsilon b]\!]$ is also true. Finally, since $[\![x' \varepsilon b]\!]$ is true, there must be yet another accessible world at which $x' \varepsilon b$ itself is true. Say this world is q. To summarize then: at q all the following are true: $[\![x \in b]\!]$, $[\![\neg(x \in \theta(b))]\!]$, $[\![x \approx x']\!]$, $[\![x' \varepsilon b]\!]$, and $x' \varepsilon b$.

Since $q \Vdash (x' \varepsilon b)$, it follows from the definition of b that x' is $\mathfrak{F}$-invariant and, for some c, $q \Vdash [\![c \approx x' \wedge c \varepsilon a]\!]$. Since $\theta^{-1} \in \mathfrak{G}$, by Theorem 3.2 and the $\mathfrak{G}$-invariance of a, $\theta^{-1}(q) \Vdash [\![\theta^{-1}(c) \approx \theta^{-1}(x') \wedge \theta^{-1}(c) \varepsilon a]\!]$. Then by definition (and Theorem 3.3), $\theta^{-1}(q) \Vdash \theta^{-1}(x') \varepsilon b$, so $\theta^{-1}(q) \Vdash [\![\theta^{-1}(x') \varepsilon b]\!]$. Now $q \Vdash [\![x' \varepsilon \theta(b)]\!]$, by Theorem 3.2 again, which is impossible since $q \Vdash [\![\neg(x' \in \theta(b))]\!]$.

We have now established that b is $\mathfrak{F}$-invariant.

(2) Finally, the proof that $p \Vdash [\![a \approx b]\!]$ is almost identical with that of Lemma 4.4 of Chapter 17. The only thing that needs to be added to the earlier proof is the observation that the d that occurs in it is $\mathfrak{F}$-invariant (since $[\![d \varepsilon f]\!]$ is true at some world). We leave it to you to check the details. ∎

Now an analog of Theorem 4.5 from Chapter 17.

Theorem 5.2. *For any $f \in \mathcal{D}_\mathfrak{F}^\mathcal{G}$ there is some $g \in \mathcal{D}_\mathfrak{F}^\mathcal{G}$ such that*

$$[\![(\forall x)(x \in g \equiv x \subseteq f)]\!]$$

is valid in $\mathcal{M}_\mathfrak{F}$.

Proof For some ordinal α, $f \in R_{\alpha+1}^\mathcal{G} \cap \mathcal{D}_\mathfrak{F}^\mathcal{G}$. Let g be defined by

$$\langle p, a \rangle \in g \Leftrightarrow a \in R_{\alpha+1}^\mathcal{G} \cap \mathcal{D}_\mathfrak{F}^\mathcal{G} \text{ and } p \Vdash [\![a \subseteq f]\!].$$

Directly from the definition, $g \in R_{\alpha+2}^\mathcal{G}$, but showing it is $\mathfrak{F}$-invariant is more involved.

If $\langle p, a \rangle \in g$, a is required to be $\mathfrak{F}$-invariant, so we have part (2) of the invariance definition. As usual it is part (1) that requires some work.

Since f is $\mathfrak{F}$-invariant, for some group $\mathfrak{G}$ in the sequence, f is $\mathfrak{G}$-invariant. We show that g is also $\mathfrak{G}$-invariant, and this will complete the proof. And as in the previous lemma, this can be reduced to showing the validity, for any $\theta \in \mathfrak{G}$, of $[\![g \subseteq \theta(g)]\!]$.

Let θ be an arbitrary member of $\mathfrak{G}$, and suppose $[\![g \subseteq \theta(g)]\!]$ is not valid—we derive a contradiction. Just as in the previous proof, it follows from our supposition that there is some world p at which all the following are true, for some a and a': $[\![a \in g]\!]$,

$[\![\neg(a \in \theta(g))]\!]$, $[\![a \approx a']\!]$, $[\![a' \, \varepsilon \, g]\!]$, and $a' \, \varepsilon \, g$. Also, since $p \Vdash (a' \, \varepsilon \, g)$ then $\langle p, a' \rangle \in g$, so $p \Vdash [\![a' \subseteq f]\!]$.

$\theta \in \mathfrak{G}$, so $\theta^{-1} \in \mathfrak{G}$ as well. Since $p \Vdash [\![a' \subseteq f]\!]$, $\theta^{-1}(p) \Vdash [\![\theta^{-1}(a') \subseteq \theta^{-1}(f)]\!]$, or by the invariance of f, $\theta^{-1}(p) \Vdash [\![\theta^{-1}(a') \subseteq f]\!]$. Then by the previous lemma, for some $b \in R^{\mathcal{G}}_{\alpha+1} \cap \mathcal{D}^{\mathcal{G}}_{\mathfrak{F}}$, $\theta^{-1}(p) \Vdash [\![\theta^{-1}(a') \approx b]\!]$, so also $\theta^{-1}(p) \Vdash [\![b \subseteq f]\!]$. By definition of g, $\theta^{-1}(p) \Vdash [\![b \, \varepsilon \, g]\!]$, so by Theorem 3.2 again, $p \Vdash [\![\theta(b) \, \varepsilon \, \theta(g)]\!]$. But also $p \Vdash [\![a' \approx \theta(b)]\!]$, and it follows that $p \Vdash [\![a' \in \theta(g)]\!]$. This is a contradiction since both $[\![\neg(a \in \theta(g))]\!]$ and $[\![a \approx a']\!]$ are true at p.

We now know that $g \in \mathcal{D}^{\mathcal{G}}_{\mathfrak{F}}$. It remains to show it behaves like a powerset in $\mathcal{M}_{\mathfrak{F}}$, and for this the proof of Theorem 4.5 from Chapter 17 carries over directly. We leave it to you to check the details. ∎

Next we turn to the axiom of substitution, and again our proof is very much like the earlier one in Chapter 17, §6. There is one significant change, however. In that earlier section we showed the translate of the axiom scheme of substitution was valid in $\mathcal{M}$ using a modified version of the scheme, called *collection*, a version that dropped the 'functionality' requirement. That is to say, we worked with a version having a weaker hypothesis—a version that is equivalent, in the presence of the axiom of well-foundedness. Here it is more convenient to use the standard version—we need the stronger hypothesis. So for this section, the axiom scheme of substitution has the usual form: for any formula $\varphi(x, y)$ with two free variables,

$$(\forall f)\{(\forall x)[x \in f \supset (\exists! y)\varphi(x, y)] \supset (\exists g)(\forall x)[x \in f \supset (\exists y)(y \in g \wedge \varphi(x, y))]\}.$$

In this we use $(\exists! y)\varphi(x, y)$—read "there is exactly one y such that $\varphi(x, y)$"—as an abbreviation for

$$(\exists y)\varphi(x, y) \wedge (\forall y_1)(\forall y_2)[\varphi(x, y_1) \wedge \varphi(x, y_2) \supset y_1 = y_2]$$

where, in turn, $y_1 = y_2$ abbreviates $(\forall w)(w \in y_1 \equiv w \in y_2)$.

We have to be a little careful here concerning the role of equality. We have been using $\approx$ in the modal setting all along, and we are quite familiar with its properties, but the language we are considering for the present family of models, those of the form $\mathcal{M}_{\mathfrak{F}}$, only has $\in$ as a relation symbol. This is why we introduced $=$ as an abbreviation above. Using it, the axiom of extensionality can be stated in the equivalent form:

$$(\forall x)(\forall y)(\forall z)[x = y \supset (x \in z \equiv y \in z)].$$

It is straightforward to show this version of equality has the appropriate properties.

Proposition 5.3. *Suppose $f, g \in \mathcal{D}^{\mathcal{G}}_{\mathfrak{F}}$, and X and X' are closed formulas of the language $\mathcal{L}_C$, differing in that X' has an occurrence of g in a location where X has an occurrence of f. If $p \Vdash_{\mathfrak{F}} [\![f = g]\!]$ then $p \Vdash_{\mathfrak{F}} [\![X]\!] \equiv [\![X']\!]$.*

We will need a simple case of this in the argument below—we leave the proof to you (the translate of the extensionality axiom will be needed).

§5. VERIFYING THE AXIOMS   277

Now we begin on the substitution schema, and we closely follow the earlier argument from Chapter 17, §6, which we recommend you go back and re-read. The following is the analog of Proposition 6.1 from that chapter.

Proposition 5.4. *Assume* $p \Vdash_{\mathfrak{F}} \llbracket (\forall x)[x \in f \supset (\exists! y)\varphi(x,y)] \rrbracket$ *for some* $p \in \mathcal{G}$ *and* $f \in \mathcal{D}^{\mathcal{G}}_{\mathfrak{F}}$. *Then for some* $g^* \in \mathcal{D}^{\mathcal{G}}_{\mathfrak{F}}$, $p \Vdash_{\mathfrak{F}} \llbracket (\forall x)[x \in f \supset (\exists y)(y \in g^* \wedge \varphi(x,y))] \rrbracket$.

Proof We start out more-or-less as we did in Chapter 17, with the following relation between u and v:

$$u = \langle q, a \rangle \text{ for some } q \in \mathcal{G} \text{ and } a \in \mathcal{D}^{\mathcal{G}}_{\mathfrak{F}}$$
and $v = \langle q', b \rangle$ for some q' with $q\mathcal{R}q'$ and some $b \in \mathcal{D}^{\mathcal{G}}_{\mathfrak{F}}$
and $q' \Vdash_{\mathfrak{F}} \llbracket \varphi(a, b) \rrbracket$.

This relation is definable over $\mathcal{M}$ by a formula $\psi(u, v)$—we skip the details. In showing this the requirement that $\mathfrak{F}$ is an $\mathcal{M}$-set is used.

Let

$$F = \{\langle q, a \rangle \mid q \in \mathcal{G}, a \in \mathcal{D}^{\mathcal{G}}_{\mathfrak{F}} \text{ and }$$
$$q \Vdash_{\mathfrak{F}} \llbracket (\forall x)[x \in f \supset (\exists! y)\varphi(x,y)] \rrbracket \text{ and }$$
$$q \Vdash \llbracket a \varepsilon f \rrbracket \}.$$

Much as before, F is an $\mathcal{M}$-set. Also, for every $u \in F$ there is some v such that $\psi(u, v)$. The argument for this is virtually the same as in Chapter 17, and we do not repeat it.

Now using the axiom of substitution in $\mathcal{M}$, there is a set g such that

$$(\forall u \in F)(\exists v \in g)\psi(u, v) \wedge (\forall v \in g)(\exists u \in F)\psi(u, v).$$

By definition, members of g are in $\mathcal{D}^{\mathcal{G}}$—indeed they are in $\mathcal{D}^{\mathcal{G}}_{\mathfrak{F}}$. Since g is a *set in* $\mathcal{M}$ then $g \in \mathcal{D}^{\mathcal{G}}$. Hence also $g^* \in \mathcal{D}^{\mathcal{G}}$ —recall that $g^* = \{\langle q, a \rangle \mid \langle p, a \rangle \in g$ for some $p \in \mathcal{G}$ such that $p\mathcal{R}q\}$. And further, by a slight variant of the argument in the proof of Proposition 6.1, if $p \Vdash_{\mathfrak{F}} \llbracket (\forall x)[x \in f \supset (\exists! y)\varphi(x,y)] \rrbracket$ then $p \Vdash_{\mathfrak{F}} \llbracket (\forall x)[x \in f \supset (\exists y)(y \in g^* \wedge \varphi(x,y))] \rrbracket$. (Actually we don't need the full strength of $(\exists! y)$ here, only that of $(\exists y)$.) So what remains is to show that g^* is $\mathfrak{F}$-invariant, and hence is in $\mathcal{D}^{\mathcal{G}}_{\mathfrak{F}}$. We give this part of the argument in detail.

If $\langle q, a \rangle \in g^*$ then $\langle p, a \rangle \in g$ for some p such that $p\mathcal{R}q$. And if $\langle p, a \rangle \in g$, for some $u \in F$, $\psi(u, \langle p, a \rangle)$, and hence $a \in \mathcal{D}^{\mathcal{G}}_{\mathfrak{F}}$—$a$ is $\mathfrak{F}$-invariant. So what is left to show is that g^* itself is $\mathfrak{G}$-invariant, for some group $\mathfrak{G}$ in the automorphism sequence $\mathfrak{F}$.

Since $f \in \mathcal{D}^{\mathcal{G}}_{\mathfrak{F}}$, for some automorphism group $\mathfrak{G}$ in $\mathfrak{F}$, f is $\mathfrak{G}$-invariant. We will show that $\llbracket g^* \approx \theta(g^*) \rrbracket$ is also valid in $\mathcal{M}$ for each $\theta \in \mathfrak{G}$, and this will complete the proof. And as in earlier arguments in this section, it is enough to show the validity in $\mathcal{M}$ of $\llbracket g^* \subseteq \theta(g^*) \rrbracket$, for all $\theta \in \mathfrak{G}$.

Suppose $\llbracket g^* \subseteq \theta(g^*) \rrbracket$ were not valid in $\mathcal{M}$, for some $\theta \in \mathfrak{G}$. Then, in the usual way, there would be some $q \in \mathcal{G}$ and some $a \in \mathcal{D}^{\mathcal{G}}$ such that $q \Vdash \llbracket a \in g^* \rrbracket$, $q \Vdash (a \varepsilon g^*)$, and $q \Vdash \llbracket \neg(a \in \theta(g^*)) \rrbracket$. We derive a contradiction from this.

Since $q \Vdash (a \,\varepsilon\, g^*)$, $\langle p, a \rangle \in g$ for some $p \in \mathcal{G}$ with $p\mathcal{R}q$, and so for some $\langle p_0, b \rangle \in F$, $\psi(\langle p_0, b \rangle, \langle p, a \rangle)$ is true. Since $\langle p_0, b \rangle \in F$, $p_0 \Vdash_{\widetilde{\mathfrak{F}}} \big[\!\!\big[(\forall x)[x \in f \supset (\exists! y)\varphi(x, y)]\big]\!\!\big]$ and $p_0 \Vdash \big[\!\!\big[b \,\varepsilon\, f\big]\!\!\big]$. Since $\psi(\langle p_0, b \rangle, \langle p, a \rangle)$, $p_0 \mathcal{R} p$ and $p \Vdash_{\widetilde{\mathfrak{F}}} \big[\!\!\big[\varphi(b, a)\big]\!\!\big]$. It follows that $q \Vdash_{\widetilde{\mathfrak{F}}} \big[\!\!\big[(\forall x)[x \in f \supset (\exists! y)\varphi(x, y)]\big]\!\!\big]$ and $q \Vdash \big[\!\!\big[b \,\varepsilon\, f\big]\!\!\big]$.

By Theorem 3.5 and the fact that f is $\mathfrak{G}$-invariant, we have $\theta^{-1}(q) \Vdash_{\widetilde{\mathfrak{F}}} \big[\!\!\big[(\forall x)[x \in f \supset (\exists! y)\varphi(x, y)]\big]\!\!\big]$. Also by Theorem 3.2, $\theta^{-1}(q) \Vdash \big[\!\!\big[\theta^{-1}(b) \,\varepsilon\, f\big]\!\!\big]$. By definition then, $\langle \theta^{-1}(q), \theta^{-1}(b) \rangle \in F$. So for some $\langle r, c \rangle \in g$, $\psi(\langle \theta^{-1}(q), \theta^{-1}(b) \rangle, \langle r, c \rangle)$, so $\theta^{-1}(q)\mathcal{R}r$ and $r \Vdash_{\widetilde{\mathfrak{F}}} \big[\!\!\big[\varphi(\theta^{-1}(b), c)\big]\!\!\big]$. Since $\langle r, c \rangle \in g$, $r \Vdash (c \,\varepsilon\, g)$, so $r \Vdash \big[\!\!\big[c \,\varepsilon\, g^*\big]\!\!\big]$, and it follows that $r \Vdash_{\widetilde{\mathfrak{F}}} \big[\!\!\big[c \in g^*\big]\!\!\big]$.

Using Theorem 3.5 again, $\theta(r) \Vdash_{\widetilde{\mathfrak{F}}} \big[\!\!\big[\varphi(b, \theta(c))\big]\!\!\big]$ and $\theta(r) \Vdash_{\widetilde{\mathfrak{F}}} \big[\!\!\big[\theta(c) \in \theta(g^*)\big]\!\!\big]$. Also since $\theta^{-1}(q)\mathcal{R}r$ then $q\mathcal{R}\theta(r)$. Since $q \Vdash \big[\!\!\big[b \,\varepsilon\, f\big]\!\!\big]$ then $\theta(r) \Vdash \big[\!\!\big[b \,\varepsilon\, f\big]\!\!\big]$, and it follows that $\theta(r) \Vdash_{\widetilde{\mathfrak{F}}} \big[\!\!\big[b \in f\big]\!\!\big]$ as well. Also $\theta(r) \Vdash_{\widetilde{\mathfrak{F}}} \big[\!\!\big[(\forall x)[x \in f \supset (\exists! y)\varphi(x, y)]\big]\!\!\big]$. Consequently $\theta(r) \Vdash_{\widetilde{\mathfrak{F}}} \big[\!\!\big[(\exists! y)\varphi(b, y)\big]\!\!\big]$. We also have that $\theta(r) \Vdash_{\widetilde{\mathfrak{F}}} \big[\!\!\big[\varphi(b, a)\big]\!\!\big]$ (since we have this at p), and so $\theta(r) \Vdash_{\widetilde{\mathfrak{F}}} \big[\!\!\big[a = \theta(c)\big]\!\!\big]$. We know that $\theta(r) \Vdash_{\widetilde{\mathfrak{F}}} \big[\!\!\big[\theta(c) \in \theta(g^*)\big]\!\!\big]$, so by substitutivity of equality, $\theta(r) \Vdash_{\widetilde{\mathfrak{F}}} \big[\!\!\big[a \in \theta(g^*)\big]\!\!\big]$. Since $\Vdash$ and $\Vdash_{\widetilde{\mathfrak{F}}}$ agree on atomic formulas of $\mathcal{L}_C^*$, it follows that $\theta(r) \Vdash \big[\!\!\big[a \in \theta(g^*)\big]\!\!\big]$, and this contradicts the fact that $q \Vdash \big[\!\!\big[\neg(a \in \theta(g^*))\big]\!\!\big]$. ∎

EXERCISES

Exercise 5.1. Complete the proof of Theorem 5.2.

Exercise 5.2. Prove Proposition 5.3.

Exercise 5.3. Suppose $a, b \in \mathcal{D}_{\widetilde{\mathfrak{F}}}^{\mathcal{G}}$, and $p \in \mathcal{G}$. Show:

$$p \Vdash \big[\!\!\big[a \approx b\big]\!\!\big] \text{ iff } p \Vdash_{\widetilde{\mathfrak{F}}} \big[\!\!\big[a = b\big]\!\!\big],$$

where $a = b$ is the abbreviation introduced in this section.

§6 AC is independent

We produce an $S4$-ZF model that shows the axiom of choice is independent. In this model there will be a domain member, T, having a countable number of members, t_0, $t_1, \ldots, t_n, \ldots$. Each t_n in turn consists of members chosen from the list $s_0, s_1, \ldots, s_n$, $\ldots$, each of which is created more-or-less the same way the non-constructible set of Chapter 18 was. And there will be enough symmetry imposed on the model so that T has no choice function. In effect, each t_n is a countable analog of a pair of Russell's socks.

A partition of the counting numbers We begin by dividing ω into disjoint families $I_0, I_1, I_2, \ldots$. The first few families are as follows.

§6. AC IS INDEPENDENT

$$I_0 = 0, 2, 5, 9, 14, \ldots$$
$$I_1 = 1, 4, 8, 13, 19, \ldots$$
$$I_2 = 3, 7, 12, 18, 25, \ldots$$
$$I_3 = 6, 11, 17, 24, 32, \ldots$$
$$I_4 = 10, 16, 23, 31, 40, \ldots$$

You may recognize the pattern of Cantor's so called *reverse diagonal* arrangement. Formally, $I_n = \{(n+k)(n+k+1)/2 + k \mid k = 0, 1, 2, \ldots\}$. As a matter of fact, all we need of this partition are the following simple items—any other partition having the same properties would do.

(1) Each member of ω is in exactly one I_n.

(2) If $k \in I_n$ then $k \geq n$.

(3) Each I_n is infinite.

The uses of this partition will be clear shortly.

Model specifics A *forcing condition* is a function p with domain ω such that:

(1) for each $n \in \omega$, $p(n) = \langle P_n, N_n \rangle$, where P_n and N_n are finite, disjoint sets of integers;

(2) $p(n) = \langle \emptyset, \emptyset \rangle$ for all but a finite number of n.

$\mathcal{G}$ is the set of all forcing conditions. For $p, q \in \mathcal{G}$, let $p\mathcal{R}q$ provided, for each $n \in \omega$, if $p(n) = \langle P_n, N_n \rangle$ and $q(n) = \langle P'_n, N'_n \rangle$, then $P_n \subseteq P'_n$ and $N_n \subseteq N'_n$.

As a matter of fact, this is the same definition as in Chapter 19, except that now forcing conditions are functions that have ω as domain, instead of ω_2. As usual, $\mathcal{R}$ is easily checked to be reflexive and transitive, and $\langle \mathcal{G}, \mathcal{R} \rangle$ is an M-set. Let $\mathcal{M} = (\mathcal{G}, \mathcal{R}, \mathcal{D}^{\mathcal{G}}, \Vdash)$ be the resulting $S4$-regular model over M.

Next, for each $n \in \omega$ let

$$s_n = \{\langle p, \hat{k} \rangle \mid p \in \mathcal{G}, p(n) = \langle P_n, N_n \rangle \text{ and } k \in P_n\}.$$

Each $s_n \in \mathcal{D}^{\mathcal{G}}$—in fact, $s_n \in R^{\mathcal{G}}_{\omega+1}$. Next, for each $n \in \omega$ let

$$t_n = \{\langle p, s_k \rangle \mid p \in \mathcal{G} \text{ and } k \in I_n\}.$$

Then each $t_n \in \mathcal{D}^{\mathcal{G}}$ as well. Finally, let

$$T = \{\langle p, t_n \rangle \mid p \in \mathcal{G} \text{ and } n \in \omega\}.$$

A few remarks before things get really serious. First, from the way the items above are characterized, we have the following.

$$p \Vdash (\hat{k} \,\varepsilon\, s_n) \Leftrightarrow p(n) = \langle P_n, N_n \rangle \text{ and } k \in P_n$$

$$p \Vdash (s_k \, \varepsilon \, t_n) \Leftrightarrow k \in I_n$$

$$p \Vdash (t_n \, \varepsilon \, T).$$

Next, for each $n, k \in \omega$, if $n \neq k$ then $\llbracket \neg(s_n \approx s_k) \rrbracket$ is valid in $\mathcal{M}$. The proof is virtually the same as that of Proposition 4.1 from Chapter 19, and we skip it here.

Finally, $\langle \mathcal{G}, \mathcal{R} \rangle$ satisfies the countable chain condition. This can be proved along the lines of the argument for Proposition 3.1 from Chapter 19. We won't make any fundamental use of this, but it is an interesting observation that, as a consequence, cardinals are well behaved in $\mathcal{M}$.

Automorphisms We now begin to impose symmetry on the model $\mathcal{M}$. It is easiest to start with the integers. A *permutation* of ω is a mapping $\theta : \omega \xrightarrow[\text{onto}]{\text{1-1}} \omega$. A permutation is said to have *finite support* if it is the identity on all but a finite set—that is, if it only moves finitely many numbers. We will call a permutation θ *special* if θ has finite support and respects our partition of ω, that is, for each n, $\theta : I_n \xrightarrow[\text{onto}]{\text{1-1}} I_n$. Now we use special permutations of ω to define the class of automorphisms we are interested in.

Definition 6.1. Let θ be a special permutation. We extend the action of θ to $\mathcal{G}$ as follows. For each $p \in \mathcal{G}$, $\theta(p)$ is the mapping given by:

$$\theta(p)(n) = p(\theta^{-1}(n)).$$

It is easy to check that, for each special permutation, θ is in fact an automorphism of $\langle \mathcal{G}, \mathcal{R} \rangle$. (In fact, the condition of being special is not needed for this.) An easy calculation using the definition shows: $\theta(p)(\theta(n)) = p(\theta^{-1}\theta(n)) = p(n)$, and consequently:

$$p \Vdash (\hat{k} \, \varepsilon \, s_n) \Leftrightarrow \theta(p) \Vdash (\hat{k} \, \varepsilon \, s_{\theta(n)}).$$

Next we introduce the particular automorphism sequence we need.

Definition 6.2. For each n let $\mathfrak{G}_n$ be the set of all special permutations of ω that are the identity on all integers $< n$.

Proposition 6.3. *Sequence $\mathfrak{F} = \mathfrak{G}_0, \mathfrak{G}_1, \ldots, \mathfrak{G}_n, \ldots$ is an automorphism sequence.*

Proof It is simple to check that each $\mathfrak{G}_n$ is a group and $\mathfrak{G}_{n+1} \subseteq \mathfrak{G}_n$, so all that really needs checking is: if $\theta \in \mathfrak{G}_0$ then θ normalizes $\mathfrak{G}_k, \mathfrak{G}_{k+1}, \mathfrak{G}_{k+2}, \ldots$, for some k. And this is easy.

θ has finite support. Let n be the biggest integer θ moves. We will show that, for each $\pi \in \mathfrak{G}_{n+1}$, $\theta^{-1}\pi\theta = \pi$, from which it follows that θ normalizes each of $\mathfrak{G}_{n+1}, \mathfrak{G}_{n+2}, \mathfrak{G}_{n+3}, \ldots$.

First, note that since n is the biggest integer θ (or θ^{-1}) moves, θ maps $\{0, 1, \ldots, n\}$ to itself. And since $\pi \in \mathfrak{G}_{n+1}$, π can change nothing below $n + 1$, and so π maps $\{n+1, n+2, \ldots\}$ to itself. The same is true for π^{-1} since it is also a member of $\mathfrak{G}_{n+1}$. Now we show that $\theta^{-1}\pi\theta = \pi$ by showing that for each integer m, $(\theta^{-1}\pi\theta)(m) = \pi(m)$.

§6. AC IS INDEPENDENT

Suppose $m \geq n + 1$. Then $\pi(m) \geq n + 1$, so $\theta^{-1}(\pi(m)) = \pi(m)$. But also $m = \theta(m)$, and so $(\theta^{-1}\pi\theta)(m) = \pi(m)$.

Suppose next that $m \leq n$. Then $\theta(m) \leq n$ so $\pi(\theta(m)) = \theta(m)$, and it follows that $(\theta^{-1}\pi\theta)(m) = (\theta^{-1}\theta)(m) = m = \pi(m)$. ∎

The real model We have an $S4$-regular model $\mathcal{M}$ over M, and an automorphism sequence $\mathfrak{F} = \mathfrak{G}_0, \mathfrak{G}_1, \ldots$ over M, so we can form the $S4$ model $\mathcal{M}_{\mathfrak{F}}$. From the work in the previous section, we know it is an $S4$-ZF model. We will show the axiom of choice fails in it, thus establishing its independence. Incidentally, the assertion that x is a cardinal can be made using a Π-formula (x is an ordinal and, for every function from x onto a smaller ordinal, the function is not 1-1). Then by Corollary 4.3 and the fact that cardinals are well behaved in $\mathcal{M}$, it follows that they are also well behaved in $\mathcal{M}_{\mathfrak{F}}$. We will make no use of this, but it is an interesting observation.

To begin, we want to show that all of s_n, t_n, and T are $\mathfrak{F}$ invariant, and hence are in the domain $\mathcal{D}_{\mathfrak{F}}^{\mathcal{G}}$, of $\mathcal{M}_{\mathfrak{F}}$. Clearly T requires no work.

Proposition 6.4. *For each $n \in \omega$, s_n is $\mathfrak{F}$-invariant.*

Proof Let n be fixed—we show s_n is $\mathfrak{F}$-invariant. And from the definition of s_n, one of the two conditions for $\mathfrak{F}$-invariance is easy. What takes a small amount of work is showing that s_n is $\mathfrak{G}$-invariant for some group $\mathfrak{G}$ in the sequence $\mathfrak{F}$. Specifically, we show s_n is $\mathfrak{G}_{n+1}$-invariant. Let $\theta \in \mathfrak{G}_{n+1}$.

(1) We first show that $p \Vdash (\hat{k}\,\varepsilon\, s_n)$ iff $p \Vdash (\hat{k}\,\varepsilon\,\theta(s_n))$. Using an earlier observation,

$$p \Vdash (\hat{k}\,\varepsilon\, s_n) \Leftrightarrow \theta^{-1}(p) \Vdash (\hat{k}\,\varepsilon\, s_{\theta^{-1}(n)}).$$

But since $\theta^{-1} \in \mathfrak{G}_{n+1}$, it is the identity below $n + 1$, so

$$\theta^{-1}(p) \Vdash (\hat{k}\,\varepsilon\, s_{\theta^{-1}(n)}) \Leftrightarrow \theta^{-1}(p) \Vdash (\hat{k}\,\varepsilon\, s_n).$$

Finally, using Theorem 3.2, and the invariance of $\hat{k}$ under all automorphisms,

$$\theta^{-1}(p) \Vdash (\hat{k}\,\varepsilon\, s_n) \Leftrightarrow p \Vdash (\hat{k}\,\varepsilon\,\theta(s_n)).$$

(2) Now the validity in $\mathcal{M}$ of $[\![s_n \approx \theta(s_n)]\!]$ is an easy consequence, and we leave it as an exercise. ∎

Proposition 6.5. *For each $n \in \omega$, t_n is $\mathfrak{F}$-invariant. In fact, t_n is $\mathfrak{G}_k$-invariant for each k.*

Proof By our definition, $p \Vdash (s_k\,\varepsilon\, t_n)$ if and only if $k \in I_n$. Invariance follows easily using the fact that each special permutation maps I_n 1-1, onto itself. ∎

Independence We now know that all of s_n, t_n and T are in $\mathcal{D}_{\mathfrak{F}}^{\mathcal{G}}$, and $\mathcal{M}_{\mathfrak{F}}$ is an $S4$-ZF model. To complete things we show that in this model T has no choice function. For this purpose, let $AC(x, y)$ be a formula in the language $\mathcal{L}_C$ asserting that x is a choice function for y. The exact details of the formula construction are not important—obviously there is such a formula and that is all we need to know.

Theorem 6.6. *For each $F \in \mathcal{D}_{\tilde{\mathfrak{F}}}^{\mathcal{G}}$, $[\![\neg AC(F,T)]\!]$ is valid in $\mathcal{M}_{\tilde{\mathfrak{F}}}$. Hence the axiom of choice fails in $\mathcal{M}_{\tilde{\mathfrak{F}}}$, and so is independent.*

Proof Suppose otherwise. It follows that for some $p \in \mathcal{G}$, $p \Vdash_{\tilde{\mathfrak{F}}} [\![AC(F,T)]\!]$—we derive a contradiction.

$F \in \mathcal{D}_{\tilde{\mathfrak{F}}}^{\mathcal{G}}$ so for some n, F is $\mathfrak{G}_n$-invariant. We will show, speaking very informally, that F has trouble choosing a member of t_n.

Since $p \Vdash_{\tilde{\mathfrak{F}}} [\![AC(F,T)]\!]$, it follows that, at p, translates of sentences are true that assert: (1) F is a function; (2) the domain of F is T; (3) for each member of T, F chooses a member of it. Now, $p \Vdash_{\tilde{\mathfrak{F}}} [\![t_n \in T]\!]$, so as a consequence, $p \Vdash_{\tilde{\mathfrak{F}}} [\![(\exists x)(x \in t_n \wedge \langle t_n, x \rangle \in F)]\!]$. (That is, there is an x in t_n that F chooses.) Then for some p' with $p\mathcal{R}p'$, and for some $a \in \mathcal{D}_{\tilde{\mathfrak{F}}}^{\mathcal{G}}$, $p' \Vdash_{\tilde{\mathfrak{F}}} [\![a \in t_n \wedge \langle t_n, a \rangle \in F]\!]$. And since $p' \Vdash_{\tilde{\mathfrak{F}}} [\![a \in t_n]\!]$, by definition, $p' \Vdash [\![a \in t_n]\!]$, so for some b, $p' \Vdash [\![a \approx b \wedge b \,\varepsilon\, t_n]\!]$. Then further, for some p'' with $p'\mathcal{R}p''$, $p'' \Vdash (b \,\varepsilon\, t_n)$, so $\langle p'', b \rangle \in t_n$, and it follows from the definition of t_n that $b = s_k$ for some $k \in I_n$. So $p' \Vdash [\![a \approx s_k]\!]$ and by Exercise 5.3, $p' \Vdash_{\tilde{\mathfrak{F}}} [\![a = s_k]\!]$.

Now to summarize the essentials so far. F is asserted by p to be a function with domain T, hence also by p'. And (by substitutivity of equality) $p' \Vdash_{\tilde{\mathfrak{F}}} [\![s_k \in t_n \wedge \langle t_n, s_k \rangle \in F]\!]$. Also $k \in I_n$, so $k \geq n$. Finally, F is $\mathfrak{G}_n$-invariant. From here to a contradiction is rather easy.

One of the requirements on forcing conditions in this section is: they are $\langle \emptyset, \emptyset \rangle$ in all but a finite number of cases. Also I_n is infinite. Then we can choose an integer m meeting the following three conditions: (1) $m \neq k$; (2) $m \in I_n$; (3) $p'(m) = \langle \emptyset, \emptyset \rangle$. Define a permutation θ by setting $\theta(m) = k$, $\theta(k) = m$, and on all other integers, θ is the identity. Obviously θ has finite support, and equally obviously, it respects our partition of the integers. Consequently it is a special permutation. And since both m and k are $\geq n$ (because both are in I_n), θ is the identity below n, so $\theta \in \mathfrak{G}_n$.

Now, $p' \Vdash_{\tilde{\mathfrak{F}}} [\![\langle t_n, s_k \rangle \in F]\!]$ so by Theorem 3.5 $\theta(p') \Vdash_{\tilde{\mathfrak{F}}} [\![\langle \theta(t_n), \theta(s_k) \rangle \in \theta(F)]\!]$. By Proposition 6.5 (and Exercise 5.3), $[\![\theta(t_n) = t_n]\!]$ is valid in $\mathcal{M}_{\tilde{\mathfrak{F}}}$. Also F is $\mathfrak{G}_n$-invariant, and $\theta \in \mathfrak{G}_n$, so $[\![\theta(F) = F]\!]$ is valid in $\mathcal{M}_{\tilde{\mathfrak{F}}}$. Finally, $\theta(s_k) = s_{\theta(k)} = s_m$. Consequently, $\theta(p') \Vdash_{\tilde{\mathfrak{F}}} [\![\langle t_n, s_m \rangle \in F]\!]$.

By definition of θ, p' and $\theta(p')$ must agree on all integers except k and m. Further, $p'(m) = \langle \emptyset, \emptyset \rangle$ while $\theta(p')(k) = p'(\theta^{-1}(k)) = p'(m) = \langle \emptyset, \emptyset \rangle$. Now let r agree with p' and $\theta(p')$ on all integers except k and m, and set $r(k) = p'(k)$, and $r(m) = \theta(p')(m)$. Then $r \in \mathcal{G}$ and it is easy to see that $p'\mathcal{R}r$ and $\theta(p')\mathcal{R}r$.

In $\mathcal{M}_{\tilde{\mathfrak{F}}}$, at r, we have both $[\![\langle t_n, s_k \rangle \in F]\!]$ (since we have this at p'), and $[\![\langle t_n, s_m \rangle \in F]\!]$ (since we have this at $\theta(p')$). And we have $[\![\neg(s_k = s_m)]\!]$ since $k \neq m$ (an easy consequence of the validity in $\mathcal{M}$ of $[\![\neg(s_k \approx s_m)]\!]$). But we also have a formula asserting that F is a function, hence single valued. This is impossible, and we have our contradiction. Such an F cannot exist. ∎

EXERCISES

Exercise 6.1. Complete the proof of Proposition 6.4 by giving the argument for part (2).

Exercise 6.2. Show that each s_n is non-constructible in the model $\mathcal{M}$.

CHAPTER 21

CONSTRUCTING CLASSICAL MODELS

§1 On countable models

Gödel showed the continuum hypothesis was not disprovable in set theory. In doing so he also created a model of set theory of interest in its own right—the class of constructible sets. We have shown the continuum hypothesis is not provable either and, while our methods are semantic, we have produced no classical models at all. Modal models have an interest of their own, but finally, classical models are the real thing. So the question remains: can we construct a natural classical model of set theory in which the continuum hypothesis is false? Cohen showed how this could be done, and his method has since been refined by others. But it is subject to a curious restriction—one that cannot be escaped. We need to assume the existence of a *countable* standard model. Before getting involved in the details, we should consider just what the strength of this assumption is.

A *model* for ZF is a relational system (A, R) (not necessarily a set) in which all the axioms of ZF are true, interpreting the formal symbol "$\in$" by "R" (see Chapter 10). It is a *standard model* if R is the actual member-of relation, $\in$, and A is transitive. (Note that the axiom of well foundedness is true in a standard model provided it is true in V.) What we will need is the existence of a countable standard model. If there is any standard model that is a set, a Löwenheim-Skolem argument can be applied to produce a countable standard model. So the basic question becomes: is there any standard model that is a set.

First of all we might try applying the M.S.T.V. Theorem 4.1, from Chapter 11. If $A_1, \ldots, A_n$ is any finite list of axioms of ZF, since these are all true over V it follows that there is a countable, transitive set T in which all of $A_1, \ldots, A_n$ are true. (In applying the M.S.T.V. theorem, take the pure sentence X to be the conjunction of $A_1, \ldots, A_n$, and take the set A to be ω.) Unfortunately ZF has an infinite family of axioms, so all this argument tells us is that we have countable sets that are not full standard models, but that "approximate" to arbitrarily large finite parts of ZF. This is important, but it does not yield the existence of a countable standard model.

By Gödel's completeness theorem for first-order logic, consistency of ZF is equivalent to the existence of a model for the axioms of ZF. This argument can be formalized in set theory itself — it is a theorem of ZF that if ZF is consistent, there is a model for ZF *that is a set*. However, such a model need not be well founded, so the Mostowski-Shepherdson mapping theorem (Theorem 4.1 of Chapter 10) cannot be applied—we cannot conclude, by such an argument, the existence of a standard model for ZF, just from the assumption of the consistency of ZF. In fact, a stronger assertion can be made—one can prove that the consistency of ZF simply does not imply the existence of a standard model that is a set. We do not give this argument here, however, since it would take us too far afield.

Assuming the existence of a standard model that is a set (equivalently, of a countable standard model) is stronger than assuming the consistency of ZF. The assumption is not necessary to establish the various independence results—we have already given proofs without making use of it. But the assumption is necessary to construct classical models. It is an assumption most platonistically inclined mathematicians would have no trouble accepting as "obviously" true. And whether true or not, its consequences are so interesting that further exploration is "forced" on us.

Assumption For the rest of this chapter, we assume there is a countable standard model in V.

Remarkably, the need for this assumption arises only at a single point in the model construction, and that happens in quite a direct way. There is nothing subtle about the matter.

Finally, we note that the existence of a countable standard model M in V is not inherently contradictory—V knows M is countable, but M does not know it. Since M is countable as far as V is concerned, there is a 1-1 function $f \in V$ that maps ω onto M, but that function is not a member of M. In effect, the fewer sets a model has, the harder it is for the model to know sets are the same size.

§2 Cohen's way

Paul Cohen introduced a method for constructing classical models, based on forcing. We sketch it briefly before turning to subsequent modifications of it which we present in more detail.

Suppose $M \in V$ is a countable standard model. Pick one of the specific $S4$-regular models over M presented earlier in this part — say the one in Chapter 18 showing independence of the axiom of constructibility. That construction was carried out in a first-order universe—assume it was done in M. Then $\mathcal{G} \in M$ and $\mathcal{R} \in M$, while $\mathcal{D}^\mathcal{G}$ and $\Vdash$ are subclasses of M. M believes $\mathcal{D}^\mathcal{G}$ is a proper class, since it isn't a member of M. But V knows M is countable, so V also knows $\mathcal{D}^\mathcal{G}$ is countable. Now consider the classical language $\mathcal{L}_C$, with only "∈" as a non-logical symbol. Even if we allow constants from $\mathcal{D}^\mathcal{G}$ to appear in formulas of $\mathcal{L}_C$ there are still only countably many formulas, as far as V is concerned. Enumerate them in V: $X_0, X_1, \ldots, X_n, \ldots$.

Next, we recall a fundamental property of forcing that we have used many times. For any formula X, if $[\![X]\!]$ is not true at p then there must be some p' accessible from p such that $[\![\neg X]\!]$ is true at p'. This is the key to the whole construction.

Define a sequence of forcing conditions as follows. p_0 is arbitrary. Having chosen p_n, consider X_n, the nth formula in our enumeration. If $[\![X_n]\!]$ is true at p_n, let $p_{n+1} = p_n$. Otherwise there must be some p' accessible from p at which $[\![\neg X_n]\!]$ is true; choose one and let it be p_{n+1}. In this way a sequence $p_0, p_1, \ldots, p_n, \ldots$ is produced in V having the property that, for any formula X of $\mathcal{L}_C$ having constants from $\mathcal{D}^\mathcal{G}$, eventually the status of X is settled by the sequence—that is, either $[\![X]\!]$ is true at all worlds of the sequence from some point on, or $\llbracket \neg X \rrbracket$ is true from some point on. Such a sequence is called *complete*.

Now if we have a complete sequence, the status of every formula of $\mathcal{L}_C$ is eventually settled. Then we have, in effect, a complete description of a classical model, and the actual construction of it is now fairly routine. In the particular $S4$-regular model we are using, if X is the axiom of constructibility $\llbracket \neg X \rrbracket$ is true at every world. It follows that any classical model produced by this method must be one in which that axiom fails. As we said earlier, we skip the details in this section because we will present an alternative version shortly. Notice, however, the role that countability played: we used it to establish the existence of a complete sequence. Once we have a sequence that eventually settles the status of every formula, countability is never needed again. The construction has its subtleties, however. The countable standard model M is a set in V. Members of the complete sequence all belong to M, but we did not establish the existence of the complete sequence within M, since M does not know, of itself, that it is countable. The existence of a complete sequence is a fact about V—the complete sequence is a V-set, not an M-set, even though all the terms of the sequence are M-sets (as well as V-sets).

This construction of Cohen is quite lovely, but there is one point on which improvement is possible. One feels that once the underlying frame, $\langle \mathcal{G}, \mathcal{R} \rangle$, has been specified, everything essential has been done. All else follows inexorably, with no more creativity on our part. But to define a complete sequence, and hence to construct a classical model, we must make a detour through a formal language (not to mention the full construction of a modal model). Now while we use language to specify what we are talking about, objects of discourse exist independently of language. One ought to be able to show the existence of a classical model in which the axiom of constructibility fails without using a formal language in such a central way. Such a modification of Cohen's construction has indeed been developed, and it is what we present next.

§3 Dense sets, filters, and generic sets

Cohen's method of constructing classical models has been altered by abstracting away details of formal language, and also details of sequences. We present each of these in turn.

When using an $S4$ model $\mathcal{M} = (\mathcal{G}, \mathcal{R}, \mathcal{D}^{\mathcal{G}}, \Vdash)$, instead of talking about a sentence X, we can talk about the set of worlds at which X is true. This, of course, replaces language constructs with special subsets of $\mathcal{G}$, but is there any simple way of saying which ones? Well, it turns out we don't need to say which ones exactly—we can be overgenerous in this matter and still have sufficient structure to work with. For $S4$-regular models we are not generally interested in all modal formulas, but only those of the form $\llbracket X \rrbracket$ for some classical X. Any such formula is $\square \lozenge Y$ for some Y—what can be said about the set of worlds at which such a formula is true?

Suppose first that $\square \lozenge Y$ is valid in $\mathcal{M}$. Pick an arbitrary world $p \in \mathcal{G}$. Since $p \Vdash \square \lozenge Y$, and $\mathcal{R}$ is reflexive, $p \Vdash \lozenge Y$, so for some p' with $p\mathcal{R}p'$, $p' \Vdash Y$. Conversely, suppose for every world $p \in \mathcal{G}$ there is some p' accessible from p at which Y is true. Let q be an arbitrary world in $\mathcal{G}$—we claim $\square \lozenge Y$ is true at q. The argument is simple. Pick any world p accessible from q. By our assumption, there is a world p' accessible from p at which Y is true, hence $\Diamond Y$ is true at p. Since p was any world accessible from q, $\Box \Diamond Y$ must be true at q, and since q was also arbitrary, $\Box \Diamond Y$ must be valid in $\mathcal{M}$.

Definition 3.1. A set $D \subseteq \mathcal{G}$ is *dense in* $\langle \mathcal{G}, \mathcal{R} \rangle$ provided, for every $p \in \mathcal{G}$ there is some $p' \in D$ such that $p\mathcal{R}p'$.

Then $\Box \Diamond Y$ is valid in $\mathcal{M}$ iff the set of worlds at which Y is true is dense in $\langle \mathcal{G}, \mathcal{R} \rangle$. Thus corresponding to each valid formula of the form $\Box \Diamond Y$ is some dense set, though not every dense set corresponds to a formula. The notion of denseness is more generous than we need, but it does avoid any explicit connection with a formal language. Notice that the definition of denseness only depends on the frame $\langle \mathcal{G}, \mathcal{R} \rangle$, and makes no mention of the domain, $\mathcal{D}^{\mathcal{G}}$, or the forcing relation $\Vdash$.

In what follows we will replace the direct use of formulas with applications of properties of dense sets, thus altering the original notion of completeness in Cohen's complete sequences. But we want to abstract on the notion of sequence as well. And this is quite a well-known abstraction, one that arose in topology many years ago: sequences are replaced by *filters*, a more natural construct when *partial* orderings are involved.

Definition 3.2. A non-empty subset $G \subseteq \mathcal{G}$ is a *filter* in $\langle \mathcal{G}, \mathcal{R} \rangle$ if:

(1) for every $p, q \in G$ there exists some $r \in G$ such that $p\mathcal{R}r$ and $q\mathcal{R}r$;

(2) for every $q \in G$, if $p\mathcal{R}q$ then $p \in G$.

Think of members of $\mathcal{G}$ as pieces of information, pieces that may or may not agree with each other. A filter G is a subset of $\mathcal{G}$ that is coherent and can be thought of as a collective attempt to describe something. Item (1) says that it must be possible to fit together any two pieces of information in G—a coherency condition. Suppose we assume that G contains not just some pieces of information describing an object, but *all* of them. Then if $q \in G$, q is a partial description of that object, and item (2) says that any weakening of the description q must also apply correctly, and so is part of G as well.

Examples

(1) If we have a sequence in $\mathcal{G}$, $p_0, p_1, \ldots, p_n, \ldots$, with $p_n \mathcal{R} p_{n+1}$, it is trivial to check that $\{p_0, p_1, \ldots, p_n, \ldots\}$ satisfies item (1) of the filter definition.

(2) Again suppose we have a sequence $p_0, p_1, \ldots, p_n, \ldots$, in $\mathcal{G}$, with $p_n \mathcal{R} p_{n+1}$. Let $G = \{q \in \mathcal{G} \mid q\mathcal{R}p_n \text{ for some } p_n \text{ in the sequence}\}$. G satisfies both conditions, hence is a filter.

(3) Take for $\mathcal{G}$ all subsets of some set S; $\mathcal{G} = \mathcal{P}(S)$, and let $\mathcal{R}$ be the subset relation. Then $\langle \mathcal{G}, \mathcal{R} \rangle$ is a partial ordering, and so is an $S4$-frame. Choose some member $T \in \mathcal{G}$, and let G be the collection of all finite subsets of T. It is easy to see that G is a filter. In an intuitive sense, G "approximates" to T and to nothing else. This notion of approximation can be made precise—we do not do so since we will not need it.

We will use filters instead of sequences in this chapter. And we have seen that dense sets generalize valid formulas—each valid formula of the form $[\![X]\!]$ determines a dense set, though not conversely. Then the generalization of a complete sequence is quite direct: it is a filter that has one or more members in common with each of a sufficiently rich family of dense sets. These generalizations are called *generic*, rather than *complete*.

Definition 3.3. Let M be a set and $\langle \mathcal{G}, \mathcal{R} \rangle$ a frame. A set $G \subseteq \mathcal{G}$ is $\langle \mathcal{G}, \mathcal{R} \rangle$-*generic over M* if:

(1) G is a filter in $\langle \mathcal{G}, \mathcal{R} \rangle$;

(2) if $D \in M$ is dense in $\langle \mathcal{G}, \mathcal{R} \rangle$ then $D \cap G \neq \emptyset$.

EXERCISES

Exercise 3.1. Show that any filter in $\langle \mathcal{G}, \mathcal{R} \rangle$ can be extended to a maximal filter (maximal with respect to inclusion).

Exercise 3.2. Let $\mathcal{M} = (\mathcal{G}, \mathcal{R}, \mathcal{D}^\mathcal{G}, \Vdash)$ be an $S4$ model with $p \in \mathcal{G}$. A set $D \subseteq \mathcal{G}$ is called *dense below p* if, for every world accessible from p there is a world accessible from it that is in D. Show that $p \Vdash \Box \Diamond X$ if and only if the set of worlds at which X is true is dense below p.

§4 When generic sets exist

Generic sets will play a fundamental role throughout this chapter, so it would be nice to know some exist. In showing this we need to assume there is a countable standard model—it is the only place such an assumption is needed.

Theorem 4.1. *Assume M is a countable standard model in V and $\langle \mathcal{G}, \mathcal{R} \rangle \in M$. Then there is a set (that is, a member of V) that is $\langle \mathcal{G}, \mathcal{R} \rangle$-generic over M. More specifically, for any $r \in \mathcal{G}$ there is a set G that is $\langle \mathcal{G}, \mathcal{R} \rangle$-generic over M such that $r \in G$.*

Proof Since M is countable (with respect to V) we can enumerate the members of M that are dense in $\langle \mathcal{G}, \mathcal{R} \rangle$: $D_0, D_1, \ldots, D_n, \ldots$. Now define a sequence, $p_0, p_1, \ldots, p_n, \ldots$, in $\mathcal{G}$ as follows.

$p_0 = r$.

Given p_n, since D_n is dense in $\langle \mathcal{G}, \mathcal{R} \rangle$ it has a member p' such that $p_n \mathcal{R} p'$. Choose one and let it be p_{n+1}.

Now let $G = \{q \in \mathcal{G} \mid q \mathcal{R} p_n \text{ for some } n\}$. This has r as a member (because $r = p_0$ and $\mathcal{R}$ is reflexive). And it overlaps every member of M that is dense in $\langle \mathcal{G}, \mathcal{R} \rangle$ since $p_{n+1} \in G \cap D_n$. What remains is to show it is a filter.

Suppose $q \in G$ and $p \mathcal{R} q$. Since $q \in G$, for some n, $q \mathcal{R} p_n$. But then $p \mathcal{R} p_n$ (since $\mathcal{R}$ is transitive), so $p \in G$.

Suppose $p, q \in G$. Then for some n and k, $p \mathcal{R} p_n$ and $q \mathcal{R} p_k$. Say we have $k \leq n$—then $p_k \mathcal{R} p_n$, so $p \mathcal{R} p_n$ and $q \mathcal{R} p_n$, where $p_n \in G$.

Thus G is a filter, and the proof is complete. ∎

Generic sets have generally replaced Cohen's complete sequences in the literature. To conclude this section, then, we show there is a clear sense in which they are a natural extension. First we recall Definition 5.2 from Chapter 18. It says that members $p, q \in G$ are *compatible* if they have a common extension, that is, for some $r \in G$, $p\mathcal{R}r$ and $q\mathcal{R}r$; they are *incompatible* if they are not compatible.

Lemma 4.2. *Let M be a standard model with $\langle \mathcal{G}, \mathcal{R} \rangle \in M$, let G be $\langle \mathcal{G}, \mathcal{R} \rangle$-generic over M, and let $S \in M$ be an arbitrary subset of $\mathcal{G}$. Then either:*

(1) for some $p \in G$, $p \in S$, or

(2) for some $p \in G$, p is incompatible with every member of S.

Proof Let

$$A = \{q \in \mathcal{G} \mid p\mathcal{R}q \text{ for some } p \in S\}$$
$$B = \{q \in \mathcal{G} \mid q \text{ is incompatible with every member of } S\}.$$

We first show that $A \cup B$ is dense in $\langle \mathcal{G}, \mathcal{R} \rangle$. Suppose $p \in \mathcal{G}$; we must show that some world accessible from p is in $A \cup B$. If $p \in B$, we are done, since $p\mathcal{R}p$. If $p \notin B$, p is compatible with some member of S, say with q. Then for some r, $p\mathcal{R}r$ and $q\mathcal{R}r$. Since $q\mathcal{R}r$, $r \in A$, and since $p\mathcal{R}r$, there is a world accessible from p in A, hence in $A \cup B$. Thus $A \cup B$ is dense. Since M is a first-order universe with $\langle \mathcal{G}, \mathcal{R} \rangle \in M$, it follows easily that $A \cup B \in M$.

Since G is $\langle \mathcal{G}, \mathcal{R} \rangle$-generic over M, for some p, $p \in G$ and $p \in A \cup B$. Suppose first that $p \in B$. Then $p \in G$ and p is incompatible with every member of S, so we are in case (2). Otherwise, $p \in A$. Then for some $q \in S$, $q\mathcal{R}p$. But $p \in G$, and G is a filter, so $q \in G$ and we are in case (1). ∎

Now recall that the characteristic of a complete sequence in Cohen's sense is: for each formula X of $\mathcal{L}_C$, either the sequence "settles on" $[\![X]\!]$ or on $[\![\neg X]\!]$. The following says generic sets have a similar property.

Proposition 4.3. *Again let M be a standard model with $\langle \mathcal{G}, \mathcal{R} \rangle \in M$, and suppose G is $\langle \mathcal{G}, \mathcal{R} \rangle$-generic over M. For any formula X of $\mathcal{L}_C$, exactly one of:*

(1) for some $p \in G$, $p \Vdash [\![X]\!]$, or

(2) for some $p \in G$, $p \Vdash [\![\neg X]\!]$.

Proof First, we cannot have both items (1) and (2) since any two members of G are compatible, by definition of generic. Next, to show we must have at least one of (1) or (2), we observe the following validities: $[\![X]\!] \equiv \Box[\![X]\!]$ and $[\![\neg X]\!] \equiv \Box\neg\Box[\![X]\!]$. So it is enough to show that for each formula Y, either $p \Vdash \Box Y$ for some $p \in G$, or $p \Vdash \Box\neg\Box Y$ for some $p \in G$.

Let $S = \{q \in \mathcal{G} \mid q \Vdash \Box Y\}$. The proof now divides into two cases, following the previous lemma.

Case (1), for some $p \in G$, $p \in S$. In this case, $p \Vdash \Box Y$ by definition of S.

Case (2) for some $p \in G$, p is incompatible with every member of S. Then for any q such that $p\mathcal{R}q$ it must be that $q \notin S$, so $q \not\Vdash \Box Y$, or $q \Vdash \neg \Box Y$. But then $p \Vdash \Box \neg \Box Y$.
∎

§5 Generic extensions

We can now outline the construction of classical models in broad strokes—the details will be presented in this and the following sections. For these sections we make the following

General assumptions $M \subseteq V$ is a standard model and $\mathcal{M} = (\mathcal{G}, \mathcal{R}, \mathcal{D}^{\mathcal{G}}, \Vdash)$ is an $S4$-regular modal model over M. We know that if M is a countable set in V, $\langle \mathcal{G}, \mathcal{R} \rangle$-generic sets must exist. We explicitly assume their existence when necessary.

We will show that if G is $\langle \mathcal{G}, \mathcal{R} \rangle$-generic over M, there is a smallest standard model, denoted $M[G]$, that extends M and has G as a member. We will also show that, except in trivial cases, this is a proper extension. And we will show there is a close connection between truth at the possible worlds of the modal model $\mathcal{M}$ and truth in the classical model $M[G]$—enough to easily let us produce an $M[G]$ in which the continuum hypothesis is false.

Way back in Chapter 16 we pointed out there were two basic approaches to independence proofs. One possibility was to change the logic—we have done that, using the modal logic $S4$. The other was to change the meaning of inner. This is what we are discussing now. The model $M[G]$ is not generally a submodel of M—it extends it. But the $S4$-regular model $(\mathcal{G}, \mathcal{R}, \mathcal{D}^{\mathcal{G}}, \Vdash)$ is available within M, and it tells us things about $M[G]$, so M can "discuss" $M[G]$ even though $M[G]$ is not a submodel.

The following picture illustrates, in a general way, the situation we are about to examine in detail.

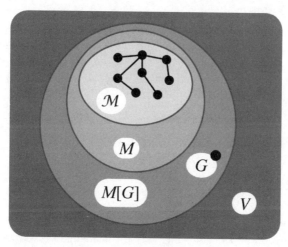

In this, we assume G is a generic set in V. Both M and $M[G]$ are submodels of the real universe V, with $M[G]$ extending M. Inside M is the $S4$ model $\mathcal{M}$, with

its many possible worlds and accessibility relation, and which, as we will see in the next section, tells us facts about $M[G]$. Finally, G is a member of $M[G]$, though not generally a member of M itself.

If we are to establish things about formulas without constants, along the way formulas with constants will come up. We will show that the members of $\mathcal{D}^{\mathcal{G}}$ can be thought of as "names" for the sets in $M[G]$, and the sets in $M[G]$ are exactly the sets having such names. The details start now.

Definition 5.1. For each $f \in \mathcal{D}^{\mathcal{G}}$, and for each set $G \subseteq \mathcal{G}$, we define a set f_G, by recursion on $\mathcal{G}$-rank, as follows:

$$f_G = \{g_G \mid p \Vdash \llbracket g \, \varepsilon \, f \rrbracket \text{ for some } p \in G\}.$$

This definition looks like it uses the truth machinery of the modal model $\mathcal{M}$, and we are trying to get away from such things. In fact, it is easy to do so. We use a notion that previously appeared in Exercise 3.2, and which we now introduce more formally.

Definition 5.2. A set $D \subseteq \mathcal{G}$ is called *dense below* p if, for every $q \in \mathcal{G}$ such that $p \mathcal{R} q$ there is some $r \in D$ such that $q \mathcal{R} r$.

Now we leave it to you to verify that

$$g_G \in f_G \text{ iff } \{q \mid \langle q, g \rangle \in f\} \text{ is dense below } p \text{ for some } p \in G.$$

Thus the mapping from f to f_G can be characterized using only features of the frame $\langle \mathcal{G}, \mathcal{R} \rangle$. Now, for each set G this maps each f in the modal domain $\mathcal{D}^{\mathcal{G}}$ to a corresponding set f_G. We can think of f as a name for f_G. Eventually we will want to take G to be generic, but even without this assumption we can show that the counterparts in $\mathcal{D}^{\mathcal{G}}$ of members of M in effect name themselves.

Proposition 5.3. *For each $x \in M$, $(\hat{x})_G = x$.*

Proof The argument is by a simple transfinite induction on rank (not $\mathcal{G}$-rank). Suppose the proposition is known for all sets of lower rank than x.

(1) Suppose $a \in (\hat{x})_G$. Then a is of the form b_G, where $p \Vdash \llbracket b \, \varepsilon \, \hat{x} \rrbracket$, for some $p \in G$. By definition of $\hat{x}$, $b = \hat{y}$ for some $y \in x$, so by the induction hypothesis, $y = (\hat{y})_G = b_G = a$, so $a \in x$.

(2) Suppose $a \in x$. Then for every $p \in \mathcal{G}$, $p \Vdash \llbracket \hat{a} \, \varepsilon \, \hat{x} \rrbracket$, so $(\hat{a})_G \in (\hat{x})_G$. But by the induction hypothesis, $(\hat{a})_G = a$, so $a \in (\hat{x})_G$. ∎

Definition 5.4. For any set $G \in V$, $M[G] = \{f_G \mid f \in \mathcal{D}^{\mathcal{G}}\}$.

This defines a class $M[G]$ for each G, generic or not—it is the class of sets that have names in $\mathcal{D}^{\mathcal{G}}$. We will show that if G is $\langle \mathcal{G}, \mathcal{R} \rangle$-generic over M then this characterization of $M[G]$ is equivalent to the one we gave at the beginning of this section—the smallest transitive ZF model extending M and containing G. The following makes a start in showing this.

Proposition 5.5. $M \subseteq M[G]$.

§5. GENERIC EXTENSIONS

Proof If $x \in M$ then $\hat{x} \in \mathcal{D}^{\mathcal{G}}$, so $(\hat{x})_G \in M[G]$ and by Proposition 5.3, $x \in M[G]$. ∎

Before showing $G \in M[G]$ we need a preliminary result, of interest for its own sake.

Lemma 5.6. *Suppose G is $\langle \mathcal{G}, \mathcal{R} \rangle$ generic over M, and $p \in G$. If D is dense below p then some member of D is in G.*

Proof Suppose D and G have no members in common. Then by Lemma 4.2 there must be some $q \in G$ that is incompatible with every member of D. Now, both p and q are in G, which is a filter, so for some $r \in G$, $p\mathcal{R}r$ and $q\mathcal{R}r$. Since D is dense below p, and $p\mathcal{R}r$, there must be some $r' \in D$ such that $r\mathcal{R}r'$. But then $q\mathcal{R}r'$, and this is impossible since q is incompatible with every member of D. ∎

We also recall a notion from Chapter 17, §6, which we repeat for convenience. For each $g \in \mathcal{D}^{\mathcal{G}}$, a new member g^* of $\mathcal{D}^{\mathcal{G}}$ is defined as follows.

$$g^* = \{\langle q, f \rangle \mid \langle p, f \rangle \in g \text{ for some } p \in \mathcal{G} \text{ such that } p\mathcal{R}q\}.$$

It is easy to check that the following items hold for this notion.

(1) $p \Vdash (a \,\varepsilon\, f) \Rightarrow p \Vdash (a \,\varepsilon\, f^*)$

(2) $p \Vdash (a \,\varepsilon\, f^*) \Leftrightarrow q \Vdash (a \,\varepsilon\, f)$ for some q such that $q\mathcal{R}p$

(3) $p \Vdash (a \,\varepsilon\, f) \Rightarrow p \Vdash \llbracket a \,\varepsilon\, f^* \rrbracket$.

Now for a rather cute trick. If $p \in \mathcal{G}$ then $p \in M$ so $\hat{p} \in \mathcal{D}^{\mathcal{G}}$. Define g as follows:

$$g = \{\langle p, \hat{p} \rangle \mid p \in G\}.$$

Then $g \in \mathcal{D}^{\mathcal{G}}$ and for $p \in G$, $p \Vdash (x \,\varepsilon\, g)$ if and only if $x = \hat{p}$. Now we have the following.

Proposition 5.7. *If G is $\langle \mathcal{G}, \mathcal{R} \rangle$-generic then $(g^*)_G = G$, and so $M[G]$ has G as a member.*

Proof We present the argument in two parts.

(1) We show $(g^*)_G \subseteq G$. Suppose $h_G \in (g^*)_G$. Then $p \Vdash \llbracket h \,\varepsilon\, g^* \rrbracket$ for some $p \in G$, that is, $p \Vdash \square \Diamond (h \,\varepsilon\, g^*)$. By Exercise 3.2, the set of members of $\mathcal{G}$ at which $(h \,\varepsilon\, g^*)$ is true is dense below p, so by Lemma 5.6, for some $q \in G$, $q \Vdash (h \,\varepsilon\, g^*)$. Then by the properties of g^*, there must be some r such that $r\mathcal{R}q$ and $r \Vdash (h \,\varepsilon\, g)$. But G is generic, and hence is a filter, so $r \in G$ (since $q \in G$). Now by definition of g, it must be that $h = \hat{r}$. Then by Proposition 5.3, $h_G = (\hat{r})_G = r$, so $h_G \in G$.

(2) We show $G \subseteq (g^*)_G$. Suppose $p \in G$. Then $\langle p, \hat{p} \rangle \in g$, so $p \Vdash (\hat{p} \,\varepsilon\, g)$ and consequently $p \Vdash \llbracket \hat{p} \,\varepsilon\, g^* \rrbracket$. Then by definition, $(\hat{p})_G \in (g^*)_G$ so by Proposition 5.3 again, $p \in (g^*)_G$. ∎

We now know that $M[G]$ is an extension of M, and contains G provided G is generic. In subsequent sections we will show $M[G]$ is also a well founded first-order universe. We conclude this section by showing that $M[G]$ is often a *proper* extension of M. The result depends on special features of the frame $\langle \mathcal{G}, \mathcal{R} \rangle$, and these conditions are met by the frames used for the various independence results we have established.

Proposition 5.8. *Suppose the frame $\langle \mathcal{G}, \mathcal{R} \rangle$ meets the condition that, for any member p of $\mathcal{G}$ there are members q and r, both accessible from p, that are incompatible. Then if G is $\langle \mathcal{G}, \mathcal{R} \rangle$-generic over M, $G \notin M$, and hence $M[G]$ is a proper extension of M.*

Proof Assume the hypothesis, and suppose that $G \in M$. Since $\mathcal{G} \in M$ then the difference, $\mathcal{G} - G$ is also in M. If we show this difference is dense we have a contradiction since G, being generic, must have a member in common with $\mathcal{G} - G$, and this is impossible. So we show the difference is dense.

Suppose $p \in \mathcal{G}$—we show there is a world q accessible from p that is in $\mathcal{G} - G$. Since $p \in \mathcal{G}$, by hypothesis there must be q and r that are incompatible, with $p \mathcal{R} q$ and $p \mathcal{R} r$. Suppose we had both $q \in G$ and $r \in G$. Since G is a filter, for some $s \in G$, $q \mathcal{R} s$ and $r \mathcal{R} s$, and this contradicts the incompatibility of q and r. So not both can be in G, say $q \notin G$. Then $q \in \mathcal{G} - G$, so there is a world accessible from p in $\mathcal{G} - G$. ∎

EXERCISES

Exercise 5.1. Show that $f_G \cup g_G = (f \cup g)_G$, even if G is not generic.

§6 The truth lemma

For this section the general assumptions stated at the beginning of the previous section are still in force. We will soon show a central result of the subject, one relating truth in $M[G]$ with truth at worlds of M, provided G is generic. We state the result properly, prove some preliminary results, and finally present a proof of the theorem itself.

Theorem 6.1 (Truth lemma). *Assume that G is $\langle \mathcal{G}, \mathcal{R} \rangle$-generic over M; let $\varphi(x_1, \ldots, x_n)$ be a formula of $\mathcal{L}_C$, with all free variables shown, and let $c_1, \ldots, c_n \in \mathcal{D}^{\mathcal{G}}$. Then, $\varphi((c_1)_G, \ldots, (c_n)_G)$ is true in $M[G]$ if and only if $p \Vdash [\![\varphi(c_1, \ldots, c_n)]\!]$ for some $p \in G$.*

Note that since $\mathcal{M}$ is an $S4$-ZF model, it follows immediately that $M[G]$ is a well founded first-order universe. But before we get to enjoy this important consequence we need a proof, and for this we need a few items first.

We are interested in formulas of the classical language $\mathcal{L}_C$, but the characterization of $\in$ in modal models makes use of $\approx$, so we begin with an investigation of that.

Proposition 6.2. *Suppose G is $\langle \mathcal{G}, \mathcal{R} \rangle$-generic over M and $f, g \in \mathcal{D}^{\mathcal{G}}$.*

(1) For each ordinal α of M, if $p \Vdash [\![f \approx_\alpha g]\!]$ for some $p \in G$ then $f_G = g_G$ in $M[G]$.

(2) $p \Vdash [\![f \approx g]\!]$ for some $p \in G$ iff $f_G = g_G$ in $M[G]$.

§6. THE TRUTH LEMMA

Proof We start with part (1), and its proof is by transfinite induction on α. Not surprisingly, the cases where α is 0 or a limit ordinal are trivial, so we concentrate on the successor ordinal case. So, assume (1) is true at α for any members of $\mathcal{D}^{\mathcal{G}}$, and also assume $p \Vdash \llbracket f \approx_{\alpha+1} g \rrbracket$ for some $p \in G$. We show $f_G \subseteq g_G$; the inclusion in the other direction has a similar proof.

Members of f_G are of the form a_G, so suppose $a_G \in f_G$; we show $a_G \in g_G$. Since $a_G \in f_G$, by definition, $q \Vdash \llbracket a \varepsilon f \rrbracket$ for some $q \in G$. Also for some $p \in G$, $p \Vdash \llbracket f \approx_{\alpha+1} g \rrbracket$ and so $p \Vdash \llbracket (\forall x)(x \varepsilon f \supset (\exists y)(y \varepsilon g \wedge y \approx_\alpha x)) \rrbracket$. Now G is a filter and $p, q \in G$, so there is some $r \in G$ such that both $p\mathcal{R}r$ and $q\mathcal{R}r$. It follows that at r, $\llbracket (\exists y)(y \varepsilon g \wedge y \approx_\alpha a) \rrbracket$ is true, equivalently, $\square\Diamond(\exists y)\llbracket y \varepsilon g \wedge y \approx_\alpha a \rrbracket$ is true at r.

By Exercise 3.2, the subset of $\mathcal{G}$ at which $(\exists y)\llbracket y \varepsilon g \wedge y \approx_\alpha a \rrbracket$ is true is dense below r, so by Lemma 5.6, some member s of G must be in this set. Then for some $b \in \mathcal{D}^{\mathcal{G}}$, $\llbracket b \varepsilon g \wedge b \approx_\alpha a \rrbracket$ is true at s.

Since $s \in G$ and $\llbracket b \approx_\alpha a \rrbracket$ is true at s, by the induction hypothesis, $b_G = a_G$. And since $\llbracket b \varepsilon g \rrbracket$ is true at s, $b_G \in g_G$, and so $a_G \in g_G$. This completes the proof of part (1).

One direction of part (2) is an immediate consequence of part (1), the implication from left to right. We concentrate on the other direction. Suppose $f_G = g_G$ in $M[G]$; we show $p \Vdash \llbracket f \approx g \rrbracket$ for some $p \in G$. For this, it is convenient to introduce three subsets of $\mathcal{G}$.

$$\begin{aligned} A &= \{p \in \mathcal{G} \mid p \Vdash \llbracket f \approx g \rrbracket\} \\ B &= \{p \in \mathcal{G} \mid p \Vdash (\exists x)\llbracket x \varepsilon f \wedge (\forall y)(y \varepsilon g \supset \neg(x \approx y)) \rrbracket\} \\ C &= \{p \in \mathcal{G} \mid p \Vdash (\exists x)\llbracket x \varepsilon g \wedge (\forall y)(y \varepsilon f \supset \neg(x \approx y)) \rrbracket\}. \end{aligned}$$

We first claim that $A \cup B \cup C$ is dense in $\langle \mathcal{G}, \mathcal{R} \rangle$. Let $p \in \mathcal{G}$; we show there is a world accessible from p that is in $A \cup B \cup C$. Now, if $p \in A$, we are done. If not, by Proposition 2.12 of Chapter 17 either $\llbracket (\forall x)[(x \varepsilon f) \supset (\exists y)(y \varepsilon g \wedge y \approx x)] \rrbracket$ or $\llbracket (\forall x)[(x \varepsilon g) \supset (\exists y)(y \varepsilon f \wedge y \approx x)] \rrbracket$ is not true at p, say the first. Then for some q with $p\mathcal{R}q$, $q \Vdash \llbracket \neg(\forall x)[(x \varepsilon f) \supset (\exists y)(y \varepsilon g \wedge y \approx x)] \rrbracket$, equivalently, $q \Vdash \llbracket (\exists x)(x \varepsilon f \wedge (\forall y)(y \varepsilon g \supset \neg(y \approx x)) \rrbracket$. Then for some r with $q\mathcal{R}r$, $r \Vdash (\exists x)\llbracket x \varepsilon f \wedge (\forall y)(y \varepsilon g \supset \neg(x \approx y)) \rrbracket$, and so there is a world accessible from p that is in B. Likewise if the second alternative were the case, there would be a world accessible from p that is in C. It follows that $A \cup B \cup C$ is dense. And clearly, it is in M.

Since G is generic, it meets every dense set in M, consequently for some $p \in G$, $p \in A \cup B \cup C$. If we show we cannot have $p \in B$ or $p \in C$, we are done. Well, suppose we did have $p \in B$ (the other case is similar, of course). Then for some $c \in \mathcal{D}^{\mathcal{G}}$, $p \Vdash \llbracket c \varepsilon f \wedge (\forall y)(y \varepsilon g \supset \neg(c \approx y)) \rrbracket$. Since $p \Vdash \llbracket c \varepsilon f \rrbracket$, then $c_G \in f_G$. By assumption, $f_G = g_G$, so $c_G \in g_G$. Then, for some $q \in G$, $q \Vdash \llbracket c \varepsilon g \rrbracket$. Since both p and q are in G, which is a filter, for some $r \in G$ both $p\mathcal{R}r$ and $q\mathcal{R}r$. Then at r, $\llbracket c \varepsilon g \rrbracket$, since we have this at q, and we have $\llbracket (\forall y)(y \varepsilon g \supset \neg(c \approx y)) \rrbracket$, since we have this at p. Then we have

$[\![c \, \varepsilon \, g \supset \neg(c \approx c)]\!]$, and it follows that we have $[\![\neg(c \approx c)]\!]$, and this is impossible. ∎

With this out of the way, we turn to the proof of the truth lemma itself.

Proof of Theorem 6.1 The proof is by induction on the complexity of φ. We generally suppress explicit display of constants from $\mathcal{D}^{\mathcal{G}}$ to make reading easier.

Ground case: Suppose first that $p \Vdash [\![a \in b]\!]$ for some $p \in G$. Then for some c, $p \Vdash [\![c \approx a]\!]$ and $p \Vdash [\![c \, \varepsilon \, b]\!]$. By definition, $c_G \in b_G$, and by Proposition 6.2, $c_G = a_G$, and consequently $a_G \in b_G$.

Conversely, suppose $a_G \in b_B$. By definition, $p \Vdash [\![a \, \varepsilon \, b]\!]$ for some $p \in G$, and it follows immediately that $p \Vdash [\![a \in b]\!]$.

Conjunction case: Suppose the result is known for both X and Y. If $(X \wedge Y)$ is true in $M[G]$ then both X and Y are true in $M[G]$. Then for some $p \in G$, $p \Vdash [\![X]\!]$, and for some $q \in G$, $q \Vdash [\![Y]\!]$. Since G is generic, hence a filter, for some $r \in G$, $p\mathcal{R}r$ and $q\mathcal{R}r$. Then both $[\![X]\!]$ and $[\![Y]\!]$ are true at r, hence so is $[\![X \wedge Y]\!]$. This establishes one direction—the other is simpler and is omitted.

Negation case: Suppose the result is known for X. If $\neg X$ is true in $M[G]$, X is not true in $M[G]$, so by the induction hypothesis, $p \not\Vdash [\![X]\!]$ for every $p \in G$. Then by Proposition 4.3, there must be some $q \in G$ such that $q \Vdash [\![\neg X]\!]$. This is one direction—again the other is similar.

Existential case: Suppose the result is known for $\psi(x)$, and for some $p \in G$, $p \Vdash [\![(\exists x)\psi(x)]\!]$. Then $p \Vdash \Box\Diamond(\exists x)[\![\psi(x)]\!]$, so by Exercise 3.2 the set of worlds at which $(\exists x)[\![\psi(x)]\!]$ is true is dense below p. By Lemma 5.6, $(\exists x)[\![\psi(x)]\!]$ must be true at some member q of G. But then, for some $c \in \mathcal{D}^{\mathcal{G}}$, $[\![\psi(c)]\!]$ is true at q, so by the induction hypothesis, $\psi(c_G)$ is true in $M[G]$, and hence so is $(\exists x)\psi(x)$. Once again, the converse direction is simpler. □

§7 Conclusion

We have now established all the basic results, and it is time to summarize. Suppose we assume M is a countable standard model in which the axioms of well foundedness and choice are true, and $\langle \mathcal{G}, \mathcal{R} \rangle$ is a frame that is a member of M. According to Theorem 4.1 there is a set G that is $\langle \mathcal{G}, \mathcal{R} \rangle$-generic over M. (In fact, we can arrange things so that any given member of $\mathcal{G}$ is in G.) From now on assume G is generic.

In §5 we constructed $M[G]$, and it is easy to see from its definition that it is transitive. We showed in Proposition 5.5 that $M[G]$ extends M, contains G (Proposition 5.7), and under some circumstances properly extends M, since G will not be in M (Proposition 5.8).

We know that every member of $M[G]$ has a name in $\mathcal{D}^{\mathcal{G}}$, and the truth lemma (Theorem 6.1) relates truth in $M[G]$ with truth at the worlds of $\mathcal{M} = \langle \mathcal{G}, \mathcal{R}, \mathcal{D}^{\mathcal{G}}, \Vdash \rangle$. Since $\mathcal{M}$ has been shown to be an $S4$-ZF model in Chapter 17, it follows that $M[G]$ is a well founded first-order universe. It is not hard to show that the ordinals of $M[G]$

§7. CONCLUSION

correspond to the members of $\mathcal{D}^{\mathcal{G}}$ that M "thinks" are ordinals, and these correspond to the ordinals of M. Thus $M[G]$ and M have the same ordinals. Also, $M[G]$ is countable (in V) since its members have names in M, and M is countable. Thus $M[G]$ is itself a countable standard model having the same ordinals as M. (Sometimes this situation is informally described as: $M[G]$ is wider than M, but not longer.)

Finally, suppose N is an arbitrary standard model that extends M and contains G. It can be shown that being a member of $\mathcal{D}^{\mathcal{G}}$ is absolute over N, and also that the mapping from $f \in \mathcal{D}^{\mathcal{G}}$ to f_G is absolute over N. Then the construction of $M[G]$, which we carried out in V, can be carried out just as well in N, so $M[G] \subseteq N$. Since N is arbitrary, it follows that $M[G]$ must be the *smallest* standard model extending M, containing G. This provides a "forcing-free" characterization of $M[G]$. Indeed, Exercise 7.1 shows that one can reverse things, and use the classical models $M[G]$ to characterize the forcing relation itself (though forcing more-or-less as we defined it is still needed to establish basic properties).

If we start with the set-up of Chapter 19, say, showing that the continuum hypothesis is independent, and apply the construction above, we get a *classical* model $M[G]$ (though countable) in which the continuum hypothesis is false.

EXERCISES

Exercise 7.1. Given the notation and assumptions of this section, show that for a closed formula X of $\mathcal{L}_C$, $[\![X]\!]$ is valid in the modal model $\mathcal{M}$ if and only if X is true in every generic extension $M[G]$ (that is, true in $M[G]$ for every G that is $\langle \mathcal{G}, \mathcal{R} \rangle$-generic over M).

CHAPTER 22

FORCING BACKGROUND

§1 Introduction

Gödel and Cohen created very different universes. Gödel's universe is long but thin—L contains all the ordinals, but otherwise it contains nothing more than it has to. (Recall that L is the smallest first-order universe containing the ordinals.) Cohen's universes are short and fat. Since they are countable, they do not contain all the ordinals but, apart from that, they contain considerably more than the minimum: nonconstructible sets, mappings between $\mathcal{P}(\omega)$ and ω_2, and so on. Sparseness versus richness—two extremes.

The Gödel universe L turns out to be an interesting object in its own right; it is more than just a tool for establishing consistency results. In one direction ordinary recursion theory, on the natural numbers, was generalized to other initial segments of the ordinals, and it turned out that L and the L_α sequence play an essential role in this. The subject, which became known as α-*recursion theory* or *admissible set recursion theory* began with abstracts of Kripke (1964) and the dissertation of Platek (1966), and quickly produced deep results uniting previously disparate ideas drawn from set theory and from computability theory. An elegant summation can be found in (Barwise 1975). In another direction the so-called "fine structure" of the constructible universe was investigated, where not just the contents of L, but the relative complexity of these contents, are examined (Jensen & Karp 1971, Jensen 1972). These two lines of development are, in fact, deeply intertwined.

Our methodology for dealing with basic facts about L has expanded as well, since Gödel's original work. The use of Δ_0 and Σ_1 formulas arose from (Lévy 1965), for instance, and their specific introduction in turn helped make it possible to see connections between computability theory and the investigation of L. The Tarski-Vaught theorem, the Mostowski-Shepherdson mapping theorem (Mostowski 1969), and other fundamental results have enriched our toolkit for the investigation of the notion of constructibility.

Our understanding of Cohen's universes, too, is considerably deeper than it was. But here the basic methodology has changed remarkably since Cohen's original work. Things are not just better understood; the whole approach looks different. Indeed, almost immediately after Cohen's original results were obtained, it became clear that forcing is a robust cluster of methods rather than a single item. No two general treatments of forcing seem to agree on the details—not even on the definition of forcing itself—though a general family resemblance is clear. In this chapter we present a brief sketch of the history of forcing. It is a complex history, and we do not pretend to give a definitive presentation—just a general outline.

§2 Cohen's version(s)

As we have presented forcing, it generalizes the R_α sequence — the well founded universe. But as Cohen originally conceived it, forcing generalized Gödel's construction of L. Instead of starting with the empty set, one started with some collection of given sets, then proceeded essentially as Gödel did, producing a universe of sets constructible from these. Forcing determined the properties of the initial sets, and consequently the properties of the resulting model. As in Gödel's work, first-order definable subsets play a fundamental role. Curiously enough Cohen, like Gödel, presented his results first in papers in the Proceedings of the National Academy of Sciences (Cohen 1963, Cohen 1964), then in a monograph (Cohen 1966). Reversing things however, Cohen's monograph follows the approach of Gödel's Proceedings articles while Cohen's Proceedings papers generalize Gödel's monograph construction of L. Still, the two approaches of Cohen are equivalent, as are Gödel's two presentations.

A more fundamental issue is not which approach to constructibility is generalized, but what kind of thing one takes forcing to be. Cohen's first, unpublished, approach was purely syntactic. Forcing was taken to be a relation between forcing conditions and formulas; it was shown that all axioms of set theory were forced by all forcing conditions; forcing preserved logical consequence and hence all theorems of set theory were also forced by all forcing conditions; but there were forcing conditions that did not force the continuum hypothesis. All proofs involved syntactic arguments about formulas. As Cohen said in (Cohen 1966):

> "Although this point of view may seem like a tedious way of avoiding models, it should be mentioned that in our original approach to forcing this syntactical point of view was the dominant point of view, and models were later introduced as they appeared to simplify the exposition. The peculiar role of the countability of M is here entirely avoided."

In a sense, our modal approach derives directly from Cohen's syntactic version. Classical models are not constructed and countability assumptions are not needed. But instead of thinking purely in syntactic terms, we have brought in non-classical semantics. This kind of non-classical treatment began in (Grzegorczyk 1964) and was treated extensively in (Fitting 1969).

Even after classical models were introduced, Cohen's definition of forcing was of considerable complexity, and required formulas to be in prenex form (all quantifiers in front). Dana Scott showed how this could be avoided, leading to a smoother and more appealing definition—essentially the one appearing in Cohen's monograph.

Incidentally, Cohen's syntactic approach is not the only way of avoiding the need for a countable standard model. Another method, due independently to several people, first appeared in (Lévy 1971), and is as follows. If the continuum hypothesis, say, were provable in set theory, only a finite number of axioms would be needed for the proof. Let S be the finite set consisting of these axioms, plus a few other sentences, such as those requiring transitivity. Since S is finite, it can be proved in set theory that if S is consistent, it has a countable model M. Now carry out the Cohen construction in this model M. The point is, we do not need to *assume* M exists; we only need to assume set theory is consistent, and then the existence of M follows. In the approach of the previous chapter, M was a countable model for the entire collection of set theory axioms—an infinite collection—and the existence of such a countable model, assuming consistency, is no longer provable in set theory. Using the Lévy approach, independence results follow from consistency assumptions alone, and classical models are still constructed, though they are not models for the entire of ZF, but only of a 'sufficently large' finite portion of it.

§3 Boolean valued models

In classical logic, formulas take their values in the space {*true, false*}, the simplest non-trivial Boolean algebra. It is an easy generalization to allow formulas to take their values in other complete Boolean algebras (completeness makes possible the interpretation of quantifiers—never mind the details here). Rasiowa & Sikorski (1968) showed how the use of general Boolean algebras, together with a theorem of theirs concerning the existence of mappings between them, could be used to prove the completeness of first-order logic in an elegant way.

Scott & Solovay (1971) realized that the Rasiowa-Sikorski approach to first-order logic extended to the entire of set theory. The additional degree of freedom given by the choice of Boolean algebra made it possible to construct models, called *Boolean valued models*, having a wide variety of features, and in this way alternative proofs of the various independence results were given. Incidentally, we note that Rasiowa and Sikorski's theorem about the existence of mappings (or some equivalent alternative) comes in, and this requires countability assumptions similar to those that appear in Cohen's approach.

In Chapter 20, proving the independence of the axiom of choice, automorphisms were used. Their theory can be generalized to mappings between, instead of just within, structures. In this way one can show there are close connections between Cohen style and Scott-Solovay style approaches to the independence results. Suitable mappings exist between the two kinds of structures. A presentation of this can be found in (Kunen 1980).

What turned out to be of more fundamental importance in the Scott-Solovay approach was not the use of Boolean algebras, but the fact that they did not generalize the constructible hierarchy, as Cohen had, but the R_α sequence instead. They explicitly gave a natural Boolean-valued version of R_α. This showed that the emphasis on features of the constructible heirarchy was somehow misplaced.

§4 Unramified forcing

As we remarked above, Scott and Solovay made the R_α rather than the L_α sequence the basis of Boolean valued forcing. Shoenfield (1971) retrofitted this idea to Cohen's approach, making it much more elegant and general. Gödel's L_α sequence is often called the "ramified hierarchy," so avoiding its use in forcing is naturally called "unramified forcing." This is now the common version of forcing.

The move to unramified forcing amounts to a simplification and a generalization of forcing techniques. It is a simplification because both the machinery of forcing and proofs of facts concerning it become much less complex. But also, if one carries out the unramified forcing construction within the universe L of constructible sets, it can be shown that the end result is equivalent to forcing as Cohen created it. Thus it is a true generalization as well.

Finally, connections between the Boolean valued approach and the Cohen approach lead to something like the version of forcing we presented. Suppose we start with a partially ordered set of forcing conditions. These can be embedded into a Boolean algebra, as a dense subset. The Boolean valued model of set theory that can be constructed using this Boolean algebra has close connections with the Cohen model based on the original forcing conditions. But more, suppose we use the Boolean algebra, allowing all the algebraic methods developed by Scott and Solovay, but we confine our attention to the dense subset corresponding to the original forcing conditions. These are the only generalized truth values we are actually interested in anyway. Doing this gives an approach to forcing that closely resembles the modal $S4$ version we presented. This is the route taken in Jech (1986) (see also (Jech 1978)). One can think of the use of a modal language as "syntactic sugar" added to this approach, though the modal version can also be understood quite independently as well, of course.

§5 Extensions

The invention of forcing inaugurated an intense period of activity in set theory, and soon many longstanding open problems were shown to be independent. This called for much ingenuity in extending and modifying the method. The most significant such extension was probably "iterated forcing," introduced by Solovay & Tennenbaum (1970). Suslin's hypothesis is that a particular family of facts concerning the real line uniquely determines it (up to isomorphism). This hypothesis, which is more commonly stated in an equivalent version concerning trees, turns out to be consistent with and independent of the usual axioms of set theory. It was to show its consistency that iterated forcing was invented. We do not give details, only the broadest outline.

Suppose we start with a countable standard model M and, as in the previous chapter, produce a generic extension $M[\mathcal{G}_0]$. This will again be a countable standard model, so in a similar way we can produce a generic extension of it, $M[\mathcal{G}_0][\mathcal{G}_1]$, and so on. In this way a chain of countable standard models can be created with considerable freedom, since the entire machinery of forcing comes into play at each stage. Now, the real innovation of Solovay & Tennenbaum (1970) was to extend this process through a limit stage. Doing so provided enough power to show the Suslin hypothesis was consistent. Since then others have considerably extended the technique, proving a number of interesting but relatively specialized results. There is a good, extensive treatment in (Kunen 1980).

Remarkably, in addition to set-theoretic applications, forcing was quickly applied in other areas of logic as well. Kripke (1971) applied Boolean valued forcing to prove results about Boolean algebras themselves. Abraham Robinson applied it to get results in model theory (Robinson 1970, Robinson 1971). It was applied to recursion theory (Feferman 1964, Gandy & Sacks 1967, Sacks 1971). (For a more accessible appli-

§5. EXTENSIONS

cation, see Chapter 20 in (Boolos & Jeffrey 1989).) And finally, a category-theoretic approach to the whole subject was developed—see (Kock & Reyes 1977, Mac Lane & Moerdijk 1992) for a summary and further references.

We have only covered the basic applications of forcing. Work is still continuing, and articles using forcing appear every year. It has become a standard tool. For a thorough survey through the date of its publication, see (Kunen 1980). For a sketchier overview, see the article in the Handbook on the subject, (Burgess 1977). For the Boolean valued approach, see (Bell 1977).

BIBLIOGRAPHY

Aczel, P. (1988), *Non-Well-Founded Sets*, number 14 *in* 'CSLI Lecture Notes', Center for the Study of Language and Information, Menlo Park, CA.
Barwise, J. (1975), *Admissible Sets and Structures*, Springer Verlag, Berlin.
Barwise, J., ed. (1977), *Handbook of Mathematical Logic*, Studies in Logic and the Foundations of Mathematics, North-Holland, Amsterdam.
Bell, J. L. (1977), *Boolean-Valued Models and Independence Proofs in Set Theory*, 2 edn, Oxford University Press, New York.
Boolos, G. (1979), *The Unprovability of Consistency*, Cambridge University Press, Cambridge.
Boolos, G. (1993), *The Logic of Provability*, Cambridge University Press, Cambridge.
Boolos, G. S. & Jeffrey, R. C. (1989), *Computability and Logic*, 3 edn, Cambridge University Press, Cambridge.
Bull, R. A. & Segerberg, K. (1984), *Handbook of Philosophical Logic*, Vol. 2, D. Reidel Publishing Company, chapter Basic Modal Logic, pp. 1–88.
Burgess, J. P. (1977), *Forcing*, pp. 403–452. In (Barwise 1977).
Chellas, B. F. (1980), *Modal Logic, an introduction*, Cambridge University Press, Cambridge.
Cohen, P. J. (1963), 'The independence of the continuum hypothesis, I', *Proceedings of the National Academy of Sciences, U.S.A.* **50**, 1143–1148.
Cohen, P. J. (1964), 'The independence of the continuum hypothesis, II', *Proceedings of the National Academy of Sciences, U.S.A.* **51**, 105–110.
Cohen, P. J. (1966), *Set Theory and the Continuum Hypothesis*, W. A. Benjamin, New York.
Cowen, R. H. (1971), 'Superinductive classes in class-set theory', *Notre Dame Journal of Formal Logic* **12**(1), 62–68.
Feferman, S. (1964), 'Some applications of the notions of forcing and generic sets', *Fundamenta Mathematicae* **56**, 325–345.
Fitting, M. C. (1969), *Intuitionistic Logic Model Theory and Forcing*, North-Holland, Amsterdam.
Fitting, M. C. (1993), Basic modal logic, *in* D. M. Gabbay, C. J. Hogger & J. A. Robinson, eds, 'Handbook of Logic in Artificial Intelligence and Logic Programming', Vol. 1, Oxford University Press, pp. 368–448.
Fraenkel, A. A. (1922), 'Der begriff "definit" und die unabhängigkeit des auswahlaxioms', *Sitzungsberichte der Preussischen Akademie der Wissenschaften, Physikalisch-mathematische Klasse* pp. 253–257. English translation in (van Heijenoort 1967), pp 284 – 289.
Gandy, R. O. & Sacks, G. E. (1967), 'A minimal hyperdegree', *Fundamenta Mathematicae* **61**, 215–223.
Gödel, K. (1938), 'The consistency of the axiom of choice and of the generalized continuum hypothesis', *Proceedings of the National Academy of Sciences, U.S.A.* **24**, 556–557. Reprinted in (Gödel 1986–1993), Volume II.
Gödel, K. (1939), 'Consistency proof for the generalized continuum hypothesis', *Proceedings of the National Academy of Sciences, U.S.A.* **25**, 220–224. Reprinted in (Gödel 1986–1993), Volume II.
Gödel, K. (1940), *The consistency of the axiom of choice and of the generalized continuum hypothesis with the axioms of set theory*, Vol. 3 of *Annals of mathematics studies*, Princeton University Press, Princeton, N.J. Reprinted in (Gödel 1986–1993), Volume II.
Gödel, K. (1947), 'What is Cantor's continuum problem?', *American mathematical monthly* **54**, 515–525. Reprinted in (Gödel 1986–1993), Volume II.
Gödel, K. (1986–1993), *Collected Works*, Oxford University Press, New York and Oxford. Three volumes.
Grzegorczyk, A. (1964), 'A philosophically plausible formal interpretation of intuitionistic logic', *Indag. Math.* **26**, 596–601.
Halmos, P. R. (1960), *Naive Set Theory*, D. Van Nostrand, Princeton, N.J.
Hausdorff, F. (1914), *Grundzüge der Mengenlehre*, Leipzig.
Hughes, G. E. & Cresswell, M. J. (1968), *An Introduction to Modal Logic*, Methuen, London.
Jech, T. J. (1978), *Set Theory*, Academic Press, New York.
Jech, T. J. (1986), *Multiple Forcing*, Cambridge University Press, Cambridge, UK.
Jensen, R. B. (1972), 'The fine structure of the constructible hierarchy', *Annals of Math. Logic* **4**, 229–308.
Jensen, R. B. & Karp, C. (1971), 'Primitive recursive set functions'. In (Scott 1971).
Kelley, J. L. (1975), *General Topology*, Springer-Verlag, New York. Reprint of Van Nostrand, 1955.
Kock, A. & Reyes, G. E. (1977), *Doctrines in categorical logic*, pp. 283–313. In (Barwise 1977).

Kripke, S. (1964), 'Transfinite recursion on admissible ordinals I, II (abstracts)', *Journal of Symbolic Logic* **29**, 161–162.

Kripke, S. (1971), 'On the application of boolean-valued models to solutions of problems in boolean algebra'. In (Scott 1971).

Kunen, K. (1980), *Set Theory, An Introduction to Independence Proofs*, Studies in Logic and the Foundations of Mathematics, North-Holland, Amsterdam.

Kuratowski, K. (1922), 'Une méthode d'ëlimination des nombres transfinis des raissonnements mathématiques', *Fundamenta Mathematicae* **3**, 76–108.

Lévy, A. (1965), 'A hierarchy of formulas in set theory', Memoirs of the American Mathematical Society. Vol. 57.

Lévy, A. (1971), 'On the logical complexity of several axioms of set theory'. In (Scott 1971).

Mac Lane, S. & Moerdijk, I. (1992), Reals and forcing with an elementary topos, *in* Y. N. Moschovakis, ed., 'Logic from Computer Science', Springer-Verlag, New York, pp. 373–385.

Marczewski, E. (1947), 'Séparabilité at multiplication cartésienne des espaces topologiques', *Fundamenta Mathematicae* **34**, 127–143.

Mendelson, E. (1987), *Introduction to Mathematical Logic*, 3rd edn, Wadsworth and Brooks/Cole Advanced Books and Software.

Moschovakis, Y. N. (1994), *Notes on Set Theory*, Springer-Verlag, New York.

Mostowski, A. (1969), *Constructible sets with applications*, North-Holland, Amsterdam.

Platek, R. (1966), Foundations of recursion theory, PhD thesis, Stanford University, Stanford, CA.

Quine, W. V. O. (1937), 'New foundations for mathematical logic', *American Mathematical Monthly* **44**, 70–80. Reprinted in (Quine 1953).

Quine, W. V. O. (1940), *Mathematical Logic*, Harvard University Press, Boston. Revised edition, 1951.

Quine, W. V. O. (1953), *From a Logical Point of View*, Harvard University Press, Boston.

Rasiowa, H. & Sikorski, R. (1968), *The Mathematics of Metamathematics*, 2 edn, PWN — Polish Scientific Publishers, Warsaw.

Robinson, A. (1970), 'Forcing in model theory', *Symposia Math.* **5**, 69–82.

Robinson, A. (1971), Infinite forcing in model theory, *in* J. E. Fenstad, ed., 'Proceedings of the 2nd Scandinavian Logic Symposium', North-Holland, Amsterdam, pp. 317–340.

Rosser, J. B. (1969), *Simplified Independence Proofs*, Academic Press, New York.

Sacks, G. E. (1971), Forcing with perfect closed sets, pp. 331–355. In (Scott 1971).

Schoenfield, J. (1954), 'A relative consistency proof', *Journal of Symbolic Logic* **19**, 21–28.

Scott, D. S., ed. (1971), *Axiomatic Set Theory I*, Vol. 13 of *Symposium in Pure Mathematics*, American Mathematical Society, Providence, R.I. (Held in Los Angeles, 1967).

Scott, D. S. & Solovay, R. M. (1971), Boolean-valued models for set theory. In (Scott 1971).

Shoenfield, J. R. (1971), Unramified forcing, pp. 357–381. In (Scott 1971).

Sierpiński, W. (1947), 'L'hypothèse généralisée du continu et l'axiome du choix', *Fundamenta Mathematicae* **34**, 1–5.

Sion, M. & Willmott, R. (1962), 'On a definition of ordinal numbers', *Am. Math. Monthly* **39**, 381–386.

Smullyan, R. M. (1965), 'On transfinite recursion', *Transactions of the N.Y. Academy of Sciences* **28**, 175–185.

Smullyan, R. M. (1967), 'More on transfinite recursion', *Transactions of the N. Y. Academy of Sciences* **30**, 175–185.

Solovay, R. M. & Tennenbaum, S. (1970), 'Iterated Cohen extensions and Souslin's problem', *Ann. Math.* **94**, 201–245.

Tarski, A. (1930), 'Über einige fundamentale begriffe der metamathematik', *Sprawozdania z posiedzeń Towarzystwa Naukowego Warszawskiego, Wydział III* **23**, 22–29. English translation in (Tarski 1956), 30–37.

Tarski, A. (1955), 'A lattice-theoretical theorem and its applications', *Pacific Journal of Mathematics* **5**, 285–309.

Tarski, A. (1956), *Logic, semantics, metamathematics: papers from 1923 to 1938*, Oxford. Translated into English and edited by J.H. Woodger.

van Heijenoort, J. (1967), *From Frege to Gödel*, Harvard University Press, Cambridge, MA.

Vaught, R. L. (1995), *Set Theory, an Introduction*, 2 edn, Birkhauser, Boston.

Wang, H. (1949), 'On Zermelo's and von Neumann's axioms for set theory', *Proceedings of the National Academy of Sciences, USA* **35**, 150–155.

Zermelo, E. (1904), Proof that every set can be well-ordered, pp. 138–141. Translation in (van Heijenoort 1967).

Zorn, M. (1935), 'A remark on method in transfinite algebra', *Bull. Amer. Math. Soc.* **41**, 667–670.

INDEX

A-closed, 145
a-formula, 185
A-rank, 132
a-sentence, 186
absolute, 158
    downwards, 179
    over K, 158
    upwards, 179
abstraction principle, 11
accessible, 209
admissible set recursion theory, 299
almost all, 269
α recursion theory, 299
α-sequence, 84
anti-symmetric, 47
arithmetic
    cardinal, 118
    ordinal, 88
Aussonderungs, 12, 170
automorphism, 268
    group, 268
    sequence, 269
axiom
    D, 99, 104
    E, 103
    F, 103
    of choice, 8, 244
    of class formation, 16
    of collection, 106
    of constructibility, 179, 190, 249
    of empty set, 18, 170, 235
    of extensionality, 15, 170, 232
    of infinity, 13, 32, 170, 235
    of power set, 23, 170, 235
    of replacement, 13, 82
    of separation, 16, 170
    of substitution, 13, 82, 170, 239
    of union, 21, 170, 235
    of unordered pairs, 19, 170, 235
    of well foundedness, 171, 235

B-closed, 197
backward path, 52
Barcan formula, 211
basic universe, 27
Bernstein-Schröder theorem, 108
Boolean algebra, 301
Boolean operations, 22
Boolean valued model, 301
bound occurrence, 141
bound up, 197
bounded, 38

c.c.c., *see* countable chain condition
Cantor's theorem, 7, 8
cardinal
    exponentiation, 121
    number, 71, 107, 116
    product, 118
    sum, 118
cardinality, 9, 107
    higher, 107
    lower, 107
    same, 107
Cartesian product, 24
chain, 52
choice function, 59
class, 13
class existence theorem, 201
class-set theory, 9
closed, 186

308   INDEX

closed relative to, 66
closed under chain unions, 52
code, 147, 183
coding, 183
collapsing cardinals, 261
commits on, 260
comparability theorem, 82
comparable, 8, 48, 108
    under inclusion, 36
compatible, 254, 290
complete sequence, 287
components, 127
composition, 79
conservative extension, 202
consistency, 62
constructible
    class, 163
    set, 155, 156
continuous, 151
continuum hypothesis, 9, 193
    generalized, 9, 193
continuum problem, 9
correspondence
    1-1, 3, 25
    many-one, 25
countable, 111
countable chain condition, 254
countably infinite, 4
counting theorem, 83
cover, 174
Cowen's Theorem, 68
Cowen's theorem, 66

D.I.P., 53
D.S.P., 53
definable
    first-order, 147
    over K, 147
define, 147, 174, 188
definition by finite recursion, 43
degree, 143
dense, 288
    below, 289, 292
denumerable, 4, 42, 111
descending $\in$-chains, 100

describable, 149
differ on, 260
distinguished subclass, 173
domain, 24, 209
double induction, *see* induction, double
double induction principle, 35
double superinduction, *see* superinduction, double

$\in$-connected, 105
$\in$-cycles, 100
$\in$-induction, 101
$\in$-isomorphic, 135
$\in$-isomorphism, 135
elementary equivalence, 144
elementary subsystem, 144
embodies, 239
empty set, 4, 11
equinumerous, 107
equivalence relation, 213
extended to, 61
extending operation, 84
extensional, 128
extensionality, 128
extraordinary, 11

$\mathfrak{F}$-invariant, 269
field, 47
filter, 288
fine structure, 299
finite, 4, 42, 111
    character, 61
    support, 280
first-order definable, *see* definable, first-order
first-order property, 15
first-order swelled, *see* swelled, first-order
first-order universe, 169
    well founded, 169
first-order Zermelo universe, *see* Zermelo universe, first order
first-order Zermelo-Fraenkel universe, *see* Zermelo-Fraenkel universe, first-order

INDEX

fixed point, 39
fixed point lemma, 109
forces, 210, 215
    weakly, 215
forcing, 207
forcing condition, 223, 252, 260, 279
formula, 141
    A, 142
    Δ_0, 159
    Σ_0, 159
    atomic, 141
    pure, 142
    with constants, 142, 209
frame, 209
    augmented, 209
free occurrence, 141
Frege set theory, 11
function, 25
function-like, 181
Fundamental Theorem of Cardinal Arithmetic, 115
fundamental theorem of cardinal arithmetic, 114

g-inductive, 35
$\mathfrak{G}$-invariant, 268
g-ordered, 68
Γ-rank, 130
G-rank, 224
g-set, 66
g-tower, 54
    slow, 58
generalized induction, *see* induction, generalized
generalized transfinite recursion, *see* transfinite recursion, generalized
generic, 289
Gödel's isomorphism theorem, 196
greatest, 37, 48

Hartog's function, 113
Hartog's theorem, 112
Henkin closure condition, 145
Henkin-closed, 145
Henkin-closure, 146

hereditary cardinality, 195

identity function, 80
immediate extension, 64
immediately in, 224
incompatible, 254, 290
individual, 183
induction, 29
    complete, 34
    double, 35
    generalized, 129
inductive, 32
    minimally, 35
    under g, *see* g-inductive
infinite, 3, 4, 42
initial element, 99, 128
inner model, 207
intersection, 21
into, 25
intuitionistic logic, 215
inverse, 79
isomorphism, 80, 129
iterated forcing, 302

K axiom, 211
Knaster-Tarski theorem, 109
Kripke semantics, 209

Löwenheim's theorem, 144
larger, 107
law of additive absorption, 115
law of multiplicative absorption, 116
least, 37, 48
Levy's basis theorem, 195
limit
    cardinal, 198
    element, 50
    ordinal, 72
Lindenbaum construction, 65
linear ordering, 48
lower section, 51, 80
    proper, 51

M-set, 223
M.S.T.V., 150
map, 25

mathematical induction, *see* induction
maximal, 61
maximal principles, 61
    Hausdorff, 63
    Kuratowski, 61
    Tukey-Teichmüller, 61
minimally superinductive, *see* superinductive, minimally
ML, 10
modal logic, 209
modal model, 210
model of *NBG*, 202
model of *ZF*, 202
model, modal, *see* modal model
modus ponens, 220
monadic, 186
monotone, 109
    increasing, 151
Montague-Levy reflection theorem, 151, 152
Morse-Kelley set theory, 13
Mostowski-Shepherdson mapping, 127, 134, 135
    theorem, 134
Mostowski-Shepherdson-Tarski-Vaught theorem, 150
moves backwards, 81

natural number, 3, 71
necessitation rule, 211
nest, 36
NF, 10
non-classical logic, 207
non-denumerable, 4, 42, 111
non-trivial, 260
normalizes, 269
number
    cardinal, *see* cardinal number
    natural, *see* natural number, 29, 32
    ordinal, *see* ordinal number

PROBLEM 0-rank, 92
On-sequence, 84
onto, 25

operation, 25
order, 156
ordered pair, 20
ordinal hierarchy, 92
ordinal number, 71
    Robinson's definition, 105
ordinal sequence, 84
ordinary, 11, 16

Π-formula, 273
partial ordering, 47, 213
Peano postulates, 29
permutation, 280
permutation group, 268
Platonist, 10
possible world, 223
possible world semantics, 209
power, 9
power set, 7
predecessor, 49
preordering, 213
Principle E, 64
progressing, 36
    slowly, 44, 58
    strictly, 55
proper subset, *see* subset, proper
proper well ordering, *see* well ordering, proper
pseudo-constructible, 196
pseudo-construction, 196

ramified, 301
range, 24
rank, 91
realist, mathematical, 10
reflection principles, 141
reflects, 145
    completely, 145
    with respect to φ, 145
reflexive, 47
reflexivity axiom, 214
regular, 228
relation, 24
    single-valued, 25
relational structure, 212

relational system, 127
    proper, 127
relativization, 199
relevant, 175
replacement theorem, 212
restriction, 48
reverse diagonal, 279

Σ-formula, 180
Σ-relation, 180
S 4, 213
    ZF model, 221
    model, 213
    regular model, 228
S 5, 213
sandwich principle, 37, 54
satisfiable, 144
Schröder-Bernstein theorem, 8
sentence, 186
    over A, 142
separation, 15
separation principle, 12
sequence, 43
set, 15
Sierpiński's Theorem, 124
Sierpiński's theorem, 121
similar, 107
singleton, 19
size, 107
Skolem-Löwenheim theorem, 144
slowly progressing, *see* progressing, slowly
smaller, 8, 107
special, 66, 280
stabilize, 245
stable, 180
standard model, 202, 285
    countable, 286
strictly progressing, *see* progressing, strictly
subclass, 15
subformula, 141
subgroup, 269
subset
    proper, 7
substitution, 142, 186
    function, 187

successor, 49
successor ordinal, 72
supercomplete, 17
superinduction
    double, 53
    proof by, 53
superinductive, 52
    minimally, 53
swelled, 17, 169
    first-order, 169

T-closed, 76
Tarski-Vaught theorem, 145
    class version, 148
Theorem G, 194
Theorem GI, 196
Theorem K, 196
transfinite induction, 50
transfinite recursion, 84, 87
    generalized, 133
transitive, 17, 47
transitive closure, 195
transitivity axiom, 214
trichotomy, 113
trichotomy principle, 41
true
    at a world, 210
    over A, 145
true at p, 210
truth, 187
    in relational systems, 143
truth lemma, 294
type, 61, 185

uncountable, 4, 111
universal class, 15
universal generalization, 220
unordered pair, 11, 20
unramified, 301
urelements, 267

valid, 211
valuation, 187
valuation function, 188
variable, 141

NBG, 9
Von Neumann's principle, 103

weakly forces, *see* forces, weakly
well founded, 99, 104, 128
well ordered, 37
well ordering, 48
    fundamental theorem of, 82
    proper, 80
    slow, 57
well ordering principle, 41
Well ordering theorem, Zermelo, 60

x-special, *see* special

Zermelo set theory, 12
Zermelo universe, 27, 32
    first-order, 198
Zermelo-Fraenkel, *see* ZF
Zermelo-Fraenkel set theory, 13, 170
Zermelo-Fraenkel universe, 82
    first-order, 169
        well founded, 169
ZF, 9, 13
ZF-axioms, 170
ZFS, 13
Zorn's lemma, 63

LIST OF NOTATION

$\mathcal{P}$, 7
$A < B$, 8
$A \leq B$, 8
$\{x\}$, 8
ω, 9
$\neg$, 11
$\wedge$, 11
$\vee$, 11
$\supset$, 11
$\equiv$, 11
$\forall$, 11
$\exists$, 11
$=$, 11
$x \in y$, 11
$\{x \mid P(x)\}$, 11
$\emptyset$, 11
$\{a, b\}$, 11
$\cup a$, 11
ω, 11
$x \in y$, 15
$x \notin y$, 15
$\subseteq$, 15
V, 15
iff, 17
A_1, 17
A_2, 18
$\emptyset$, 18
A_3, 18
$\{a\}$, 19
$\{a, b\}$, 19
A_4, 19
$\langle a, b \rangle$, 20
$\cup A$, 21
A_5, 21
$\cap A$, 21
$\cup$, 22

$\cap$, 22
$-$, 22
$\mathcal{P}$, 23
A_6, 23
$A \times B$, 24
Dom, 24
Ran, 24
$F(x)$, 25
x^+, 31
ω, 32
A_7, 32
$<$, 40
$\leq$, 40
$>$, 40
$\geq$, 40
$\upharpoonright$, 48
$L_<(x)$, 50
$L_\leq(x)$, 50
On, 71
F'', 79
$\upharpoonright$, 79
R^{-1}, 79
fg, 79
$f \circ g$, 79
seg, 80
A_8, 82
R_α, 91
R_Ω, 91
$\omega \cdot 2$, 96
A_8^*, 106
$\cong$, 107
$\leq$, 107
$\prec$, 107
(H), 113
$\aleph_0$, 116
$\aleph_1$, 116

LIST OF NOTATION

$\aleph_2$, 116
$\aleph_\alpha$, 116
$\overline{\overline{A}}$, 116
On^2, 119
(A, R), 127
$p(x)$, 130
Γ_α, 130
$\neg$, 141
$\wedge$, 141
$\exists$, 141
$\vee$, 142
$\supset$, 142
$\equiv$, 142
$\forall$, 142
α_a^x, 142
φ_a^x, 142
$\mathcal{F}$, 147
$\ulcorner \varphi \urcorner$, 147
$\mathcal{T}$, 148
$\mathcal{T}_n$, 148
$\mathcal{T}_n^K$, 148
$\mathcal{F}$, 155
L_α, 156
L, 156
M_α, 156
Num, 159
sf Fun, 159
Δ_0, 159
$K_{i,j}^n$, 167
$K_{i,\overline{a}}^n$, 167
$K_{\overline{a},i}^n$, 167
$K_{\overline{a},\overline{b}}^n$, 167
L^n, 167
$L_{i,j}^n$, 167
$L_{i,\overline{a}}^n$, 167
Const(x), 179
$\mathcal{F}(x)$, 183
$\ulcorner \varphi \urcorner$, 183
Neg(x), 184
Con(x, y), 184
$\mathsf{E}_i(x)$, 184
$\Pi(a)$, 184
$c(a)$, 184
$\mathcal{E}_n^a$, 184

$\mathcal{E}^a$, 184
$t(\varphi)$, 185
$\mathcal{E}_b(a)$, 185
$\overline{\mathcal{E}}_b(a)$, 185
$M(a)$, 186
$S(a)$, 186
$Sub(\varphi, v_i, b)$, 186
Sub_a, 186
$\mathsf{Sub}_a(x, i, b)$, 186
$V_a(x)$, 187
V_a, 187
Def(x, y, a), 188
$\mathcal{L}(x, y)$, 189
Const(x), 189
$L(x)$, 189
c^*, 193
GCH, 193
CH, 193
$V = L$, 194
HC(x), 195
$o(K)$, 197
φ^L, 199
$\Box$, 209
$\Diamond$, 209
$\langle \mathcal{G}, \mathcal{R} \rangle$, 209
$\langle \mathcal{G}, \mathcal{R}, \mathcal{D} \rangle$, 209
$\langle \mathcal{G}, \mathcal{R}, \mathcal{D}, \mathcal{V} \rangle$, 210
$p \Vdash_M X$, 210
$p \Vdash X$, 210
$S5$, 213
$S4$, 213
$[\![A]\!]$, 216
$R_\alpha^\mathcal{G}$, 224
$\mathcal{D}^\mathcal{G}$, 224
$\mathcal{V}^\mathcal{G}$, 224
$x \, \varepsilon \, y$, 224
$\mathcal{L}_C$, 225
$\mathcal{L}_M$, 225
$\approx$, 225
$\approx_\alpha$, 225
$\mathcal{L}_C^*$, 225
$p \Vdash X$, 225
$f \approx g$, 228
$f \in g$, 228
$\hat{x}$, 233

Next_Equal(x, y, z, w), 239
Equal(x, y, z), 240
A_8^*, 242
g^*, 242
Cardinal(α), 253
Up(x, y), 262
Op(x, y), 262
$\theta_1 H \theta_2$, 269
$\mathcal{D}_{\widetilde{\mathfrak{F}}}^{\mathcal{G}}$, 269
$\Vdash_{\widetilde{\mathfrak{F}}}$, 269
$\mathcal{M}_{\widetilde{\mathfrak{F}}}$, 270
f_G, 292
$M[G]$, 292

CATALOG OF DOVER BOOKS

Mathematics

FUNCTIONAL ANALYSIS (Second Corrected Edition), George Bachman and Lawrence Narici. Excellent treatment of subject geared toward students with background in linear algebra, advanced calculus, physics and engineering. Text covers introduction to inner-product spaces, normed, metric spaces, and topological spaces; complete orthonormal sets, the Hahn-Banach Theorem and its consequences, and many other related subjects. 1966 ed. 544pp. 6 1/8 x 9 1/4. 0-486-40251-7

DIFFERENTIAL MANIFOLDS, Antoni A. Kosinski. Introductory text for advanced undergraduates and graduate students presents systematic study of the topological structure of smooth manifolds, starting with elements of theory and concluding with method of surgery. 1993 edition. 288pp. 5 3/8 x 8 1/2. 0-486-46244-7

VECTOR AND TENSOR ANALYSIS WITH APPLICATIONS, A. I. Borisenko and I. E. Tarapov. Concise introduction. Worked-out problems, solutions, exercises. 257pp. 5 5/8 x 8 1/4. 0-486-63833-2

AN INTRODUCTION TO ORDINARY DIFFERENTIAL EQUATIONS, Earl A. Coddington. A thorough and systematic first course in elementary differential equations for undergraduates in mathematics and science, with many exercises and problems (with answers). Index. 304pp. 5 3/8 x 8 1/2. 0-486-65942-9

FOURIER SERIES AND ORTHOGONAL FUNCTIONS, Harry F. Davis. An incisive text combining theory and practical example to introduce Fourier series, orthogonal functions and applications of the Fourier method to boundary-value problems. 570 exercises. Answers and notes. 416pp. 5 3/8 x 8 1/2. 0-486-65973-9

COMPUTABILITY AND UNSOLVABILITY, Martin Davis. Classic graduate-level introduction to theory of computability, usually referred to as theory of recurrent functions. New preface and appendix. 288pp. 5 3/8 x 8 1/2. 0-486-61471-9

AN INTRODUCTION TO MATHEMATICAL ANALYSIS, Robert A. Rankin. Dealing chiefly with functions of a single real variable, this text by a distinguished educator introduces limits, continuity, differentiability, integration, convergence of infinite series, double series, and infinite products. 1963 edition. 624pp. 5 3/8 x 8 1/2. 0-486-46251-X

METHODS OF NUMERICAL INTEGRATION (SECOND EDITION), Philip J. Davis and Philip Rabinowitz. Requiring only a background in calculus, this text covers approximate integration over finite and infinite intervals, error analysis, approximate integration in two or more dimensions, and automatic integration. 1984 edition. 624pp. 5 3/8 x 8 1/2. 0-486-45339-1

INTRODUCTION TO LINEAR ALGEBRA AND DIFFERENTIAL EQUATIONS, John W. Dettman. Excellent text covers complex numbers, determinants, orthonormal bases, Laplace transforms, much more. Exercises with solutions. Undergraduate level. 416pp. 5 3/8 x 8 1/2. 0-486-65191-6

RIEMANN'S ZETA FUNCTION, H. M. Edwards. Superb, high-level study of landmark 1859 publication entitled "On the Number of Primes Less Than a Given Magnitude" traces developments in mathematical theory that it inspired. xiv+315pp. 5 3/8 x 8 1/2.
0-486-41740-9

CATALOG OF DOVER BOOKS

CALCULUS OF VARIATIONS WITH APPLICATIONS, George M. Ewing. Applications-oriented introduction to variational theory develops insight and promotes understanding of specialized books, research papers. Suitable for advanced undergraduate/graduate students as primary, supplementary text. 352pp. $5^{3}/_{8}$ x $8^{1}/_{2}$. 0-486-64856-7

MATHEMATICIAN'S DELIGHT, W. W. Sawyer. "Recommended with confidence" by *The Times Literary Supplement*, this lively survey was written by a renowned teacher. It starts with arithmetic and algebra, gradually proceeding to trigonometry and calculus. 1943 edition. 240pp. $5^{3}/_{8}$ x $8^{1}/_{2}$. 0-486-46240-4

ADVANCED EUCLIDEAN GEOMETRY, Roger A. Johnson. This classic text explores the geometry of the triangle and the circle, concentrating on extensions of Euclidean theory, and examining in detail many relatively recent theorems. 1929 edition. 336pp. $5^{3}/_{8}$ x $8^{1}/_{2}$. 0-486-46237-4

COUNTEREXAMPLES IN ANALYSIS, Bernard R. Gelbaum and John M. H. Olmsted. These counterexamples deal mostly with the part of analysis known as "real variables." The first half covers the real number system, and the second half encompasses higher dimensions. 1962 edition. xxiv+198pp. $5^{3}/_{8}$ x $8^{1}/_{2}$. 0-486-42875-3

CATASTROPHE THEORY FOR SCIENTISTS AND ENGINEERS, Robert Gilmore. Advanced-level treatment describes mathematics of theory grounded in the work of Poincaré, R. Thom, other mathematicians. Also important applications to problems in mathematics, physics, chemistry and engineering. 1981 edition. References. 28 tables. 397 black-and-white illustrations. xvii + 666pp. $6^{1}/_{8}$ x $9^{1}/_{4}$. 0-486-67539-4

COMPLEX VARIABLES: Second Edition, Robert B. Ash and W. P. Novinger. Suitable for advanced undergraduates and graduate students, this newly revised treatment covers Cauchy theorem and its applications, analytic functions, and the prime number theorem. Numerous problems and solutions. 2004 edition. 224pp. $6^{1}/_{2}$ x $9^{1}/_{4}$. 0-486-46250-1

NUMERICAL METHODS FOR SCIENTISTS AND ENGINEERS, Richard Hamming. Classic text stresses frequency approach in coverage of algorithms, polynomial approximation, Fourier approximation, exponential approximation, other topics. Revised and enlarged 2nd edition. 721pp. $5^{3}/_{8}$ x $8^{1}/_{2}$. 0-486-65241-6

INTRODUCTION TO NUMERICAL ANALYSIS (2nd Edition), F. B. Hildebrand. Classic, fundamental treatment covers computation, approximation, interpolation, numerical differentiation and integration, other topics. 150 new problems. 669pp. $5^{3}/_{8}$ x $8^{1}/_{2}$.
0-486-65363-3

MARKOV PROCESSES AND POTENTIAL THEORY, Robert M. Blumental and Ronald K. Getoor. This graduate-level text explores the relationship between Markov processes and potential theory in terms of excessive functions, multiplicative functionals and subprocesses, additive functionals and their potentials, and dual processes. 1968 edition. 320pp. $5^{3}/_{8}$ x $8^{1}/_{2}$. 0-486-46263-3

ABSTRACT SETS AND FINITE ORDINALS: An Introduction to the Study of Set Theory, G. B. Keene. This text unites logical and philosophical aspects of set theory in a manner intelligible to mathematicians without training in formal logic and to logicians without a mathematical background. 1961 edition. 112pp. $5^{3}/_{8}$ x $8^{1}/_{2}$. 0-486-46249-8

CATALOG OF DOVER BOOKS

A TREATISE ON ELECTRICITY AND MAGNETISM, James Clerk Maxwell. Important foundation work of modern physics. Brings to final form Maxwell's theory of electromagnetism and rigorously derives his general equations of field theory. 1,084pp. $5^{3}/_{8}$ x $8^{1}/_{2}$. Two-vol. set.　　　　　　Vol. I: 0-486-60636-8　Vol. II: 0-486-60637-6

MATHEMATICS FOR PHYSICISTS, Philippe Dennery and Andre Krzywicki. Superb text provides math needed to understand today's more advanced topics in physics and engineering. Theory of functions of a complex variable, linear vector spaces, much more. Problems. 1967 edition. 400pp. $6^{1}/_{2}$ x $9^{1}/_{4}$.　　　　　　0-486-69193-4

INTRODUCTION TO QUANTUM MECHANICS WITH APPLICATIONS TO CHEMISTRY, Linus Pauling & E. Bright Wilson, Jr. Classic undergraduate text by Nobel Prize winner applies quantum mechanics to chemical and physical problems. Numerous tables and figures enhance the text. Chapter bibliographies. Appendices. Index. 468pp. $5^{3}/_{8}$ x $8^{1}/_{2}$.　　　　　　0-486-64871-0

METHODS OF THERMODYNAMICS, Howard Reiss. Outstanding text focuses on physical technique of thermodynamics, typical problem areas of understanding, and significance and use of thermodynamic potential. 1965 edition. 238pp. $5^{3}/_{8}$ x $8^{1}/_{2}$.

0-486-69445-3

THE ELECTROMAGNETIC FIELD, Albert Shadowitz. Comprehensive under- graduate text covers basics of electric and magnetic fields, builds up to electromagnetic theory. Also related topics, including relativity. Over 900 problems. 768pp. $5^{3}/_{8}$ x $8^{1}/_{4}$.

0-486-65660-8

GREAT EXPERIMENTS IN PHYSICS: FIRSTHAND ACCOUNTS FROM GALILEO TO EINSTEIN, Morris H. Shamos (ed.). 25 crucial discoveries: Newton's laws of motion, Chadwick's study of the neutron, Hertz on electromagnetic waves, more. Original accounts clearly annotated. 370pp. $5^{3}/_{8}$ x $8^{1}/_{2}$.　　　　　　0-486-25346-5

EINSTEIN'S LEGACY, Julian Schwinger. A Nobel Laureate relates fascinating story of Einstein and development of relativity theory in well-illustrated, nontechnical volume. Subjects include meaning of time, paradoxes of space travel, gravity and its effect on light, non-Euclidean geometry and curving of space-time, impact of radio astronomy and space-age discoveries, and more. 189 b/w illustrations. xiv+250pp. $8^{3}/_{8}$ x $9^{1}/_{4}$.　　0-486-41974-6

THE VARIATIONAL PRINCIPLES OF MECHANICS, Cornelius Lanczos. Philosophic, less formalistic approach to analytical mechanics offers model of clear, scholarly exposition at graduate level with coverage of basics, calculus of variations, principle of virtual work, equations of motion, more. 418pp. $5^{3}/_{8}$ x $8^{1}/_{2}$.　　　　　　0-486-65067-7

Paperbound unless otherwise indicated. Available at your book dealer, online at www.doverpublications.com, or by writing to Dept. GI, Dover Publications, Inc., 31 East 2nd Street, Mineola, NY 11501. For current price information or for free catalogues (please indicate field of interest), write to Dover Publications or log on to www.doverpublications.com and see every Dover book in print. Dover publishes more than 400 books each year on science, elementary and advanced mathematics, biology, music, art, literary history, social sciences, and other areas.